S0-BLS-322

Chiral Separations by Liquid Chromatography and Related Technologies

CHROMATOGRAPHIC SCIENCE SERIES

A Series of Textbooks and Reference Books

Editor: JACK CAZES

1. Dynamics of Chromatography: Principles and Theory, *J. Calvin Giddings*
2. Gas Chromatographic Analysis of Drugs and Pesticides, *Benjamin J. Gudzinowicz*
3. Principles of Adsorption Chromatography: The Separation of Nonionic Organic Compounds, *Lloyd R. Snyder*
4. Multicomponent Chromatography: Theory of Interference, *Friedrich Helfferich and Gerhard Klein*
5. Quantitative Analysis by Gas Chromatography, *Josef Novák*
6. High-Speed Liquid Chromatography, *Peter M. Rajcsanyi and Elisabeth Rajcsanyi*
7. Fundamentals of Integrated GC-MS (in three parts), *Benjamin J. Gudzinowicz, Michael J. Gudzinowicz, and Horace F. Martin*
8. Liquid Chromatography of Polymers and Related Materials, *Jack Cazes*
9. GLC and HPLC Determination of Therapeutic Agents (in three parts), *Part 1 edited by Kiyoshi Tsuji and Walter Morozowich, Parts 2 and 3 edited by Kiyoshi Tsuji*
10. Biological/Biomedical Applications of Liquid Chromatography, *edited by Gerald L. Hawk*
11. Chromatography in Petroleum Analysis, *edited by Klaus H. Altgelt and T. H. Gouw*
12. Biological/Biomedical Applications of Liquid Chromatography II, *edited by Gerald L. Hawk*
13. Liquid Chromatography of Polymers and Related Materials II, *edited by Jack Cazes and Xavier Delamare*
14. Introduction to Analytical Gas Chromatography: History, Principles, and Practice, *John A. Perry*
15. Applications of Glass Capillary Gas Chromatography, *edited by Walter G. Jennings*
16. Steroid Analysis by HPLC: Recent Applications, *edited by Marie P. Kautsky*
17. Thin-Layer Chromatography: Techniques and Applications, *Bernard Fried and Joseph Sherma*
18. Biological/Biomedical Applications of Liquid Chromatography III, *edited by Gerald L. Hawk*
19. Liquid Chromatography of Polymers and Related Materials III, *edited by Jack Cazes*
20. Biological/Biomedical Applications of Liquid Chromatography, *edited by Gerald L. Hawk*
21. Chromatographic Separation and Extraction with Foamed Plastics and Rubbers, *G. J. Moody and J. D. R. Thomas*
22. Analytical Pyrolysis: A Comprehensive Guide, *William J. Irwin*
23. Liquid Chromatography Detectors, *edited by Thomas M. Vickrey*
24. High-Performance Liquid Chromatography in Forensic Chemistry, *edited by Ira S. Lurie and John D. Wittwer, Jr.*
25. Steric Exclusion Liquid Chromatography of Polymers, *edited by Josef Janca*
26. HPLC Analysis of Biological Compounds: A Laboratory Guide, *William S. Hancock and James T. Sparrow*

27. Affinity Chromatography: Template Chromatography of Nucleic Acids and Proteins, *Herbert Schott*
28. HPLC in Nucleic Acid Research: Methods and Applications, *edited by Phyllis R. Brown*
29. Pyrolysis and GC in Polymer Analysis, *edited by S. A. Liebman and E. J. Levy*
30. Modern Chromatographic Analysis of the Vitamins, *edited by André P. De Leenheer, Willy E. Lambert, and Marcel G. M. De Ruyter*
31. Ion-Pair Chromatography, *edited by Milton T. W. Hearn*
32. Therapeutic Drug Monitoring and Toxicology by Liquid Chromatography, *edited by Steven H. Y. Wong*
33. Affinity Chromatography: Practical and Theoretical Aspects, *Peter Mohr and Klaus Pommerening*
34. Reaction Detection in Liquid Chromatography, *edited by Ira S. Krull*
35. Thin-Layer Chromatography: Techniques and Applications. Second Edition, Revised and Expanded, *Bernard Fried and Joseph Sherma*
36. Quantitative Thin-Layer Chromatography and Its Industrial Applications, *edited by Laszlo R. Treiber*
37. Ion Chromatography, *edited by James G. Tarter*
38. Chromatographic Theory and Basic Principles, *edited by Jan Åke Jönsson*
39. Field-Flow Fractionation: Analysis of Macromolecules and Particles, *Josef Janca*
40. Chromatographic Chiral Separations, *edited by Morris Zief and Laura J. Crane*
41. Quantitative Analysis by Gas Chromatography: Second Edition, Revised and Expanded, *Josef Novák*
42. Flow Perturbation Gas Chromatography, *N. A. Katsanos*
43. Ion-Exchange Chromatography of Proteins, *Shuichi Yamamoto, Kazuhiro Nakanishi, and Ryuichi Matsuno*
44. Countercurrent Chromatography: Theory and Practice, *edited by N. Bhushan Mandava and Yoichiro Ito*
45. Microbore Column Chromatography: A Unified Approach to Chromatography, *edited by Frank J. Yang*
46. Preparative-Scale Chromatography, *edited by Eli Grushka*
47. Packings and Stationary Phases in Chromatographic Techniques, *edited by Klaus K. Unger*
48. Detection-Oriented Derivatization Techniques in Liquid Chromatography, *edited by Henk Lingeman and Willy J. M. Underberg*
49. Chromatographic Analysis of Pharmaceuticals, *edited by John A. Adamovics*
50. Multidimensional Chromatography: Techniques and Applications, *edited by Hernan Cortes*
51. HPLC of Biological Macromolecules: Methods and Applications, *edited by Karen M. Gooding and Fred E. Regnier*
52. Modern Thin-Layer Chromatography, *edited by Nelu Grinberg*
53. Chromatographic Analysis of Alkaloids, *Milan Popl, Jan Fähnrich, and Vlastimil Tatar*
54. HPLC in Clinical Chemistry, *I. N. Papadoyannis*
55. Handbook of Thin-Layer Chromatography, *edited by Joseph Sherma and Bernard Fried*
56. Gas–Liquid–Solid Chromatography, *V. G. Berezkin*
57. Complexation Chromatography, *edited by D. Cagniant*
58. Liquid Chromatography–Mass Spectrometry, *W. M. A. Niessen and Jan van der Greef*
59. Trace Analysis with Microcolumn Liquid Chromatography, *Milos Krejcl*

60. Modern Chromatographic Analysis of Vitamins: Second Edition, *edited by André P. De Leenheer, Willy E. Lambert, and Hans J. Nelis*
61. Preparative and Production Scale Chromatography, *edited by G. Ganetsos and P. E. Barker*
62. Diode Array Detection in HPLC, *edited by Ludwig Huber and Stephan A. George*
63. Handbook of Affinity Chromatography, *edited by Toni Kline*
64. Capillary Electrophoresis Technology, *edited by Norberto A. Guzman*
65. Lipid Chromatographic Analysis, *edited by Takayuki Shibamoto*
66. Thin-Layer Chromatography: Techniques and Applications, Third Edition, Revised and Expanded, *Bernard Fried and Joseph Sherma*
67. Liquid Chromatography for the Analyst, *Raymond P. W. Scott*
68. Centrifugal Partition Chromatography, *edited by Alain P. Foucault*
69. Handbook of Size Exclusion Chromatography, *edited by Chi-san Wu*
70. Techniques and Practice of Chromatography, *Raymond P. W. Scott*
71. Handbook of Thin-Layer Chromatography: Second Edition, Revised and Expanded, *edited by Joseph Sherma and Bernard Fried*
72. Liquid Chromatography of Oligomers, *Constantin V. Uglea*
73. Chromatographic Detectors: Design, Function, and Operation, *Raymond P. W. Scott*
74. Chromatographic Analysis of Pharmaceuticals: Second Edition, Revised and Expanded, *edited by John A. Adamovics*
75. Supercritical Fluid Chromatography with Packed Columns: Techniques and Applications, *edited by Klaus Anton and Claire Berger*
76. Introduction to Analytical Gas Chromatography: Second Edition, Revised and Expanded, *Raymond P. W. Scott*
77. Chromatographic Analysis of Environmental and Food Toxicants, *edited by Takayuki Shibamoto*
78. Handbook of HPLC, *edited by Elena Katz, Roy Eksteen, Peter Schoenmakers, and Neil Miller*
79. Liquid Chromatography–Mass Spectrometry: Second Edition, Revised and Expanded, *W. M. A. Niessen*
80. Capillary Electrophoresis of Proteins, *Tim Wehr, Roberto Rodríguez-Díaz, and Mingde Zhu*
81. Thin-Layer Chromatography: Fourth Edition, Revised and Expanded, *Bernard Fried and Joseph Sherma*
82. Countercurrent Chromatography, *edited by Jean-Michel Menet and Didier Thiébaut*
83. Micellar Liquid Chromatography, *Alain Berthod and Celia García-Alvarez-Coque*
84. Modern Chromatographic Analysis of Vitamins: Third Edition, Revised and Expanded, *edited by André P. De Leenheer, Willy E. Lambert, and Jan F. Van Bocxlaer*
85. Quantitative Chromatographic Analysis, *Thomas E. Beesley, Benjamin Buglio, and Raymond P. W. Scott*
86. Current Practice of Gas Chromatography–Mass Spectrometry, *edited by W. M. A. Niessen*
87. HPLC of Biological Macromolecules: Second Edition, Revised and Expanded, *edited by Karen M. Gooding and Fred E. Regnier*
88. Scale-Up and Optimization in Preparative Chromatography: Principles and Biopharmaceutical Applications, *edited by Anurag S. Rathore and Ajoy Velayudhan*
89. Handbook of Thin-Layer Chromatography: Third Edition, Revised and Expanded, *edited by Joseph Sherma and Bernard Fried*

90. Chiral Separations by Liquid Chromatography and Related Technologies, *Hassan Y. Aboul-Enein and Imran Ali*

ADDITIONAL VOLUMES IN PREPARATION

Handbook of Size Exclusion Chromatography and Related Techniques, Second Edition, Revised and Expanded, *Chi-san Wu*

Chiral Separations by Liquid Chromatography and Related Technologies

Hassan Y. Aboul-Enein

King Faisal Specialist Hospital and Research Centre
Riyadh, Saudi Arabia

Imran Ali

National Insitute of Hydrology
Roorkee, India

MARCEL DEKKER, INC. NEW YORK • BASEL

Library of Congress Cataloging-in-Publication Data
A catalog record for this book is available from the Library of Congress.

ISBN: 0-8247-4014-9

This book is printed on acid-free paper.

Headquarters
Marcel Dekker, Inc., 270 Madison Avenue, New York, NY 10016, U.S.A.
tel: 212-696-9000; fax: 212-685-4540

Distribution and Customer Service
Marcel Dekker, Inc., Cimarron Road, Monticello, New York 12701, U.S.A.
tel: 800-228-1160; fax: 845-796-1772

Eastern Hemisphere Distribution
Marcel Dekker AG, Hutgasse 4, Postfach 812, CH-4001 Basel, Switzerland
tel: 41-61-260-6300; fax: 41-61-260-6333

World Wide Web
http://www.dekker.com

The publisher offers discounts on this book when ordered in bulk quantities. For more information, write to Special Sales/Professional Marketing at the headquarters address above.

Current printing (last digit):

10 9 8 7 6 5 4 3 2 1

PRINTED IN THE UNITED STATES OF AMERICA

Dedicated to the memory of my late parents,
Basheer Ahmed and Mehmudan Begum. *Imran Ali*

This book is dedicated to the memory of my father, Dr. Youssef Aboul-Enein,
my mother, and my wife, Najla Al-Mojadady, for her patience,
understanding, and never-ending support. *Hassan Y. Aboul-Enein*

Preface

The development of chiral resolution methods is considered an urgent need in pharmaceutical, agricultural, and other chemical industries. Perhaps most essential is the development of new chiral drugs, which are necessary because of the different physiological properties of enantiomers. Various approaches to chiral resolution have been developed and used, but the direct resolution by liquid chromatography on chiral stationary phases has proven to be one of the most effective, practical, and economical. This volume deals with the art of chiral resolution by liquid chromatography, focusing on high-performance liquid chromatography (HPLC), sub- and supercritical fluid chromatography (SFC), capillary electrochromatography (CEC), and thin-layer chromatography (TLC). These methods are examined as they are used in analysis and development of drugs, pharmaceuticals, xenobiotics, and other chiral molecules.

The book provides systematic and detailed descriptions of the numerous approaches to chiral resolution. The first chapter is an introduction to basic concepts of molecular chirality and liquid chromatography. Chapters 2 through 9 discuss the chiral resolution of various classes of chiral stationary phases. Chapter 10 deals with chiral resolution using chiral mobile phase additives. These discussions elaborate the types, structures, and properties of the chiral phases, as well as their preparations, their applications, and the future scope of chiral resolution. It is our hope that this volume will be a valuable resource for scientists, researchers, academics, and graduate students in the chromatographic sciences.

ACKNOWLEDGMENTS

I express my deep regard and gratitude to my wife, Hajan Seema Imran, who has helped me and supported me throughout this work. I also thank my dearest son, Haji al Arsh Basheer Baichain, who has inspired me throughout this difficult job. I would also like to acknowledge my other family members and relatives who have helped me directly and indirectly during this period.

I also extend my sincere thanks to Professor Vinod K. Gupta, Chemistry Department, Indian Institute of Technology, Roorkee, India, for his efforts in helping me to complete this book. His constant moral support was a great help—one that I will always remember. Finally, I would like to thank Mr. Russell Dekker of Marcel Dekker, Inc.

Imran Ali

I would like to express my thanks to the administration of King Faisal Specialist Hospital and Research Centre for their support of this work. Special thanks to Mr. Russell Dekker and the editorial staff of Marcel Dekker, Inc., for their assistance in publishing this book.

Hassan Y. Aboul-Enein

Contents

1

Introduction

1.1 CHIRALITY AND ITS OCCURRENCE

In 1883 Lord Kelvin used the term "chirality," derived from the Greek *kheir* or *chiros* for handedness [1]. Any object lacking the three elements of symmetry (i.e., plane of symmetry, center of symmetry, and axis of symmetry) and existing in two forms that are nonsuperimposable mirror images of each other is called a chiral object. From elementary particles to humans, chirality is found in a wide range of objects [2]. This observation suggests that chirality has very important and essential roles in the existence of the universe and is still a mystery. Several examples indicate the presence of chirality in our universe. In the ancient kingdoms of Upper and Lower Egypt, many mural paintings in burial chambers depict significant events in ways that emphasize their chirality [1]. Besides, out of 1168 galaxies listed in *Carnegie Atlas of Galaxies*, 540 are found to be chiral when coupled with the direction of their recession velocities [3]. The influence of the chirality is evident in plants and animals, where numerous examples of asymmetric structures can be observed. For example, helical structures of plants and animals make them asymmetric. Briefly, chirality exists almost everywhere in this universe.

1.2 CHEMICAL EVOLUTION OF CHIRALITY

Study of the chemical evolution of chirality started in 1809 with the discovery of Haüy [4], who postulated from crystal cleavage observations that a crystal and each of its constituent space-filling molecules are images of each other in overall shape. Later, in 1848, Pasteur reported the different destruction rates of the dextro- and levorotatory forms of ammonium tartrate by the mold *Penicillium glaucum* [5]. These observations could not be explained properly at that time, but in 1874 Le Bel [6] and van't Hoff [7] independently proposed that the four valences of the carbon atom are directed toward the vertices of an atom-centered tetrahedron. This finding allowed the development of the theory of the three-dimensional structure of molecules by which the phenomenon of chirality and Pasteur's discovery were explained scientifically.

1.3 CHIRALITY IN MOLECULES

Generally, the phenomenon of chirality exists in organic molecules. However, it is also found in some inorganic compounds. In some molecules, the carbon atoms remain attached to four different atoms or groups. This arrangement makes the whole molecule asymmetric. Molecules of this type differ in their three-dimensional configurations and exist in two forms, which are mirror images of each other. No matter what symmetry operation is applied to such molecule, it is never possible to superimpose the two mirror images upon each other. These mirror images are called optical isomers (having the capacity to rotate the plan polarized light) or stereoisomers or enantiomers or enantiomorphs or antipodes or chiral molecules. The existence of these different enantiomers comprises the phenomenon of stereoisomerism, or chirality. A 50 : 50 ratio of enantiomers is called a racemic mixture. In addition to central chirality, axial chirality can occur in allenes and cumulenes. In the former class, the substituents are not necessarily different, since the second double bond causes the loss of the C_3 rotational symmetry element. In the latter class, only members with an odd number of accumulated carbon atoms are potentially chiral, whereas an even number of carbon atoms results in E/Z isomerism (geometric isomerism) [8].

Another type of axial chirality is represented by atropisomers, which possess conformational chirality. As long as the ortho substituents in tetrasubstituted biaryls are large enough, the rotation around a C—C single bond will be hindered, preventing the two forms from interconverting. Finally, there exists a planar chirality that arises from the arrangement of atoms or groups of atoms relative to a stereogenic plane. However, this form of chirality is rather rare [8]. Helicity is a special form of chirality and often occurs in macromolecules such as biopolymers, proteins, and polysaccharides [9]. A helix is always chiral owing to its right-handed (clockwise) or left-handed (counterclockwise) arrangement.

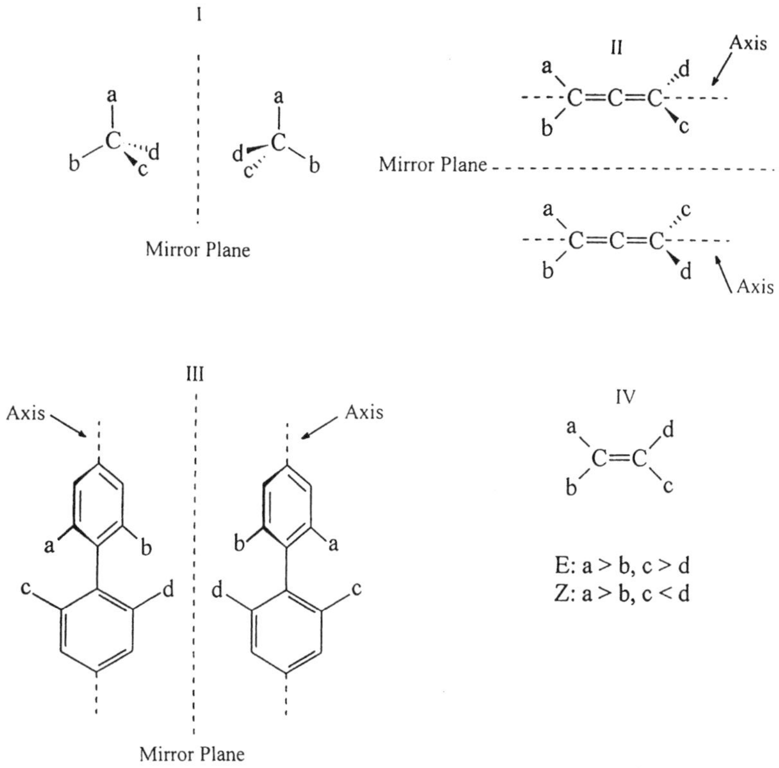

FIGURE 1 The different types of stereoisomerism: I, central chirality; II, axial chirality; III, atropisomerism; IV, E-/Z-isomerism (geometrical).

When a stereoisomer has more than one stereogenic center (e.g., n), the number of theoretically possible enantiomers can be derived from the 2^n formula. The four types of stereoisomerism are shown in Figure 1.

1.3.1 Nomenclature of Chiral Molecules

Another point connected to chirality is the nomenclature of enantiomers. In the beginning, the optical isomers were distinguished with (+) and (−) signs or *d* (*dextro*) and *l* (*levo*), indicating the direction in which the enantiomers rotate a plane of polarized light. In this nomenclature, (+) or *d* stands for a rotation to the right (clockwise), whereas (−) or *l* indicates a rotation to the left (counter-clockwise). The main drawback of such an assignment is that one cannot derive the number of chirality centers from it. Rather, it is necessary to apply the R/S

notation, which describes the absolute configuration (the spatial arrangement of the substituents) around the asymmetric carbon atom. This assignment, based on the Cahn–Ingold–Prelog (CIP) convention [10], has mostly replaced the older *D*/*L* notation, which correlates the configuration of a molecule to the configuration of *D*/*L*-glyceraldehyde according to the Fischer convention. Today, the latter designation is restricted mainly to amino acids and carbohydrates [9]. The assignment of *R* or *S* according to CIP follows the sequence rule, that is, the order of priority of the substituents on the center of chirality. It can be determined on the basis of the decrease in the atomic number of the atoms directly bonded to the center of chirality. If two or more of these atoms are identical, the next bonded atoms are considered, eventually the third bonded atoms and so on. Thus the branch containing the atoms with the highest atomic numbers has the highest priority. Atoms connected by double or triple bonds are higher in weight than two or three singly bonded atoms [9]. Atomic isotopes can be arranged in order of priority by listing them in a series from highest to lowest mass number. It is very interesting to observe that in closely related structures the designation of the absolute configuration may change, whereas the spatial arrangement of the substituents is maintained. Other consequences of chirality are manifest in the metabolic processes. Several transformations, such as prochiral to chiral, chiral to chiral, chiral to diastereoisomer, chiral to nonchiral, and chiral inversion can occur [11,12].

1.4 CHIRALITY AND ITS CONSEQUENCES IN BIOLOGICAL SYSTEMS

In a nonchiral environment, the enantiomers of a racemate possess the same physical and chemical properties. But in the early 1930s, Easson and Stedman introduced a three-point attachment model that laid the basis for the initial understanding of stereochemical differences in pharmacological activity [13]. The authors described the differences in the bioaffinity of the enantiomers to a common site on an enzyme or receptor surface, with the receptor or enzyme needing to possess three nonequivalent binding sites to discriminate between the enantiomers. The enantiomer that interacts simultaneously with all three sites is called the eutomer (active enantiomer), whereas the other, which binds to fewer than three sites at the same time, is called the distomer (inactive enantiomer) [14].

Therefore, metabolic and regulatory processes mediated by biological systems are sensitive to stereochemistry, and different responses often can be observed upon comparing the activities of a pair of enantiomers. These differences can be expressed in distribution rates, in metabolism and excretion, in antagonistic actions relative to each other, or in toxicological properties. It is very interesting to note that the inactive enantiomer may be not only ballast but also antagonistic to the action of the eutomer, or sometimes even toxic. The relation of the potency of the active enantiomer to that of the less active one is called the eudismic ratio [15,16]. Hence today the main interest in the development of new drugs is directed toward finding the eutomer. To properly work with chiral drugs in drug development, in therapeutic drug monitoring, or in clinical and forensic toxicology, the optically active pure forms of the racemic compound must be identified. The different biological properties of the enantiomers of some drugs are given in Table 1.

1.5 CHIRALITY AND DRUG DEVELOPMENT

Although the optical isomers of a racemic drug can exhibit different pharmacological activities in living systems [12,15,17–23], bioactive synthetic compounds, which comprise most of the chiral drugs, are administered as racemates [24]. Obviously, the more chiral centers present in a drug molecule, the more complex the situation. To ensure the optimum therapeutic effect, it would seem to be convenient to administer the eutomer. However, applying a single enantiomer to humans does not necessarily prevent side effects or tissue/organ damage, since, for example, the formation of harmful metabolites, as well as chiral inversion or racemization, can occur in vivo. An example of chiral inversion without negative side effects is provided by ibuprofen, where the inactive (*R*)-(−)-isomer is converted by an enzymatic mechanism into the active (*S*)-(+) form. The (*R*)-isomer can be considered to be a prodrug of its (*S*)-enantiomer [12]. A negative example is thalidomide, which was introduced to the market in the late 1960s as a sedative, in the racemic form. Even when applied in the therapeutic and harmless (+)-form, in vivo interconversion into the harmful (−)-isomer was shown to be responsible for disastrous malformations of embryos when thalidomide was prescribed for women during pregnancy [8,25,26]. In the diuretic indacrinon, however, the presence of the distomer was useful, since it promoted the efficacy of the therapeutic eutomer by antagonizing one of its side effects [27]. In addition to creating a general awareness, the thalidomide tragedy resulted in stricter controls, and the guidelines for approving new drugs were reconsidered. To protect patients from unwanted and harmful enantiomers [12] and side effects, the possibility of a different action of the individual enantiomers with regard to pharmacology and toxicology had to be taken into account.

In 1993 Witte et al. [28] and in 1994 Rauws and Groen [29] reviewed the status of the regulation of chiral medicinal products on the pharmaceutical industries in the United States, Japan, and some European countries where designated authorities are responsible for the control and approval of newly developed drugs (both chiral and nonchiral). In the United States, new drug applications are submitted to the Food and Drug Administration (FDA), which requires appro-

TABLE 1 The Different Physiological Properties of the Enantiomers of Some Drugs

Drugs	(+)-Enantiomer	(−)-Enantiomer
Barbiturates	Excitation	Sedation
Dobutamine	β_1- and β_2-adrenoceptor agonist (vasodilatation)	α_1-Adrenoceptor agonist (positive inotropic/vasoconstriction)
Fluoxetine	Selective serotonin reuptake inhibitor	Minimal effect
Ketamine	Strong anesthetic	Weak anesthetic
Fenfluramine	Selective serotonin reuptake inhibitor	Norepinephrine/dopamine reuptake inhibitor (adverse side effect)
Levodopa	Antiparkinsonian	Agranulocytosis
Methadone	Minimal effect	Strong analgesic
Methamphetamine	Central nervous system stimulant	Peripheral vasodilator
Penicillamine	Antirheumatic (Wilson's disease)	Neurotoxic
Pentazocine	Anxiety	Analgesia, respiratory depression
Propoxyphene	Analgesia	Antitussive
β-Adrenergic antagonist (e.g., propranolol)	Suppress ventricular arrhythmia without β-adrenergic blockade	Active β-adrenergic blocker
Morphine	Minimal effect	Strong analgesic
Thyroxine	Inactive	Thyroxemic effect
Verapamil	Minimal effect	Negative dromotropic; negative inotropic and chronotropic effects
Warfarin	Weak anticoagulant	Anticoagulant
Thalidomide	Sedative	Sedative, teratogenic
Picenadol	μ-Receptor agonist (analgesic)	Weak μ-receptor antagonist
Tetramisole	Minimal effect	Anthelmintic
Nonsteroidal anti-inflammatory agents (NSAIAs)	Anti-inflammatory	Minimal effect

priate information on chemistry as well as manufacturing and control data, such as methods and specifications, results of stability tests, proper labeling, pharmacological activity, pharmacokinetic profile, toxicology, and impurity limits [30]. If the drug contains a racemate, the stereochemistry and all pharmacological and toxicological data on the individual isomers must be included. For single enantiomers, a pharmacology and toxicology evaluation must be conducted, relying on the existing knowledge of the racemate. However, the application papers should include the longest repeat dose and data on reproductive toxicities [30]. Additionally, the evaluation should include data regarding the conversion of the studied eutomer into the distomer.

The FDA now may approve single enantiomers of racemates as new drugs, which will offer new marketing opportunities for racemic drugs that are about to go off-patent [31]. In 1995 Stinson [32] predicted that the production and marketing of single enantiomers (i.e., chiral drugs already approved as racemates but being redeveloped as single enantiomers) might grow by more than 30% by the end of the year 2000.

In Europe, guidelines on the quality, safety, and efficacy of medicinal products for human use are also formulated [33]. The European guidelines generally are similar to the FDA guidelines, since they are based on the same scientific knowledge [29]. In Japan, no official statements on the regulation of chiral drugs have been issued [28,34]. However, the Japanese registration authorities recommend adherence to the guidelines of other countries.

1.6 CHIRAL DRUGS AND THEIR ECONOMIC STATUS

Many of the pharmaceutical industries have started to market the optically active pure enantiomeric form of some of their drugs [19–23]. The most important such drugs are anti-inflammatory agents, analgesics, antiviral drugs, anticancer drugs, cardiovascular medications, and other pharmaceuticals used for central nervous system, dermatological, gastrointestinal, and ophthalmic conditions, and respiratory diseases [35].

The debate over racemic compounds versus enantiomers resulted in a new marketing strategy, the so-called racemic switch. A racemic switch results in the development in single enantiomer form of a drug that was first approved as a racemate. Economic interests are very important in the development of the drugs, and since the separation methods (of racemic drugs) are less costly than the asymmetric synthesis methods, most pharmaceutical companies are preparing single enantiomers by separation methods. A literature search revealed that about $300 billion was spent worldwide on drugs in 1997 [35–37]. It has also been estimated that about 28% of this amount was accounted for by single enantiomers [35]. Interestingly, sales of enantiomeric pure drugs, which were 26% of the total in 1996, had increased to 28% by 1997 [35], and the trend was continuing owing to an increased population and the development of new biological assays of racemic drugs. Briefly, all the pharmaceutical companies are competing intensely

TABLE 2 Sales (U.S. $, Millions) of Enantiomeric Intermediates and Single Enantiomer Drugs

	Enantiomeric intermediates		Bulk enantiomeric drugs	
Drugs	1999	2000	1999	2000
Anti-inflammatory/analgesics	150	156	200	223
Antiviral	794	830	983	1180
Cancer	892	1073	1783	2146
Cardiovascular	1133	2281	1889	3802
Central nervous system	1038	1142	1483	1632
Dermatology	82	85	164	170
Gastrointestinal	251	331	413	567
Ophthalmic	238	284	340	405
Respiratory	516	656	1151	1511
Other	140	170	315	426
Total	5294	7008	8721	12,062

Source: Technology Catalysts International.

with respect to the costs and the marketing of new (optically active pure form) drugs and also are demanding economical methods for preparing pure enantiomeric drugs. The sales figures for some enantiomeric intermediates and bulk enantiomeric drugs in 1999 and 2000 are given in Table 2.

1.7 CHIRALITY AND ITS CONSEQUENCE IN THE ENVIRONMENT

Many xenobiotics and pollutants are chiral, and the two enantiomers of such pollutants may have different toxicities [1]. Besides, the degradation of some achiral pollutants results in chiral toxic metabolites. Moreover, the degradation of some chiral pollutants is stereospecific in the environment. Therefore, to predict the exact toxicity of a given pollutant, the concentrations of both enantiomers is required and essential. For example, the two enantiomers of α-hexachlorocyclohexane pesticide have different toxicities. Moreover, the rates of degradation of the enantiomers of α-hexachlorocyclohexane also are different. In 1991 Kallenborn et al. [38] reported the enantioselective metabolism of α-hexachlorocyclohexane in the organs of eider ducks, while in the same year Faller et al. [39] reported the enantioselective degradation, by marine bacteria, of α-hexachlorocyclohexane. Therefore, environmental chemists are also looking for the optimum technique to determine the chiral ratio of xenobiotics in the environment. Besides, in the field of archaeology, the measurement of racemization of specific amino acids is used to determine the age of mammalian fossils [40]. Another interesting application of enantiomer separation in our environment is the determination of amino acids in meteorites [41]. Because of the different biological activities present in pairs of enantiomers, diverse groups including regulators, executives in the material industries, clinicians, nutritional experts, agriculturalists, and environmentalists are now demanding data on the ratio of enantiomers rather than total concentration of a racemate (analytical scale). Moreover, the preparative separation and isolation of pure enantiomers is especially attractive for medicinal chemists, who can compare the biological activities of such compounds with those of racemates. Besides, the fate of the chiral xenobiotics is an important issue in environmental studies. Therefore, chiral separation (resolution) at the analytical and preparative scales of racemic compounds is essential and urgently needed in the modern pharmaceutical, agricultural, and food industries, and for environmental reasons as well. The resolution of enantiomers having identical physical and chemical properties is difficult, but various methods have been developed and used for the chiral separation of different racemic compounds, drugs, and xenobiotics.

1.8 RESOLUTION OF STEREOISOMERS

The separation of a racemic compound into its enantiomers is called resolution. Various methodologies have been used for the resolution of the enantiomers on both analytical and preparative scales. The different techniques may be categorized into two classes: the classical approach, using enzymatic degradation of one of the enantiomers, and preferential crystallization. Modern technologies include spectroscopic, electrophoretic, and chromatographic methods.

1.8.1 Classical Methods

In the enzymatic method, resolution is effected by means of a biochemical process that destroys one enantiomeric form. When certain microorganisms such as yeast, molds, and bacteria are allowed to grow in a solution of racemic mixtures, they assimilate one form selectively, leaving behind the other one in solution. For example, if the ordinary mold *Penicillium glaucum* is added to a racemic solution of ammonium tartrate, the solution becomes levorotatory owing to the destruction of the dextrorotatary form [5]. The principle of crystallization is based on the formation by two enantiomers of the diastereomeric salts of the optically pure compound, and these diastereoisomeric salts can be separated easily [2,30]. In this process, the optically active resolving agent must be of high

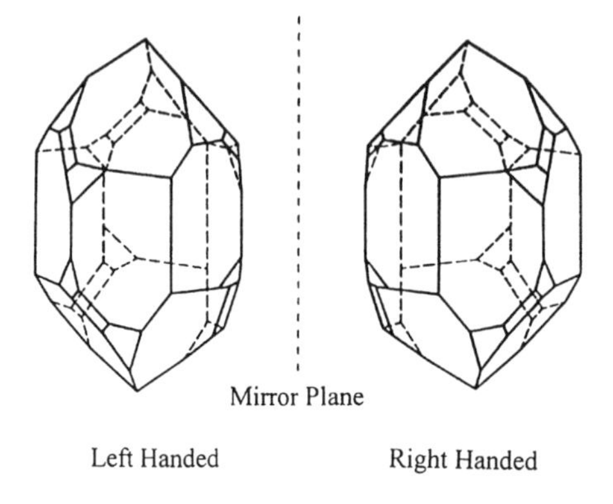

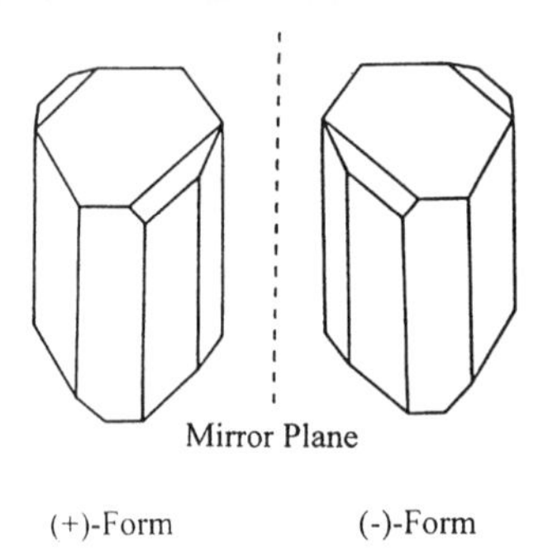

FIGURE 2 The stereostructures of quartz and sodium ammonium tartrate crystals.

optical purity. In most cases, after the desired enantiomers have been separated from the diastereomeric salts, the resolving agent is recovered and becomes available for reuse [30–33]. Mechanical methods of separation (by needle, etc.) also can be utilized for the separation of crystals of some racemic compounds, such as sodium ammonium tartrate and quartz, since their crystals are mirror images of each other (Fig. 2). These classical methods never achieved the status of routine laboratory practice owing to certain drawbacks, such as the degradation of one of the enantiomers in the enzymatic method, and the very limited number of applications of the crystallization method. Moreover, these methods cannot be used at the analytical scale, especially in natural samples.

1.8.2 Modern Methods

Nowadays, the chromatographic, electrophoretic, spectroscopic, biosensor, and membrane methods are the ones most commonly applied (Fig. 3) [19–23,42,43]. Among the spectroscopic methods, investigators use optical rotation measurements, nuclear magnetic resonance (NMR) spectroscopy, and infrared (IR) spectroscopy. The latter, as differential scanning calorimetry (DSC), can

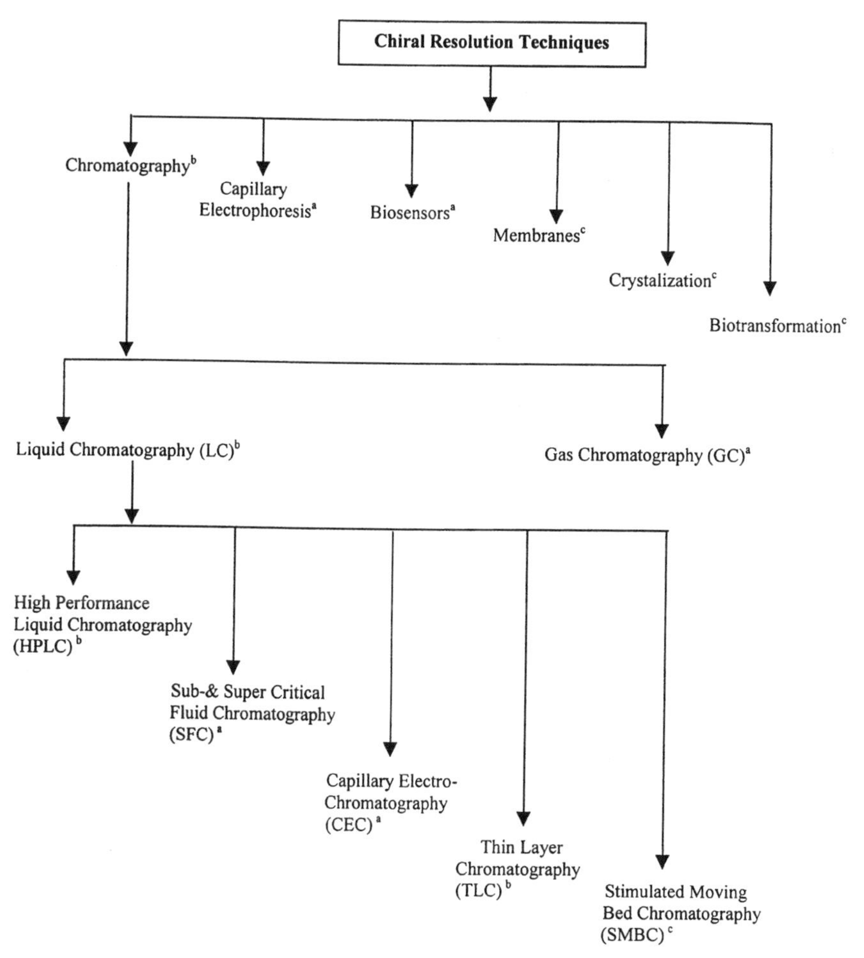

FIGURE 3 The different techniques of chiral resolution: *a*, analytical; *b*, analytical and preparative; *c*, preparative.

distinguish only racemic mixtures (±) and individual enantiomers. In NMR spectroscopy, chiral solvating agents (CSA) can be utilized to promote a change in the chemical shift of the chiral (carbon) atoms. However, the spectroscopic and DSC methods are sensitive to interference from chiral or nonchiral impurities in the sample. Additionally, they can lack sensitivity and accuracy, since the differences between the isomers and/or isomer mixtures may be very small. Moreover, these methods fail totally at the preparative scale. Chiral membranes were used to separate the enantiomers at the preparative scale in 2001 [42], but the technique is not fully developed.

Among the electrophoretic methods of chiral resolution, various forms of capillary electrophoresis such as capillary zone electrophoresis (CZE), capillary isotachophoresis (CIF), capillary gel electrophoresis (CGE), capillary isoelectric focusing (CIEF), affinity capillary electrophoresis (ACE), and separation on microchips have been used. However, in contrast to others, the CZE model has been used frequently for this purpose [44]. On the other hand, drawbacks associated with the electrophoretic technique due to lack of development of modern chiral phases have limited the application of these methods. Moreover, the electrophoretic techniques cannot be used at the preparative scale, which represents an urgent need of chiral separation science.

1.8.2.1 Chromatographic Methods

In chromatographic methods, two approaches are used: indirect and direct. The indirect chromatographic separation of racemic mixtures into their enantiomers can be achieved by derivatization of the racemic compound with a chiral derivatizing agent (CDA) resulting in the formation of diastereoisomer–salt complex. Having different physical and chemical properties, the diastereoisomers can be separated from each other by means of an achiral chromatographic method. A condition for successful derivatization is the presence of suitable functional groups in the analyte. Besides, to increase the physicochemical differentiation, derivatization should occur close to the chiral atom. Although the indirect chromatographic approach has the advantage of predetermining elution order, which can be important for the determination of optical purities, there are some limitations to this technique. The derivatization procedure is tedious and time-consuming owing to the different reaction rates of the individual enantiomers. Additionally, suitable chiral derivatizing agents in pure form can be difficult to obtain. For preparative purposes, the indirect chromatographic approach includes an additional synthetic step because after the separated diastereoisomers have been resolved in a nonchiral environment, the derivatizing agent must be cleaved off.

On the other hand, the direct chromatographic approach involves the use of the chiral selector either in the mobile phase, a so-called chiral mobile phase additive (CMPA), or in the stationary phase [i.e., the chiral stationary phase (CSP)]. In the latter case, the chiral selector is chemically bonded or coated or allowed to absorb onto a suitable solid support. Of course chiral selectors still can be used as CMPAs, but the approach is a very expensive one owing to the high amount of chiral selector required for the preparation of the mobile phase, and the large amount of costly chiral selector that is wasted (since there is very little chance of recovering this compound). Moreover, this approach is not successful in the preparative separation of the enantiomers.

In contrast, CSPs have achieved great repute in the chiral separation of enantiomers by chromatography and, today, are the tools of the choice of almost all analytical, biochemical, pharmaceutical, and pharmacological institutions and industries. The most important and useful CSPs are available in the form of open and tubular columns. However, some chiral capillaries and thin layer plates are also available for use in capillary electrophoresis and thin-layer chromatography. The chiral columns and capillaries are packed with several chiral selectors such as polysaccharides, cyclodextrins, antibiotics, Pirkle type, ligand exchangers, and crown ethers.

The chromatographic methods use gas or liquid separately as the mobile phase, hence the terms gas chromatography (GC) and liquid chromatography (LC). Gas chromatography could not be accepted as the method of choice for the chiral resolution of racemic compounds mainly because of its requirement for the conversion by derivatization of the racemic compound into a volatile species. Besides, the separated enantiomers cannot be collected for further pharmacological and other studies. Moreover, GC cannot be used at the preparative scale.

On the other hand, LC is the only and the best remaining technology for the chiral resolution of a wide variety of racemates. The main advantage of LC is its ability to determine enantiomers in both biological and environmental samples. Besides, the ease of applying LC techniques at the preparative scale contributed much to the elucidation of the pharmaceutical and toxicological properties of enantiomers in living systems. Over time, various liquid chromatographic approaches were developed and used in this work. The most important liquid chromatographic methods are high performance liquid chromatography (HPLC), sub- and supercritical fluid chromatography (SFC), capillary electrochromatography (CEC), and thin-layer chromatography (TLC).

High Performance Liquid Chromatography. Among the various liquid chromatographic techniques, HPLC remains the best modality owing to several advantages. High speed, sensitivity, and reproducible results make HPLC the method of choice in almost all laboratories. Moreover, HPLC can be used successfully at the preparative scale. There is no serious limitation of this technique. About 90% of the work in chiral resolution has been achieved by means of the HPLC mode of chromatography. Almost all the chiral selectors are available in the form of HPLC columns, owing to the wide range of applications of HPLC in chiral resolution [19–23]. A variety of mobile phases including normal, reversed, and new polar organic phases are used in HPLC. The composition of the mobile phases may be changed by the addition of various aqueous and nonaqueous solvents. A number of parameters are used in the optimization of the chiral resolution. Chapters 2 to 10 discuss in detail the methodology of chiral resolution by HPLC using different CSPs and CMPAs.

Sub- and Supercritical Fluid Chromatography. The use of supercritical fluids as the mobile phases for chromatographic separation was first reported more than 30 years ago, but most of the growth in SFC has occurred recently. A supercritical fluid exists when both temperature and pressure in the system exceed the critical values: critical temperature (T_c) and critical pressure (P_c). Critical fluids have physical properties between those of liquids and gases. Like gases, supercritical fluids are highly compressible, and properties of the fluid, including density and viscosity, can be maintained by changes in the pressure and temperature conditions. In chromatographic systems, the solute diffusion coefficients are often of a higher order of magnitude in supercritical fluids than in traditional liquids. The viscosities, however, are lower than those of liquids [45]. At temperatures below T_c and pressures above P_c, the fluid becomes a liquid. On the other hand, at temperatures above T_c and pressures below P_c, the fluid behaves as a gas. Therefore, supercritical fluids can be used as part of a mixture of liquid and gas [46]. The commonly used supercritical fluids (SFs) are carbon dioxide, nitrous oxide, and trifluoromethane [45–47]. The SF of choice is carbon dioxide, owing to its compatibility with most detectors, low critical temperature and pressure, low toxicity and environmental burden, and low costs. The main drawback of supercritical carbon dioxide as the mobile phase lies in its inability to elute more polar compounds. To improve the elution ability of CO_2, one can add organic modifiers to the relatively apolar carbon dioxide. Chiral sub- and supercritical FC have been carried out in packed and open tubular columns and capillaries [48]. The first report on chiral resolution by SFC was published in 1985 by Mourier et al. [49]. Later, several other papers and a review appeared [50–55].

Capillary Electrochromatography. Capillary electrochromatography (CEC) is a hybrid technique that works on the basic principles of capillary electrophoresis and chromatography [41]. This mode of chromatography is used on either packed or tubular capillaries/columns. The packed column approach was introduced by Pretorius et al. [60] in 1974, while open tubular CEC was presented by Tsuda et al. [61] a decade later. In 1984 Terabe et al. [62] introduced another modification in liquid chromatography, micellar electrokinetic capillary chromatography (MECC). Of course this mode also depends on the working principles of capillary electrophoresis and chromatography but adds the formation of micelles. The recent use of CEC and MECC in the chiral resolution of racemic compounds resulted in several publications [63–65]. High speed, sensitivity, lower limit of detection, and reproducible results make CEC and MECC ideal for chiral resolution. However, these methods are not used frequently because the techniques are not fully developed. Research is under way, however.

Thin-Layer Chromatography. The history of thin-layer chromatography (TLC) is very old, but its use in chiral resolution goes back only about 25 years. Most TLC enantioseparations are carried out in the indirect mode (i.e., by preparing diastereoisomers and resolving them on TLC). The derivatization of racemic mixtures and subsequently separation on silica gel or reversed-phase (RP) TLC plates represents a method of chiral resolution. There are only a few reports on direct enantiomeric separation on chiral TLC plates (i.e., using CMPAs or CSPs). Among the direct approaches, the use of CSPs is very limited. Only chiral TLC plates based on ligand exchange are commercially available for the chiral resolution of racemates. However, some reports mention the impregnation of TLC plates with suitable chiral selectors [66,67]. Several research papers and review articles on chiral resolution by TLC can be found in the literature [66–72].

Despite the development of more efficient and rapid chromatographic techniques, TLC remains an important method of chiral resolution. Low running cost, simplicity, ease of handling, and coelution of the racemate and pure enantiomers are the main advantages of TLC. The reliable, robust, and efficient methods for the resolution of enantiomers by thin-layer chromatography comprise useful additions to the techniques of chiral resolution. Moreover, TLC is the technique of choice in developing and underdeveloped countries. Its major weaknesses include relatively low resolving power, with a high limit of detection, in comparison to other liquid chromatographic approaches. The waste of the chiral selector or chiral thin-layer material is another serious drawback.

One of the classical approaches of liquid chromatography, paper chromatography, was used for chiral resolution about 50 years ago but is not part of modern practice. In paper chromatography, the stationary phase is water bonded to cellulose (paper material), which is of course chiral and hence provides a chiral surface for the enantiomers. However, some workers used chiral mobile phase additives also in paper chromatography [73,74]. In 1951 some research groups independently [73,75–77] resolved the enantiomers of amino acids. Simultaneously, numerous interesting publications on chiral resolution by paper chromatography appeared [70].

A more recently developed technique, simulated moving-bed chromatography, has been very successful for chiral separations at the preparative scale. This technique affords the separation of large amounts of enantiomers by means of developing suitable chiral selectors and mobile phases [42].

1.9 DETECTION IN LIQUID CHROMATOGRAPHY

In most of chiral resolutions by liquid chromatography, a UV detector is used, because most racemic drugs and pharmaceuticals are UV sensitive. However, conductivity, fluorescence, refractive index, and other detectors have also been used. The choice of detector depends on the properties of the racemic compound to be resolved [23,78]. Besides, chiroptical detectors, based on the principle of polarimetry [79,80] or circular dichroism [81,82], are available. In addition, liquid chromatography–mass spectrometric instruments are now being marketed for routine use [83,84]. The polarity notation of enantiomers [i.e., (+)- or (–)-] is determined by optical detector. Detection on TLC is carried out by a number of approaches. The UV-sensitive enantiomers are observed in a UV cabinet, while other non-UV-absorbing chiral molecules such as amino acids and amines are located by developing the color on the TLC plate with a suitable reagent (e.g., ninhydrin, which is used for developing color with amino acids and amines) [85]. Moreover, iodine vapors are used as the universal detection method on TLC plates. The separated enantiomers adsorb iodine vapors and turn yellow. The iodine from the spots can be removed by heating the TLC plate, and hence the pure enantiomers from the plate can be recovered and used for further pharmacological studies.

1.10 CHIRAL RECOGNITION MECHANISMS

The chiral recognition mechanisms on all the chiral selectors are quite similar except on ligand exchangers. All chiral selectors provide a chiral surface to the enantiomers, which form with the selectors temporary complexes, having different bonding energies. The enantiomers differ in their binding energies because they fit differently into the chiral selector structures. Again, the different stereoconfigurations of the enantiomers are responsible for the differences in fit on the chiral selectors. These temporary complexes are stabilized by a number of interactions. The most important interactions are hydrogen, π–π, dipole–induced dipole, ionic, inclusion complexations, and steric interactions. However, the other weaker forces such as van der Waals and charge transfer, also play a crucial role in chiral recognition mechanisms. The chiral recognition mechanism on chiral ligand exchange selectors is different from that on the other selectors. This is mainly due to the exchange of ligands through coordinate bonds on a metal ion, which is not possible on other chiral selectors. Briefly, in general, the chiral

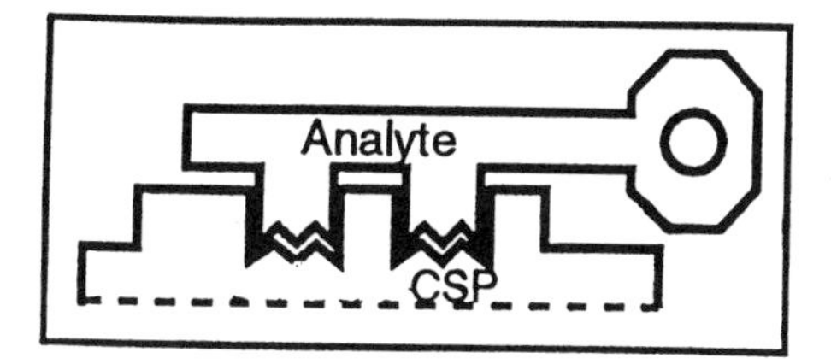

FIGURE 4 Schematic representation of the chiral recognition mechanism.

recognition mechanism on a chiral selector is based on a key-and-lock arrangement (Fig. 4). Details of the chiral recognition mechanism on each type of chiral selector are given in the chapter covering that particular type.

In view of the importance of chiral resolution and the efficiency of liquid chromatographic methods, attempts are made to explain the art of chiral resolution by means of liquid chromatography. This book consists of an introduction followed by Chapters 2 to 8, which discuss resolution chiral stationary phases based on polysaccharides, cyclodextrins, macrocyclic glycopeptide antibiotics, Pirkle types, proteins, ligand exchangers, and crown ethers. The applications of other miscellaneous types of CSP are covered in Chapter 9. However, the use of chiral mobile phase additives in the separation of enantiomers is discussed in Chapter 10.

REFERENCES

1. Kelvin L in Kallenborn R, Huhnerfuss H, Chiral Environmental Pollutants: Trace Analysis and Ecotoxicology, Springer-Verlag, Berlin (2000).
2. Hegstrom R, Kondepudi DK, Sci Am 262: 108 (1990).
3. Kondepudi D, Durand DJ, Chirality 13: 351 (2001).
4. Haüy RJ, Tableaux Comparatif des Resultats de la Crystallographie et de l'Analyse Chimique Relativement à la Classification des Mineraux, Paris, p. XVII (1809).
5. Pasteur L, C R Acad Sci 26: 535 (1848).
6. Le Bel JA, Bull Soc Chim Fr 22: 337 (1874).
7. Van't Hoff JH, Arch Neerland Sci Exactes Nat 9: 445 (1874).
8. Aitken RA, Parker D, Taylor RJ, Gopal J, Kilenyi RN, in Asymmetric Synthesis, Blackie Academic & Professional, New York (1992).
9. Hauptmann S, in Organische Chemie, Hauptmann S (Ed.), VEB Deutscher Verlag für Grundstoffindustrie (1988).
10. Cahn RS, Ingold CK, Prelog V, Experientia 12: 81 (1956).
11. Caldwell J, J Chromatogr A 694: 48 (1995).
12. Testa B, Trends Pharmacol Sci 60 (1986).
13. Easson EH, Stedman E, Biochem J 27: 1257 (1933).
14. Lehmann PA, Trends Pharmacol Sci 281 (1986).

15. Ariëns EJ, Trends Pharmacol Sci 200 (1986).
16. Waldeck B, Chirality 5: 355 (1993).
17. Simonyi M, Fitos I, Visy J, Trends Pharmacol Sci 112 (1986).
18. Cannon JG, Moe ST, Long JP, Chirality 3: 19 (1991).
19. Krstulovic AM, Chiral Separations by HPLC: Applications to Pharmaceutical Compounds, Ellis Horwood, New York (1989).
20. Allenmark S, Chromatographic Enantioseparation: Methods and Applications, 2nd ed., Ellis Horwood, New York (1991).
21. Subramanian G (Ed.), A Practical Approach to Chiral Separations by Liquid Chromatography, VCH Verlag, Weinheim, Germany (1994).
22. Aboul-Enein HY, Wainer IW (Eds.), The Impact of Stereochemistry on Drugs Development and Use, Vol. 142, John Wiley & Sons, New York (1997).
23. Beesley TE, Scott RPW, “Chiral Chromatography,” John Wiley & Sons, New York (1998).
24. Stinson SC, Chem Eng News 72: 38 (1994).
25. Blaschke G, Kraft HP, Fickentscher K, Köhler F, Drug Res 29: 1640 (1979).
26. Knoche B, Blaschke G, J Chromatogr A 666: 235 (1994).
27. Ariëns EJ, Science 259: 68 (1993).
28. Witte DT, Ensing K, Franke JP, de Zeeuw RA, World Sci 15: 10 (1993).
29. Rauws AG, Groen K, Chirality 6: 72 (1994).
30. FDA’s policy statement for the development of new stereoisomeric drugs. Chirality 4: 338 (1992).
31. Stinson SC, Chem Eng News 75: 28 (1997).
32. Stinson SC, Chem Eng News 73: 44 (1995).
33. The Rules Governing Medicinal Products in the European Union. Guidelines on the Quality, Safety and Efficacy of Medicinal Products for Human Use. Clinical Investigation of Chiral Active Substances, Pharmaceutical Press, London (1991).
34. Shindo H , Caldwell J, Chirality 3: 91 (1991).
35. Stinson SC, Chem Eng News 76: 83 (1998).
36. Stinson SC, Chem Eng News 77: 101 (1999).
37. Thayer AM, Chem Eng News 76: 25 (1998).
38. Kallenborn R, Hühnerfuss H, Köning WA, Angew Chem 103: 328 (1991).
39. Faller J, Hühnerfuss H, Köning WA, Krebber R, Ludwig P, Environ Sci Technol 25: 676 (1991).
40. Meyer VR, ACS Symp Ser 471: 217 (1991).
41. Cronin JR, Pizarello S, Science 275: 95 (1997).
42. Maier NM, Franco P, Lindner W, J Chromatogr A 906: 3 (2001).
43. Stefan RL, van Staden JF, Aboul-Enein HY, Electrochemical Sensors in Bioanalysis, Marcel Dekker, New York (2001).
44. Chankvetadze B, Capillary Electrophoresis in Chiral Analysis, John Wiley & Sons, New York, p. 353 (1997).
45. Weast RC (Ed.), Handbook of Chemistry and Physics, 54th ed., CRC Press, Cleveland (1973).
46. Berger TA, in Packed Column SFC, Smith RM (Ed.), The Royal Society of Chemistry, Chromatography Monographs, Cambridge (1995).

47. Schoenmakers PJ, in Packed Column SFC, Smith RM (Ed.), The Royal Society of Chemistry, Chromatography Monographs, Cambridge (1988).
48. Terfloth G, J Chromatogr A 906: 301 (2001).
49. Mourier PA, Eliot E, Caude RH, Rosset RH, Tambute AG, Anal Chem 57: 2819 (1985).
50. Macaudiere P, Caude M, Rosset R, Tambute A, J Chromatogr 405: 135 (1987).
51. Lee CR, Porziemsky JP, Aubert MC, Krstulovic AM, J Chromatogr 539: 55 (1991).
52. Biermanns P, Miller C, Lyon V, Wilson W, LC-GC 11: 744 (1993).
53. Pettersson P, Markides KE, J Chromatogr A 666: 381 (1994).
54. Bargmann-Leyder N, Tambute A, Caude M, Chirality 7: 311 (1995).
55. Williams KL, Sander LC, Wise SA, J Chromatogr A 746: 91 (1996).
56. Overbeke AV, Sandra P, Medvedovici A, Baeyens W, Aboul-Enein HY, Chirality 9: 126 (1997).
57. Phinney KW, Sanders LC, Wise SA, Anal Chem 70: 2331 (1998).
58. Duval R, Lévêque H, Prigent Y, Aboul-Enein HY, Biomed Chromatogr 15: 202 (2001).
59. Phinney KW, Sub- and supercritical fluid chromatography for enantiomer separations, in Chiral Separation Techniques: A Practical Approach, (Subramanian G, Ed.), p. 299, VCH Verlag, Weinheim, Germany (2001).
60. Pretorius V, Hopkins BJ, Schieke JD, J Chromatogr 99: 23 (1974).
61. Tsuda T, Nomura K, Nagakawa G, J Chromatogr 248: 241 (1982).
62. Terabe S, Otsuka K, Ichikawa K, Tsuchiya A, Ando T, Anal Chem 56: 111 (1984).
63. Schuring V, Wistuba D, Electrophoresis 20: 2313 (1999).
64. Fanali S, Catarcini P, Blaschke G, Chankvetadze, Electrophoresis 22: 3131 (2001).
65. Cohen AS, Paulus A, Karger BL, Chromatographia 24: 15 (1987).
66. Bushan R, Ali I, J Chromatogr 463 (1987).
67. Bushan R, Ali I, Chromatographia, 5, 679 (1993).
68. Günther K, Martens J, Schickedanz M, Angew Chem 96: 514 (1984).
69. Grinberg N, J Chromatogr 333: 69 (1985).
70. Günther K, Enantiomer separations, in Handbook of Thin Layer Chromatography (Sherma J, Fried B, Eds.), Marcel Dekker, New York, p. 541 (1991).
71. Aboul-Enein HY, El-Awady MI, Heard CM, Nicholls PJ, Biomed Chromatogr 13: 531 (1999).
72. Bhushan R, Martens J, Biomed Chromatogr 15: 155 (2001).
73. Bonino GB, Carassiti V, Nature 167: 569 (1951).
74. Mason M, Berg CP, J Biol Chem 195: 515 (1952).
75. Kotake M, Sakan T, Nakamura N, Senoh S, J Am Chem Soc 73: 2973 (1951).
76. Berlingozzi S, Serchi G, Adembri G, Sper Sez Chim Biol 2: 89 (1951).
77. Fujisawa Y, J Osaka City Med Center 1: 7 (1951).
78. Scott, RP, Chromatography Detectors, Chromatography Science Series, Vol 73, Marcel Dekker, New York (1996).
79. Mannschreck A, Chirality 4: 163 (1992).
80. Lloyd DK, Goodall DM, Scrivener H, Anal Chem 61: 1238 (1989).
81. Salvadori P, Bertucci C, Rosini C, Chirality 3: 376 (1991).

82. Brandl G, Kastner F, Mannschreck A, Nölting B, Andert K, Wetzel R, J Chromatogr 586: 249 (1991).
83. Bugge CJL, Crun I, Ljungqvist A, Vatankhan M, Garci DB, Warren HB, Gupta S, American Association of Pharmaceutical Scientists Annual Meeting, Seattle, WA (1996).
84. Ramos L, Bakhtiar R, Majumdar T, Hayes M, Tse F, Rapid Commun Mass Spectrom 13: 2054 (1999).
85. Stahl E, Thin Layer Chromatography, 2nd ed., Springer-Verlag, Berlin (1969).

2

Polysaccharide-Based Chiral Stationary Phases

Most of the naturally occurring polymers including polysaccharides are chiral and optically active because of their asymmetrical structures. These polymers often possess a specific conformation or higher order structure arising from the chirality that is essential for the chiral resolution of racemic compounds [1]. Therefore, polysaccharides are potentially applicable in the chiral separation of racemic compounds by liquid chromatography [2,3]. Resolution of racemates by using polysaccharides has been reported since 1951, when Kotake et al. [4], for the first time, resolved amino acids on paper chromatograms. Later on, several publications emerged on chiral resolution by means of paper chromatography and cellulose thin-layer chromatography [5]. Polysaccharide polymers such as cellulose, amylose, chitosan, xylan, curdlan, dextran, and inulin have been used for chiral resolution in liquid chromatography [6]. However, these derivatives could not be used as commercial chiral stationary phases (CSPs) because their resolution capacity was poor and there were handling problems [1]. Therefore, derivatives of these polymers have been synthesized in the last two decades [1]. Among the various polymers of polysaccharides, cellulose and amylose are the most readily available naturally occurring polymers and are suitable for chiral resolution; therefore, most of the reported chiral applications involving liquid chromatography have used these two polysaccharides [1,6]. The most useful and successful derivatives of cellulose and amylose are triesters and tricarbamate [1,6,7]. Figure 1 shows their three-dimensional structures, including the chiral grooves, of cellulose and amylose. Okamoto et al. [8] prepared several triesters and tricarbamates of cellulose and amylose derivatives and coated or bonded them onto macroporous silica gel. These derivatives represent the most universally applicable kind of chiral stationary phase. Furthermore, a search of the literature [1,6–11] revealed that polysaccharide-based chiral stationary phases in liquid chromatography are very effective and have a wide range of applications. About 80% of the racemic compounds can be resolved on these CSPs. Therefore, this chapter describes the structure, properties, preparation, applications, and mechanisms of polysaccharide-based CSPs. Attempts have also been made to describe the optimization of chiral resolution on these CSPs. Furthermore, the use of polysaccharide chiral phases in sub- and supercritical fluid chromatography, capillary electrochromatography, and thin-layer chromatography have also been discussed.

2.1 POLYSACCHARIDE DERIVATIVES

Cellulose polymer is made of glucose units and polymeric chains of *D*-(+) glucose units containing the β-1,4 linkage. The glucose units have a chair conformation with 2-OH, 3-OH, and the 5-CH_2OH groups all in equatorial positions (Fig. 2). These chains lie side by side in a linear fashion in cellulose (Fig. 1). The degree of polymerization of cellulose is not known exactly, but it

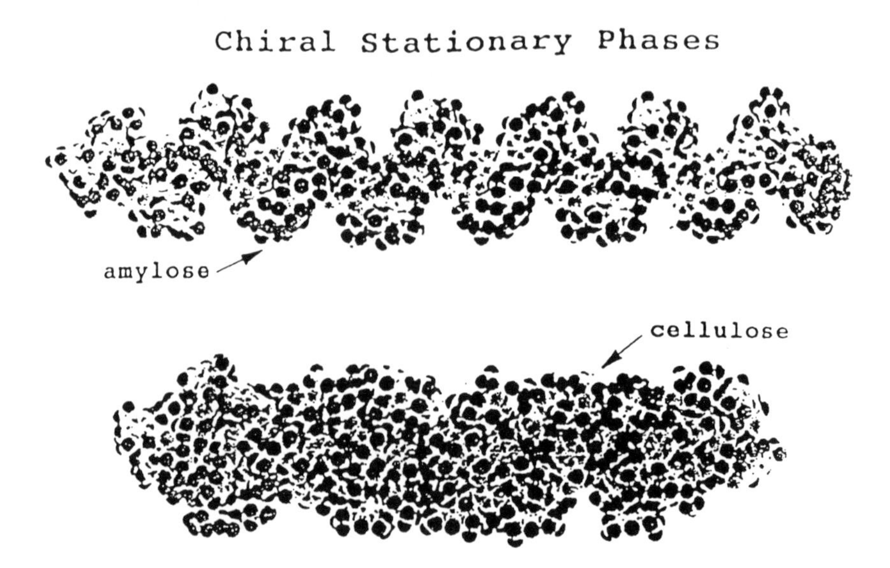

FIGURE 1 Three-dimensional structures of amylose and cellulose polymers.

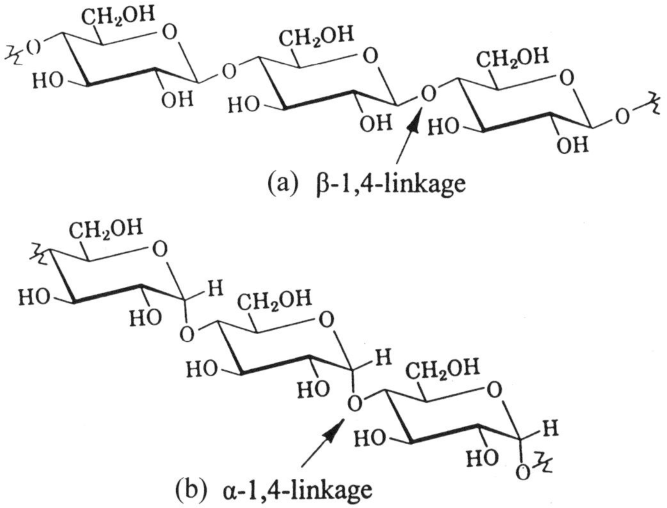

FIGURE 2 The chemical structures of (a) cellulose and (b) amylose polymers.

ranges from 200 to 14,000 units of glucose. Avicel, the most commonly used type of microcrystalline cellulose, has a degree of polymerization of about 200 to 300 glucose units [12,13]. Cellulose occurs in different forms, but the two most commonly used are native cellulose and the material obtained after reprecipitation. Amylose is made of more than 1000 glucose units, and the polymeric chain of *D*-(+)-glucose units contains the α-1,4 linkage. In amylose also, the glucose units have a chair conformation with 2-OH, 3-OH, and the 5-CH_2OH groups all in equatorial positions (Fig. 2). To increase their chiral recognition capacity, these two polysaccharides were converted into their triester and tricarbamate derivatives. Among the prepared derivatives are triacetate, tribenzoate, and triphenylcarbamate. The most commonly used derivative, and also the best choice, is triphenylcarbamate. The structures of these derivatives are discussed next.

2.1.1 Triester and Tricarbamate Derivatives of Cellulose

The properties of partially acetylated cellulose for the separation of enantiomers were recognized in 1966 by Luttringhaus and Peters [14]. But the full potential of cellulose acetate was developed by Hesse and Hagel in 1973 [15]. Since then, this stationary phase has been frequently used for the resolution of various racemic compounds. Native cellulose acetylated heterogeneously yields a crystallographic form of cellulose triacetate (CTA) known as CTA-I, which corresponds to its source, an indication that the original supramolecular structure of the starting material is preserved. However, if the process is performed in such a way that the triacetate dissolves during acetylation, the product obtained after reprecipitation is defined as CTA-II. This thermodynamically more stable form exists only as a pure polymer and has properties different from CTA-I. However, CTA-II suffers from poor chiral resolving power in comparison to CTA-I. Figure 3 shows the structures of different cellulose tribenzoate derivatives prepared by Okamoto et al. [16,17].

The effect of substituents on the phenyl groups of cellulose tribenzoate has been studied [17]. Cellulose triphenylcarbamate derivatives are the most thoroughly investigated polysaccharide phases with respect to chiral resolution and its

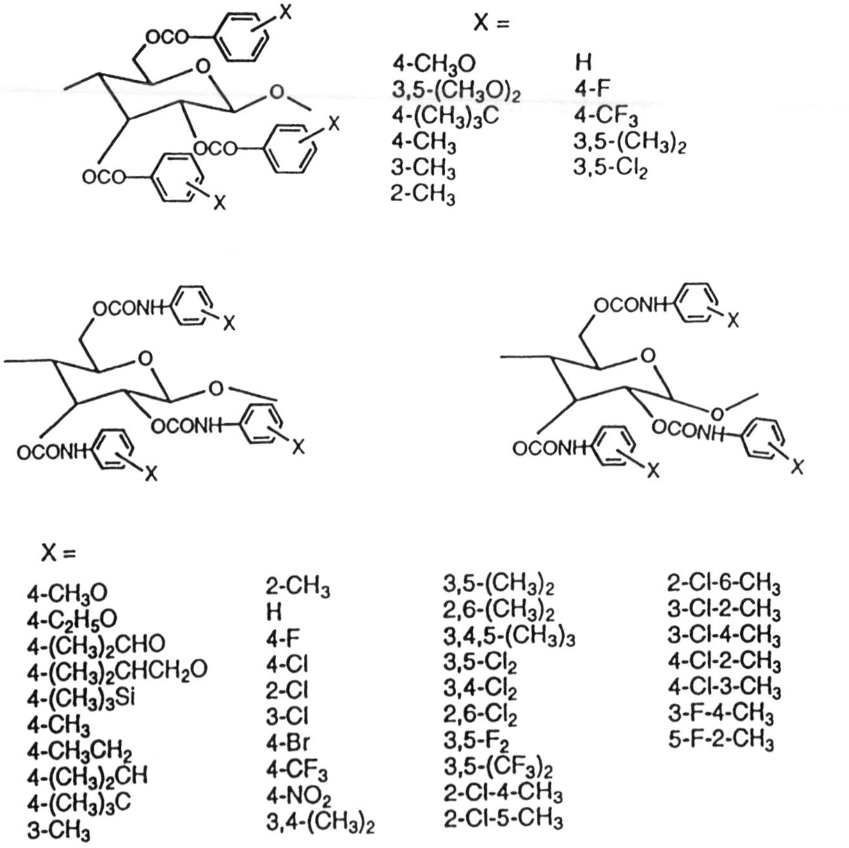

X =

$4\text{-}CH_3O$, $3,5\text{-}(CH_3O)_2$, $4\text{-}(CH_3)_3C$, $4\text{-}CH_3$, $3\text{-}CH_3$, $2\text{-}CH_3$, H, $4\text{-}F$, $4\text{-}CF_3$, $3,5\text{-}(CH_3)_2$, $3,5\text{-}Cl_2$

X =

$4\text{-}CH_3O$, $4\text{-}C_2H_5O$, $4\text{-}(CH_3)_2CHO$, $4\text{-}(CH_3)_2CHCH_2O$, $4\text{-}(CH_3)_3Si$, $4\text{-}CH_3$, $4\text{-}CH_3CH_2$, $4\text{-}(CH_3)_2CH$, $4\text{-}(CH_3)_3C$, $3\text{-}CH_3$, $2\text{-}CH_3$, H, $4\text{-}F$, $4\text{-}Cl$, $2\text{-}Cl$, $3\text{-}Cl$, $4\text{-}Br$, $4\text{-}CF_3$, $4\text{-}NO_2$, $3,4\text{-}(CH_3)_2$, $3,5\text{-}(CH_3)_2$, $2,6\text{-}(CH_3)_2$, $3,4,5\text{-}(CH_3)_3$, $3,5\text{-}Cl_2$, $3,4\text{-}Cl_2$, $2,6\text{-}Cl_2$, $3,5\text{-}F_2$, $3,5\text{-}(CF_3)_2$, $2\text{-}Cl\text{-}4\text{-}CH_3$, $2\text{-}Cl\text{-}5\text{-}CH_3$, $2\text{-}Cl\text{-}6\text{-}CH_3$, $3\text{-}Cl\text{-}2\text{-}CH_3$, $3\text{-}Cl\text{-}4\text{-}CH_3$, $4\text{-}Cl\text{-}2\text{-}CH_3$, $4\text{-}Cl\text{-}3\text{-}CH_3$, $3\text{-}F\text{-}4\text{-}CH_3$, $5\text{-}F\text{-}2\text{-}CH_3$

FIGURE 3 The chemical structures of cellulose benzoate, amylose benzoate, and carbamate derivatives. (From Ref. 32.)

mechanism. The chiral recognition power of these derivatives can be increased by adding different substituents to the phenyl rings of the carbamate derivatives [18–20].

2.1.2 Triester and Tricarbamate Derivatives of Amylose

The triester and triphenylcarbamate derivatives of amylose were prepared as in the case of cellulose, but the triester derivatives of amylose are not useful for chiral resolution. Therefore, the triester derivatives of amylose have not been studied extensively. The chemical structures of the triphenylcarbamate derivatives are shown in Fig. 3. Similarly, the chiral resolution power of amylose derivatives was studied by introducing electron-donating and electron-withdrawing groups on the phenyl ring of the amylose derivatives [21–24].

2.2 SPECTROSCOPIC STUDIES

Attempts have been made by various workers to explain the structures of these CSPs by different spectroscopic methods. The exact structures of these phases are not known owing to problems in NMR studies. Although NMR spectroscopy is powerful and comprises the main technique for revealing the structures of polysaccharide-based CSPs, these phases are soluble in solvents such as tetrahydrofuran, acetone, and pyridine. However, the presence of many interacting sites on microcrystalline celluse triacetate (MCTA) derivatives can be determined by ^{13}C NMR [25]. Further, the supramolecular structure for MCTA, having multiple interaction sites with specific surface and grooves, was determined by NMR techniques [26]. Oguni et al. [27] investigated the structure of tris(4-methylbenzoate) by ^{13}C NMR spectroscopy. Later Yashima et al. [28] were able to use NMR techniques to study the structure of cellulose tris(4-trimethylsilylphenyl) carbamate. The three-dimensional structures of cellulose- and amylase-based CSPs were determined and compared by using computational chemistry [29,30]. Vogt and Zugenmaier [31] reported that the possible structures were a 3/2 helical chain conformation for cellulose tris(phenylcarbamate) and a 4/1 helical chain conformation for amylose tris(phenylcarbamate). The amylose CSP is more helical and has well-defined grooves, making it considerably different from the corresponding cellulose analogue, which appears to be more linear and rigid.

The binding of 1,1′-bi-2-naphthol racemate with tris(5-fluoro-2-methylphenylcarbamate) was studied by mass spectroscopy and discussed by Yashima

[32]. The *S*-enantiomer bound more strongly than the *R*-enantiomer [33,34]. The relative peak intensity (RPI) of a complex of a chiral host with deuterated and nondeuterated guests has been used for detecting chiral discrimination events in chiral host–guest chemistry since 1997 [35]. A mixture of (*RS*)-1,1′-bi-2-naphthol and tris(5-fluoro-2-methylphenylcarbamate) was ionized by electron ionization (EI) from 25 to 400°C. Reconstructed ion current (RIC) profiles of (*R*)- and (*S*)-enantiomers of 1,1′-bi-2-naphthol showed different shapes, indicating different patterns of bonding. Direct chiral discrimination in EI-MS was confirmed by using partially deuterated 1,1′-bi-2-naphthol at the 3 and 3′ positions. A mixture of (*S*)-1,1′-bi-2-naphthol and (*R*)-1,1′-bi-2-naphthol-d_2 or (*S*)-1,1′-bi-2-naphthol-d_2 and (*R*)-1,1′-bi-2-naphthol in $CHCl_3$ containing tris(5-fluoro-2-methylphenylcarbamate) was directly inserted into the ion source to measure EI mass spectra.

The mass spectra at scan numbers 30–35, 80–85, and 140–145 in RIC profiles of (*S*)-1,1′-bi-2-naphthol-d_2 (M_r 288) and (*R*)-1,1′-bi-2-naphthol-d_2 (M_r 286) with tris(5-fluoro-2-methylphenylcarbamate) are shown in Figure 4, where differences in the ratios of m/z 288 to m/z 286 are clearly detected. A plot of the ratio of m/z 288 to 286 (the mean value of six scan) versus the scan number is also shown in Figure 4. Initially, the relative peak intensity of m/z 286 was larger than that of m/z 288, which increased as the sample temperature was raised. These results indicate that the molecule of M_r 288 [(*S*)-1,1′-bi-2-naphthol-d_2] vaporizes more slowly at higher temperature than the molecule of M_r 286 [(*R*)-1,1′-bi-2-naphthol-d_2]. When a mixture of (*S*)-1,1′-bi-2-naphthol-d_2 and (*R*)-1,1′-bi-2-naphthol-d_2 was used instead, the relative intensity of m/z 286 to 288 (the reciprocal of the ratio in Fig. 4) showed the same tendency.

When optically inactive polystyrene was used as adsorbent, no difference in the relative peak intensity at m/z 288 to 286 was detected. Moreover, in the resolution of (*RS*)-1,1′-bi-2-naphthol and (*RS*)-1,1′-bi-2-naphthol-d_2 on the CSP, no isotope effect was observed. These findings indicate that the difference in EI-MS spectra is due to the difference in desorption between the enantiomers from the chiral adsorbent tris(5-fluoro-2-methylphenylcarbamate). This method can be used to discriminate the chirality of other enantiomers of small molecules if they show peaks in their EI-MS spectra in the presence of chiral polymers. Similar chiral recognition was detected by negative ion fast-atom bombardment mass spectrometry [34].

It has also been reported from circular dichroism (CD) studies [36] that polysaccharide-based CSPs can induce chirality in enantiomeric guests such as (4*Z*,15*Z*)-bilirubin-Ixα (BR) (Fig. 5). Although not optically active, BR has two enantiomeric helical conformations maintained by six intramolecular hydrogen bonds between two carboxylic acid moieties and two pyrromethenone –NH– protons. These (*R*)- and (*S*)-helical conformers are in dynamic equilibrium in an achiral solution [37], but some optically active compounds can enantioselectively bind to BR to induce CD spectra in solution [38–40]. A significant induced CD

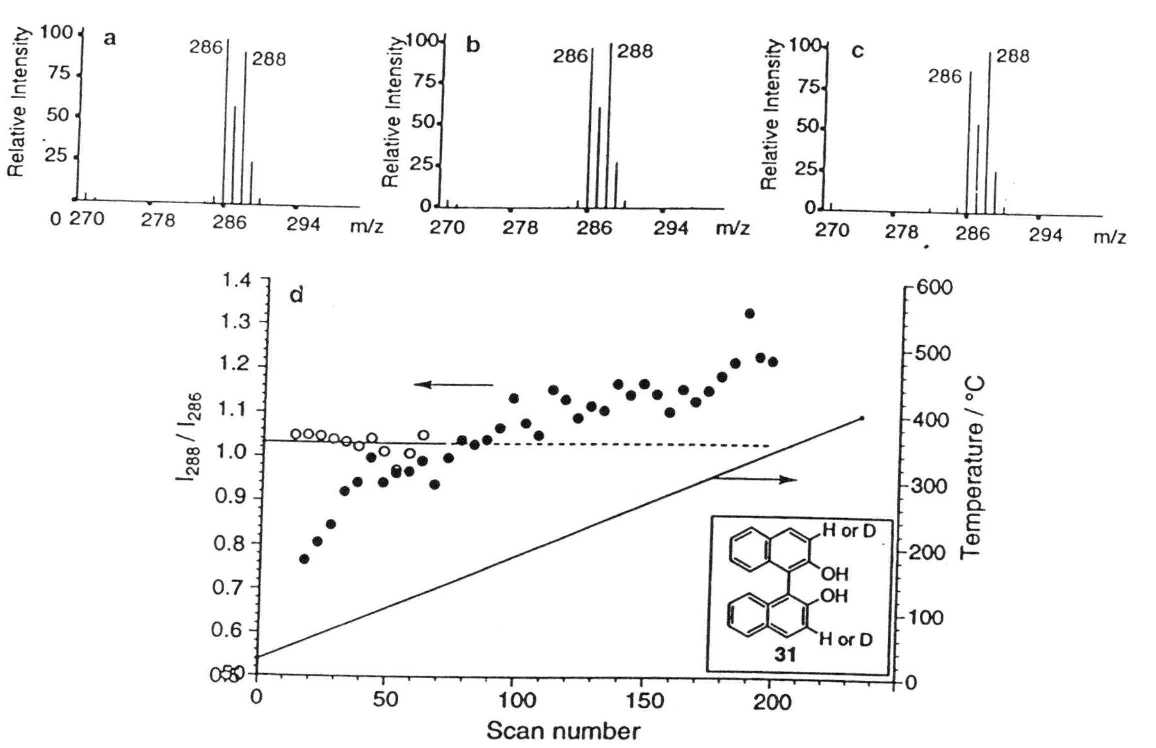

FIGURE 4 Mass spectra of a mixture of (*S*)-1,1′-bi-2-naphthol-d_2 (M_r 288) and (*R*)-1,1′-bi-2-naphthol (M_r 286) in the presence of tris(5-fluoro-2-methylphenylcarbamate) at scan numbers (a) 30–55, (b) 80–85, and (c) 140–145 in reconstructed ion current (RIC) profile. (d) The ratio of peak intensity versus scan number with (open circles) and without (solid circles) tris(5-fluoro-2-methylphenylcarbamate). (From Ref. 32.)

spectrum was observed in the UV–visible region for BR in the presence of some cellulose triphenylcarbamate (CTPC) derivatives in chloroform. Figure 5 shows typical CD spectra of BR (3.0×10^{-4} M) in chloroform in the presence of CTPC derivatives. The nature of the CD spectra is affected by the substituents of CTPC derivatives. The intramolecular hydrogen bond between phenylcarbamoyl moieties of the CTPC derivatives and BR is presumed to be the main binding force for the asymmetric transformation of BR, because addition of small amounts of 2-propanol to a chloroform solution of BR containing CTPC derivatives markedly diminishes the CD signal.

2.3 PREPARATION AND COMMERCIALIZATION

Microcrystalline cellulose triacetate (MCTA) is prepared by heterogeneous acylation of native cellulose in benzene [15,41]. Cellulose and amylose benzoates were synthesized by the reaction of corresponding benzoyl chloride in *N,N*-dimethylacetamide–lithium chloride–pyridine (15 : 1.5 : 7,v/v/v) mixture at about 100°C. These derivatives were isolated as the insoluble fractions in methanol. Macroporous silica gel was treated with a large excess of 3-aminopropyltriethoxysilane in benzene. The prepared benzoate derivatives were dissolved in chloroform or tetrahydrofuran and added to the silanized gel separately. This coating procedure was repeated two or three times and the silica gel obtained was added to hexane and packed in the desired stainless steel columns [8,17,42,43]. Phenylcarbamate derivatives of cellulose and amylose were prepared by the reaction of cellulose or amylose with an excess of the corresponding substituted phenylisocyanate in dry pyridine at 80 to 100°C. The phenylcarbamates obtained were isolated as a methanol-insoluble fraction (72–90% yield). ^{1}H-NMR data showed that hydroxyl groups of cellulose and amylose were almost quantitatively converted into the carbamate moieties [18,44,45]. The prepared derivatives of cellulose and amylose again are not useful CSPs and, therefore, these derivatives are coated or adsorbed on silica gel. In this process, the macroporous silica gel is treated with 3-aminopropylthriethoxysilane in benzene at 80°C. The polysaccharide derivatives are dissolved in tetrahydrofuran or *N,N*-dimethylacetamide and added to the above-mentioned silica gel. After the wetted silica gel has been dried under vacuum, the resulting dried material is packed in a stainless steel column of the desired dimension by a slurry method [18]. This column is used for the chiral resolution in HPLC.

First Rimbock et al. [46] and later Francotte et al. [47–50] developed beads of benzoyl cellulose, which were found to be very effective for chiral resolution at the preparative scale. Benzoyl cellulose was dissolved in a mixture of dichloromethane and heptanol and added dropwise to an aqueous solution of a surfactant (sodium lauryl sulfate), which is mechanically stirred at a well-defined speed. After addition of the benzoyl cellulose solution, stirring is maintained and the

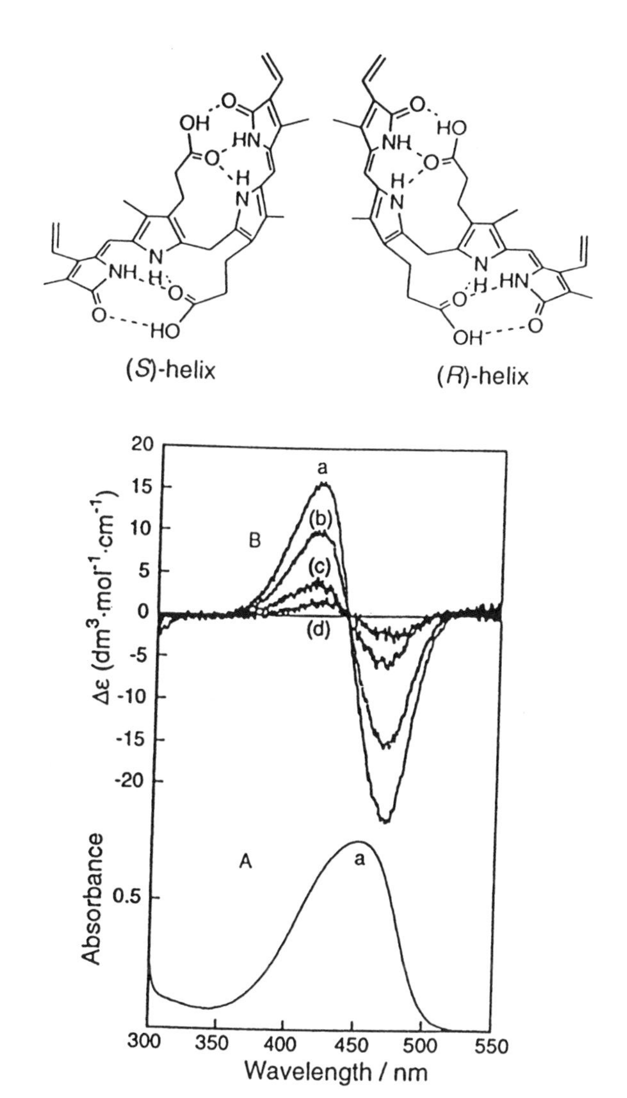

FIGURE 5 The two enantiomeric conformations of bilirubin (top) and (A) UV–visible and (B) CD spectra of bilirubin (3×10^{-4} M) in chloroform–ethanol in the presence of cellulose triphenylcarbamate derivatives (3×10^{-2} M glucose units) with the following substituents: (a) 4-Br, (b) 3-Cl, (c) 4-CH_3, and (d) 3,5-F_2. (From Ref. 32.)

emulsion obtained is heated at 40°C, to remove dichloromethane. This product is filtered and washed with sufficient amounts of water and ethanol. After washing, the product is dried at 80°C under vacuum. Recently, Castells and Carr [51] coated tris(3,5-dimethylphenylcarbamate) on microparticulate porous zirconia. The CSP that developed was tested for the chiral resolution of amino alcohol compounds such as β-adrenergic agents.

Since polysaccharide derivatives coat silica gel THF, chloroform and *N,N*-dimethylacetamide (in which most polysaccharides are soluble) cannot be used as the mobile phase, greatly limiting the choice of mobile phase. Therefore, a variety of approaches have been used to bond the CSPs developed to silica gel [52]. Okamoto el al. [53] developed an approach for stabilizing these CSPs on a silica surface by immobilization. Later this immobilization technique was improved by a number of groups [54–56], and it is now used by researchers. However, CSPs prepared by the immobilization technique show lower chiral recognition ability, particularly at higher concentrations of the chemical bond, than the corresponding coated-type CSPs. The introduction of the chemical bond may disturb the formation of the regular higher order polysaccharide structure.

Okamoto et al. [54] prepared cellulose tris(3,5-dimethylphenylcarbamate (CDMPC) and amylose tris(3,5-dimethylphenylcarbamate (ADMPC) CSPs by regioselective bonding to silica gel with 4,4′-diphenylmethane diisocyanate as a spacer. The ADMPC regioselectively bonded to silica gel when a small amount of diisocyanate was used, and the CSP that developed showed a higher resolving power. For CDMPC, the position of glucose in immobilization on silica gel affects chiral recognition hardly at all. Some racemic compounds were more efficiently resolved by using chloroform as a component of the mobile phase on the chemically bonded type of CSP. When CSPs are chemically bonded to silica gel through plural hydroxy groups of polysaccharides, there is an alteration in the higher order structure of the polymers, giving rise to a decrease in chiral recognition ability. Okamoto et al. [57] bonded amylose 3,5-dimethylphenylcarbamate to silica gel at the reducing terminal residue of amylose. More recently, Felix [58] has reviewed the regioselectively modified polysaccharide derivatives as chiral stationary phases in HPLC, describing methods of preparation of several regioselectively modified polysaccharide derivatives. The goal was to combine the effects of the selectivities of the known tris(aryl carbamate) and tris(aryl esters) of polysaccharides. The use of new substituted derivatives of polysaccharides as the chiral stationary phase in HPLC for chiral resolution was investigated.

Okamoto et al. [56] prepared CSPs of amylose phenylcarbamate bonded to silica gel by the following enzymatic methods. Amylose that had been prepared by enzymatic polymerization of α-*D*-(+)-glucose-1-phosphate dipotassium catalyzed by a phosphorylase, using two kinds of the primer derived from maltopentose, was chemically bonded to silica gel. In method I, maltopentose was first lactonized and allowed to react with (3-aminopropyl)triethoxysilane to form an amide bond. Amylose chains having the desired chain length and a narrow molecular weight distribution were then constructed by means of enzymatic polymerization. The resulting amylose bearing a trialkoxysilyl group at the terminal was allowed to react with silica gel for immobilization. In method II, maltopentose was first oxidized to form a potassium gluconate at the residual terminal. After enzymatic polymerization with the potassium gluconate, the amylose end was lactonized to be immobilized to 3-aminopropyl-silanized silica gel through amide bond formation.

The two amylose-conjugated silica gels thus obtained were treated with a large excess of 3,5-dimethylphenyl isocyanate to convert the hydroxyl groups of amylose to the corresponding carbamate residues. The CSP derived through method II was found to be superior to the CSP obtained by method I. The scheme of the preparation of these CSPs are given in Figure 6. In 2001 Okamoto's group [59] synthesized cellulose derivatives having vinyl groups and polymerized on silica gel. Silica gel first was coated with the cellulose phenylcarbamate derivative and then copolymerized under various conditions with vinyl monomers such as styrene with a radical initiator. The cellulose phenylcarbamate derivatives were efficiently immobilized on the silica surface and showed a high capacity for enantioseparation, similar to that of coated-type CSPs.

In 2001, Franco et al. [52] reviewed an alternative method using a cross-linking technique to improve the durability of polysaccharide CSPs. Kimata et al. [60] developed a method for the fixation of cellulose tris(4-vinylbenzoate) on modified silica by means of a radical copolymerization reaction. In this case there was no spacer between the matrix and the polysaccharide derivatives, both of which bore activated double bonds that were capable of being polymerized. On the other hand, the γ-aminopropyl silica gel used as a matrix was treated with acryloyl chloride. Cellulose was fully derivatized with 4-vinylbenzoyl chloride. After the cellulose derivative had been coated onto the modified silica, a suspension of the resulting material in heptane was heated in the presence of a radical initiator. The CSP obtained from the above-described process was stable in THF and dichloromethane solvents. For comparison, the coated CSP of the same material was copolymerized. The bonded CSP showed lower chiral recognition capacity, but its higher stability in a variety of solvents was proved. The procedure of this technique is presented in Figure 7.

As noted earlier, not all the prepared derivatives of cellulose and amylose are useful CSPs. Therefore, some of the useful derivatives of cellulose and amylose were selected and commercialized. The cellulose and amylose CSPs were commercialized by Daicel Chemical Industries of Tokyo. About 20 derivatives of cellulose and amylose are commercially available, as shown in Figure 8, with their chemical and trade names. The trade name of the cellulose and amylose derivatives are Chiralcel and Chiralpak, respectively. An R added to the end of a Chiralcel or Chiralpak trade name denotes a reversed-phase nature CSP. These CSPs are available in stainless steel columns of different dimensions and with different particle sizes. Generally the CSPs are available in the following range of column sizes and particle sizes (in parentheses): 25 cm × 0.46 cm

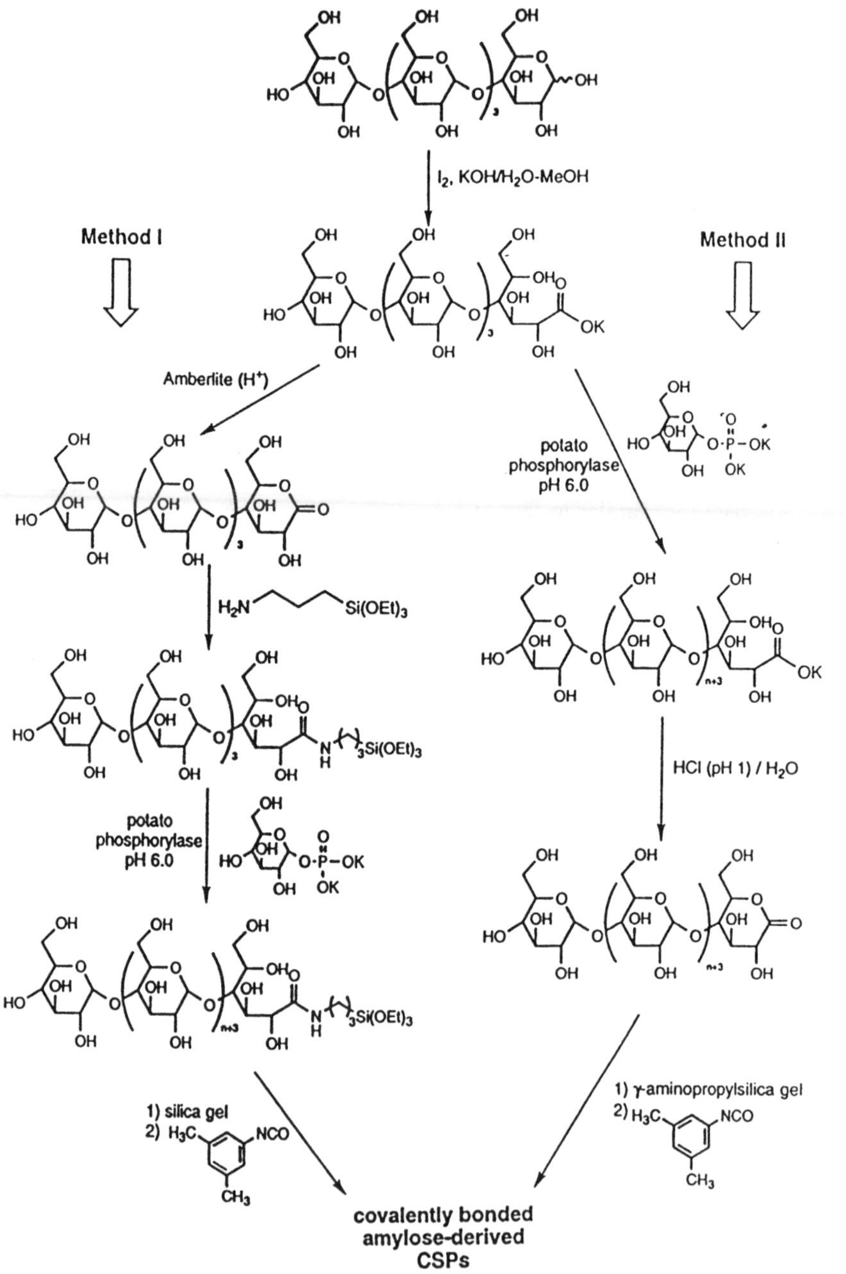

I_2, KOH/H_2O-MeOH

Method I

Method II

Amberlite (H^+)

H_2N–$(CH_2)_3$–$Si(OEt)_3$

potato phosphorylase pH 6.0

1) silica gel
2) 3,5-dimethylphenyl isocyanate (H_3C, NCO, CH_3)

potato phosphorylase pH 6.0

HCl (pH 1) / H_2O

1) γ-aminopropylsilica gel
2) 3,5-dimethylphenyl isocyanate (H_3C, NCO, CH_3)

covalently bonded amylose-derived CSPs

FIGURE 6 Preparation of amylose-derived CSPs in which the chiral selector is fixed to the matrix by the reducing terminal residues. (From Ref. 52.)

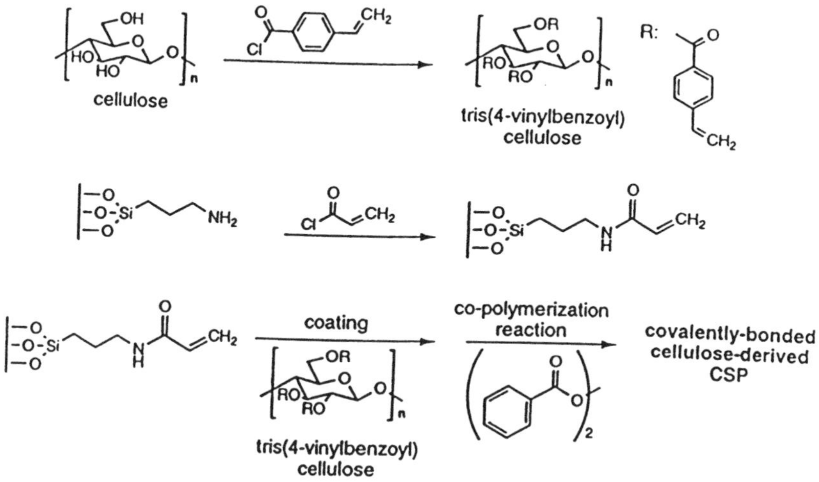

FIGURE 7 The polymerization of tris(4-vinylbenzoate) of cellulose onto acrylamidopropylsilica gel. (From Ref. 52.)

(5 μm), 15 cm × 0.46 cm (10 μm), 15 cm × 0.46 cm (5 μm), and 25 cm × 0.46 cm (10 μm).

2.4 SELECTIVITIES

Although about 20 polysaccharide-based CSPs have been commercialized and much work on enantioresolution has been carried out on these phases, it remains very difficult to predict the best CSP for the chiral resolution of a particular compound. It has been observed that most of the resolved racemic compounds contain aromatic rings or groups such as carbonyl, sulfinyl, nitro, amino, and benzoyl. However, some reports have been published on the chiral resolution of nonaromatic racemates on polysaccharide CSPs [61]. As in the case of other CSPs, polysaccharide-based CSPs do not require a certain combination of functional groups. However, only one group can afford a satisfactory separation. Presumably some chiral space (e.g., a concavity or ravine existing on a polysaccharide derivative) could enable such a separation [62].

It is not easy to determine the differences in structural selectivity among all possible derivatives. A rationalization based on an electronic effect [18] and the length of the acyl substituent on cellulose were attempted [42]. CTA-I (microcrystalline cellulose triacetate) is very specific and can be used for resolution of racemic compounds both having aromatic rings and carbonyl groups. On the other hand, a CTA-II (cellulose triacetate) CSP has a different selectivity. Both

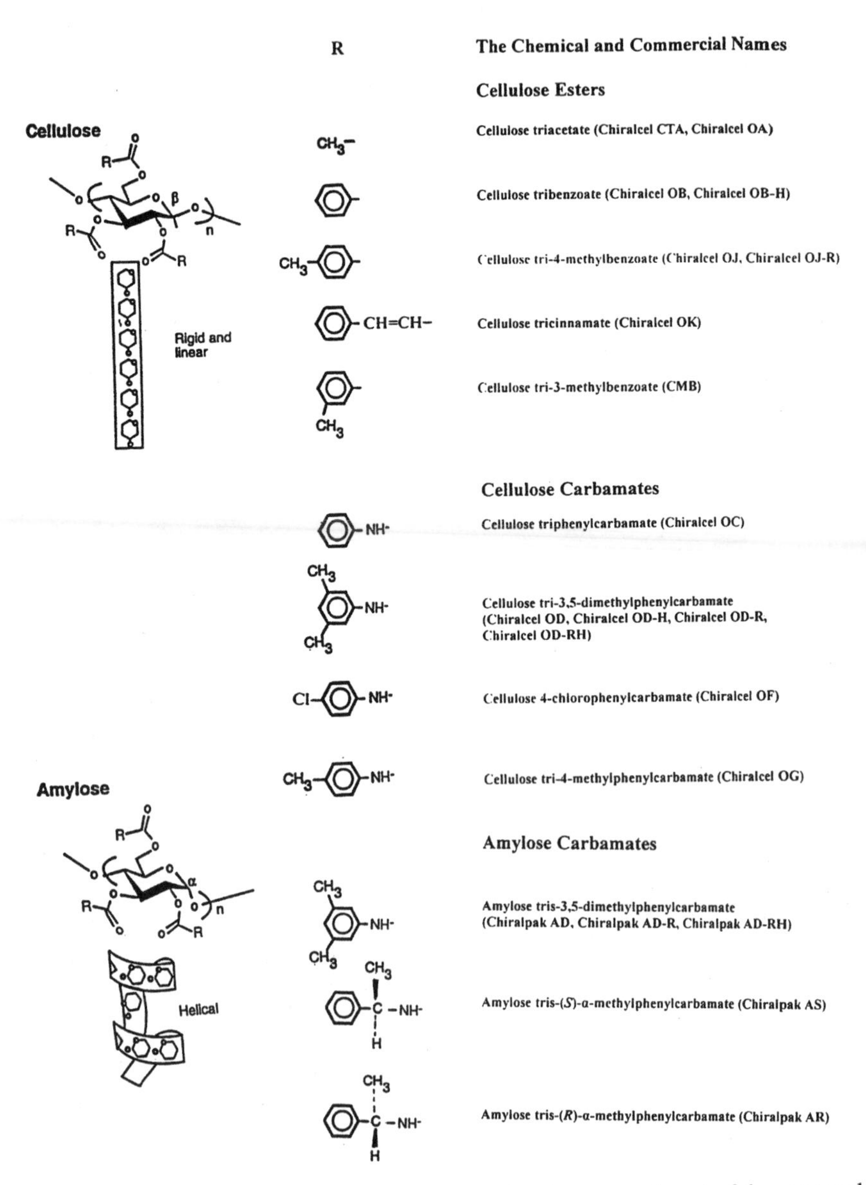

FIGURE 8 Chemical structures and chemical and trade names of most of the commonly used polysaccharide-based CSPs.

CSPs (CTA-I and CTA-II) have inverse selectivity for Tröger's base and *trans*-1,2-diphenyloxirane racemates. These characteristics of CTA CSPs are responsible for good chiral resolution of small cyclic carbonyl compounds [42]. In 2001 Aboul-Enein and Ali [63] observed the reversed order of elution of nebivolol on a Chiralpak AD column when ethanol and 2-propanol were used separately as the mobile phases. Table 1 presents selectivity data for the polysaccharide-based CSPs. Okamoto et al. [42] observed that the introduction of a methyl group at the para position of cellulose tribenzoate results in a dramatic shift of the structural selectivity toward aromatic compounds with larger skeletons, and its selectivity was rather similar to that of cellulose tricinnamate.

2.5 APPLICATIONS

2.5.1 Analytical Separations

A single CSP cannot be used for the chiral resolution of all racemic compounds. Therefore, different CSPs were used for the chiral resolution of different racemates. To make this part easy and clear, Table 1 includes the names of 20 CSPs and their most frequent applications. However, some other interesting applications are possible. Upon screening about 510 racemic compounds described in the literature, we observed that 229 of them resolved completely and 86 partially on cellulose tris(3,5-dimethylphenylcarbamate), and the rest not at all. For amylose tris(3,5-dimethylphenylcarbamate) CSP, we screened 384 racemic compounds and observed that 107 resolved completely and 102 partially. Clearly, cellulose and amylose tris(3,5-diphenylcarbamate) CSPs have the ability to resolve about 80% of the racemic compounds investigated.

Various aliphatic and aromatic compounds have been resolved on cellulose triacetate (CTA) CSPs at both the analytical and the preparative scale. Hesse and Hagel in 1973 reported the complete resolution of Tröger's base on a 16 cm column filled with MCTA [15]. Francotte et al. [50] extensively studied the relationship between the crystallinity of CTA and resolution power; CTA-I showed the best resolution power, whereas CTA-II exhibited poor resolution. Moreover, enhancement of the crystallinity of CTA-I by annealing resulted in lower resolving power for CTA-I, and the chromatographic behavior was rather similar to that of CTA-II [50]. Further, Francotte et al. [64,65] established that heterogeneous acetylation may provide a supramolecular structure for CTA-I having multiple interaction sites with specific surfaces and grooves inside the matrix that may be responsible for high chiral recognition for a wide range of enantiomers. The existence of such a supramolecular structure for CTA was also proposed as a result of NMR studies [26]. When CTA-I is coated on silica gel, the CSP is called Chiralcel OA, and its properties are different from those of CTA-I. The new coated CSP offers more advantages, including greater column efficiency,

TABLE 1 Various Polysaccharide-Based Commercial CSPs

Trade name[a]	Chemical name	Applications
Cellulose CSPs		
Chiralcel OB	Cellulose trisbenzoate	Small aliphatic and aromatic compounds
Chiralcel OB-H[b]	Cellulose trisbenzoate	Small aliphatic and aromatic compounds
Chiralcel OJ	Cellulose tris(4-methyl benzoate)	Aryl methyl esters, aryl methoxy esters
Chiralcel OJ-R[b]	Cellulose tris(4-methyl benzoate)	Aryl methyl esters, aryl methoxy esters
Chiralcel CMB	Cellulose tris(3-methylbenzoate)	Aryl esters and arylalkoxy esters
Chiralcel OC	Cellulose trisphenylcarbamate	Cyclopentenones
Chiralcel OD	Cellulose tris(3,5-dimethylphenylcarbamate)	Alkaloids, tropines, amines, β-adrenergic blockers
Chiralcel OD-H[b]	Cellulose tris(3,5-dimethylphenylcarbamate)	Alkaloids, tropines, amines, β-adrenergic blockers
Chiralcel OD-R[c]	Cellulose tris(3,5-dimethylphenylcarbamate)	Alkaloids, tropines, amines, β-adrenergic blockers
Chiralcel OD-RH[d]	Cellulose tris(3,5-dimethylphenylcarbamate)	Alkaloids, tropines, amines, β-adrenergic blockers
Chiralcel OF	Cellulose tris(4-chlorophenylcarbamate)	β-Lactams, dihydroxypryidines, alkaloids
Chiralcel OG	Cellulose tris(4-methylphenylcarbamate)	β-Lactams, alkaloids
Chiralcel OA	Cellulose triacetate on silica gel	Small aliphatic compounds
Chiralcel CTA	Cellulose triacetate, microcrystalline	Amides, biaryl compounds
Chiralcel OK	Cellulose triscinnamate	Aromatic compounds
Amylose CSPs		
Chiralpak AD	Amylose tris(3,5-dimethylphenylcarbamate)	Alkaloids, tropines, amines, β-adrenergic blockers
Chiralpak AD-R[e]	Amylose tris(3,5-dimethylphenylcarbamate)	Alkaloids, tropines, amines, β-adrenergic blockers
Chiralpak AD-RH[b]	Amylose tris(3,5-dimethylphenylcarbamate)	Alkaloids, tropines, amines, β-adrenergic blockers
Chiralpak AR	Amylose tris(*R*)-1-phenylethylcarbamate	Alkaloids, tropines, amines
Chiralpak AS	Amylose tris(*S*)-1-methylphenylcarbamate	Alkaloids, tropines, amines

[a] Columns supplied by Daicel Chemical Industries, Tokyo, Japan. Dimensions are column size 25 cm × 0.46 cm, particle size 10 μm, except as noted.
[b] Column size 25 cm × 0.46 cm, particle size 5 μm.
[c] Column size 15 cm × 0.46 cm, particle size 10 μm.
[d] Column size 15 cm × 46 cm, particle size 5 μm.

greater durability, and wider choice of mobile phase. X-ray studies indicate that Chiracel OA is almost amorphous rather than crystal [25], indicating that microcrystallinity is not essential for chiral resolution. It has also been observed that the chiral recognition of a CSP greatly depends on the conditions of preparation of that CSP, such as the type of coating solvent and the molecular weight of the cellulose [25,66].

Nevertheless, cellulose triacetate derivatives have certain limitations. An ethanol–water mixture may cause CTA to swell by up to 40% of its volume. Other solvents (e.g., acetone, acetonitrile, dimethylformamide, chlorinated alkanes) cannot be used because they dissolve cellulose triacetate more or less completely. A gel is formed when THF is used as the mobile phase. Furthermore, in pure 1,4-dioxane, or in a 50 : 50 v/v mixture of dimethoxyethane and toluene-1,4-dioxane, cellulose triacetate swells strongly. It has been also observed that CTA swells more in lower alcohols than in ethers or alkanes. Mannschreck et al. [67] found that in general the capacity factors of the solutes investigated increased when methanol, ethanol, or propanol, respectively, was used as the eluent. Similar results were found by Rizzi [68] and also by Isaksson et al. [69]. Thus, since the degree to which CTA will swell depends on the kind of solvent used, care must be taken when changing from one solvent to another. If shrinkage of the packing bed occurs, a void volume is created on top of the column, resulting in decreased column efficiency. Therefore, it is advisable to use a column equipped with a movable piston to enable compensation for packing bed volume changes. A few examples of the enantiomeric resolution on CTA are shown in Figure 9.

The chiral recognition capacities of cellulose tribenzoate (CTB) derivatives was explored by Okamoto et al. [8,17]. The effect of the substituents on the phenyl ring of cellulose tribenzoate (Chiralcel OB) has also been systematically studied [17]. Alkyl, halogen, trifluoromethyl, and methoxy groups were selected as the substituents. Resolution ability was greatly affected by the inductive effect of such substituents. Benzoate derivatives having electron-donating substituents such as methyl groups showed better chiral recognition ability than those having electron-withdrawing substituents, such as halogens. However, the most heavily electron-donating methoxy group was not suitable because of the high polarity of the substituent itself. Among the benzoate celluloses, tris(4-methylbenzoate) (Chiralcel OJ) exhibited very high chiral recognition for various racemic compounds, including drugs, and appears to be a practically useful CSP. Francotte and coworkers resolved large quantities of many drugs on CSPs derived from CTB [49,50,65,70]. They claimed that the benzoyl cellulose beads showed complementary chiral recognition to the CTA phase [49]. The CTB derivative beads show almost the same enantioseparation ability as the corresponding coated-type CTB derivatives on silica gel. Figure 10 shows the chromatograms of chiral resolution on cellulose tribenzoate CSPs.

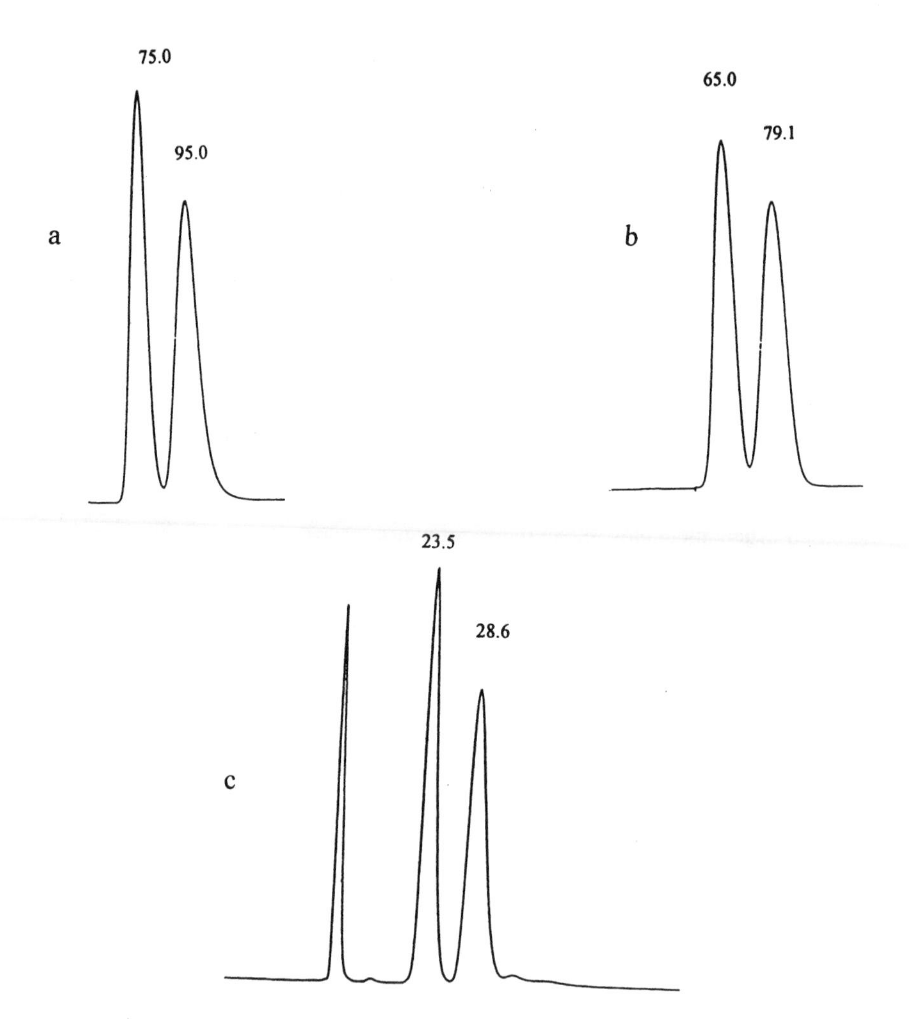

FIGURE 9 Chromatograms of enantiomeric resolution on cellulose triacetate CSPs. (a) *N*-Benzoylalanine methyl ester, with hexane–ethanol (80 : 20, v/v) as the mobile phase. (b) Benzoin with hexane–2-propanol–water (70 : 27 : 3, v/v/v) as the mobile phase. (c) Mandelic acid with ethanol as the mobile phase. (From Ref. 9.)

The chiral recognition ability of a series of cellulose phenylcarbamate derivatives has been evaluated extensively [18,43]. The introduction of an electron-donating methyl group or an electron-withdrawing halogen at the 3- and/or 4-position improved the resolution ability, but substituents at position 2 showed poor chiral resolution capacity. The derivatives with heteroatom substi-

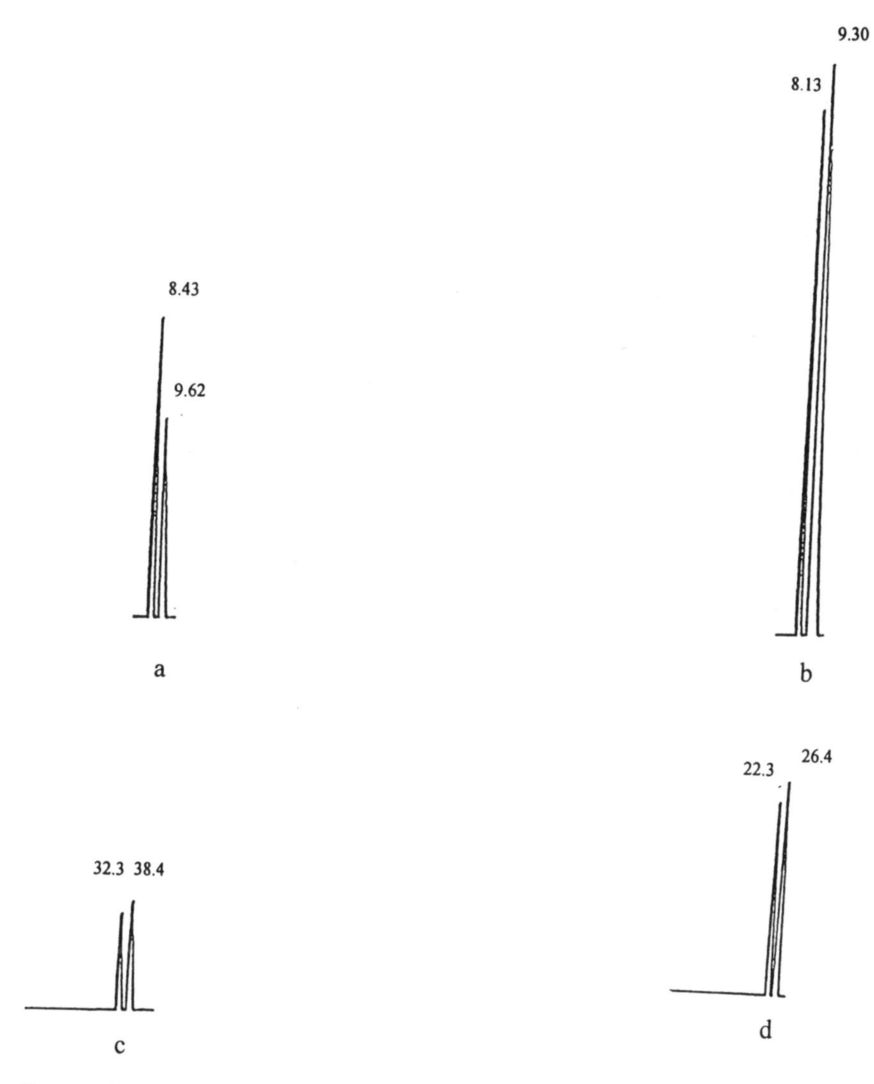

FIGURE 10 Chromatograms of enantiomeric resolution on cellulose tribenzoate CSPs. (a) (±)-*threo*-Methylphenidate on Chiralcel OB CSP, with hexane–ethanol–methanol–trifluoroacetic acid (480 : 9.75 : 9.75 : 0.5, v/v/v/v) as the mobile phase containing 0.2 mM benzoic acid. (b) (±)-*threo*-methylphenidate on Chiralcel OB CSP, with hexane–ethanol–methanol–trifluoroacetic acid (480 : 9.75 : 9.75 : 0.5, v/v/v/v) as the mobile phase containing 0.2 mM phenol. (c) Ic aromatase inhibitor on Chiralcel OJ-R CSP, with acetonitrile–water (50 : 50, v/v) as the mobile phase. (d) IIc aromatase inhibitor on Chiralcel OJ-R CSP, with acetonitrile–water (50 : 50, v/v) as the mobile phase. For structures of Ic and IIc aromatase inhibitors, see later (Fig. 18). (From Refs. 98, 100.)

tuents, such as methoxy and nitro groups, show poor chiral recognition [18]. Racemic compounds can interact with polar substituents far from a chiral glucose residue. Therefore, bulky alkoxy substituents such as isopropoxy and isobutoxy improve resolving power [23]. Phenylcarbamate derivatives having both an electron-donating methyl group and electron-withdrawing substituents (chloro or fluoro groups), on the phenyl moieties were found to exhibit high enantioseparation for many racemates. For example, cellulose 3,4- or 3,5-chloromethylphenylcarbamate showed particularly high chiral recognition ability [19–21]. The chromatograms of the chiral resolution on cellulose tricarbamate are shown in Figure 11.

Similarly, the chiral resolution power of amylose CSPs was improved by introducing methyl or chloro groups on the phenyl moieties [21,22]. However, in contrast to the cellulose derivatives, amylase tris(4-methylphenylcarbamate) [23] and tris(5-chloro-2-methylphenylcarbamate) [24] showed high chiral recognition. In general, amylose derivatives were better CSPs than the cellulose derivatives. Their superiority may be due to the structure of amylose, which is more helical than that of cellulose [31]. Among many of the amylose derivatives prepared so far, 3,5-disubstituted derivatives such as 3,5-dimethyl- and 3,5-dichlorophenylcarbamate show particularly interesting and effective optical resolving abilities for a variety of racemic compounds [18,21,71]. For ready reference on the nature of the resolution, Figure 12 shows chromatograms on amylose triphenylcarbamate CSPs.

Tricarbamates are the most important cellulose and amylose derivatives used for chiral resolution, and derivatives of both types have comparable chiral recognition capabilities. However, some compounds could not be resolved completely on amylose carbamate CSPs. For example, both the calcium antagonist nicardipine and nitredipine were completely resolved on cellulose tris(4-*tert*-butylphenylcarbamate) [72], whereas amylose carbamate could not resolve these compounds. Thus this is a respect in which the two CSPs (cellulose and amylose derivatives) differ in their chiral recognition abilities. Some enantiomers that elute in reversed order on CDMPC may be resolved ADMPC, and vice versa [43]. The two phases (CDMPC and ADMPC) were compared for performance in the enantioseparation of a series of amidotetralines [73] and chiral sulfoxide [74], and complementary chromatographic behavior was observed. Cellulose derivatives substituted with *meta* and *para* (fluoro and methyl) groups showed higher resolving power than *ortho*- and *meta*-substituted derivatives. On the other hand, amylose derivatives having the substituents at the *ortho* position also showed characteristic high chiral recognition, comparable to that afforded by the *meta* and *para* substituents derivatives. The elution order was not influenced by the change of substituents in either cellulose or amylose derivatives [45]. Similarly, Chankvetadze et al. [24] reported that *ortho*-substituted phenylcarbamate derivatives of amylose showed high chiral recognition abilities, while

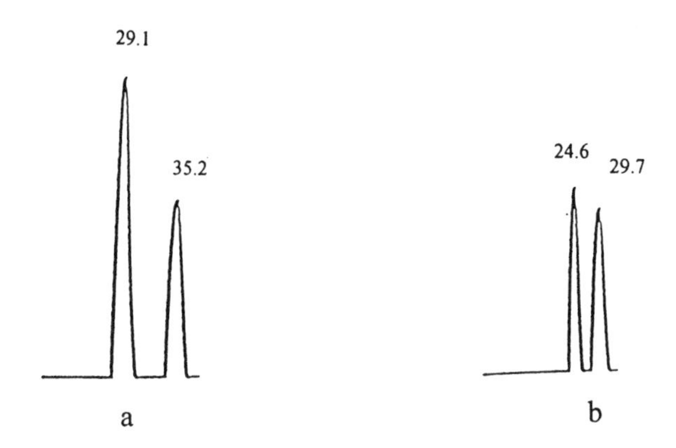

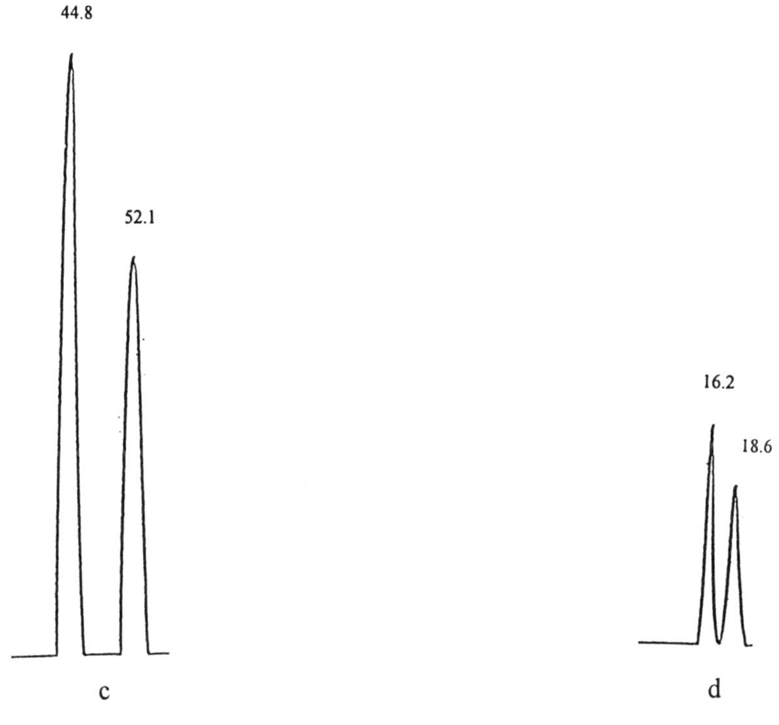

FIGURE 11 Chromatograms of enantiomeric resolution on cellulose tricarbamate CSPs. (a) Econozole. (b) Miconazole. (c) Sulconazole on a Chiralcel OF CSP, with hexane–2-propanol–diethylamine (425 : 74 : 1, v/v/v) as the mobile phase. (d) (±)-*threo*-Methylphenidate on a Chiralcel OD CSP, with hexane–ethanol–methanol–trifluoroacetic acid (480 : 9.75 : 9.75 : 0.5, v/v/v/v) as the mobile phase. (From Refs. 100, 101.)

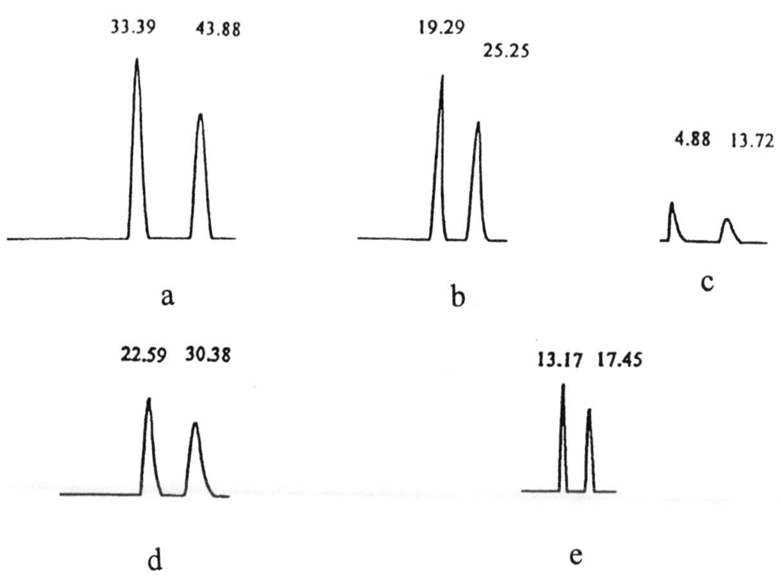

FIGURE 12 Chromatograms of enantiomeric resolution of nebivolol on amylose carbamate CSPs. (a)–(c) Chiralpak AD with (a) ethanol, (b) 1-propanol, and (c) 2-propanol. (d, e) Chiralpak AD-RH with (d) ethanol and (e) 1-propanol, separately, as the mobile phase. (From Ref. 63.)

cellulose phenylcarbamate derivatives with *ortho* substituents have poor chiral resolving power. The superiority of 5-chloro-2-methylphenylcarbamate over the corresponding dimethyl and dichlorophenylcarbamate derivatives of amylose was demonstrated. The role of the –NH– residue of the carbamate moiety and methyl and chloro groups in chiral recognition was elucidated by IR and ^{1}H NMR spectroscopy. In another study Okamoto et al. [44] observed that fluoro, chloro, bromo, and iodo groups at position 4 on the phenyl ring resulted in higher chiral recognition than was reported for the corresponding cellulose derivatives. The applications of chiral resolution on polysaccharide-based CSPs are summarized in Table 2 [1,6,7,9–116].

2.5.2 Preparative Separations

Analytical methods can be transferred easily to the preparative scale in polysaccharide-based CSPs, and these compounds have been used in preparative chromatography. Cellulose triacetate, which has good loading capacity, has been used for preparative separation at the industrial scale [70,85]. Moreover, the low cost of cellulose triacetate synthesis makes CTA ideal for preparative chromatography.

TABLE 2 The Mobile Phases Most Commonly Used with Polysaccharide-Based CSPs

Solvents[a]	Ratios (v/v)	Racemates	CSPs
Cellulose CSPs			
Hexane–2-PrOH	95 : 5	Tri- and tetrazole aromatase inhibitors	Chiralcel OD
Hexane–2-PrOH–DEA	50 : 50 : 0.4, 20 : 80 : 0.4 and 15 : 85 : 0.4	β-Adrenergic blockers	Chiralcel OD
Hexane–EtOH–DEA	80 : 20 : 0.2	Naftopidil	Chiralcel OD
Cyclohexane–MeOH–EtOH	2 : 95 : 5	Aminoglutethimide	Chiralcel OD
Hexane–EtOH–DEA	90 : 10 : 0.4	β-Adrenergic blockers	Chiralcel OD
–	–	Triacylglycerol	Chiralcel OD
Hexane–alcohols	Different ratios	Tetralin	Chiralcel OD-H
Hexane–1-PrOH	95 : 5, 97 : 3	Aromatic amides	Chiralcel OB
Hexane–1-PrOH–MeCN	96 : 3 : 1	Aromatic amides	Chiralcel OB
Hexane–2-PrOH	90 : 10	Aromatic alcohols	Chiralcel OB
–	–	α-Hydroxy-3-phenoxybenzene-acetonitrile and its *n*-butyl ester	Chiralcel OJ
Hexane–2-PrOH–DEA	425 : 74 : 1	Imidazole antifungal agents	Chiralcel OD, OJ, OB, OK, OC, and OF
Hexane–2-PrOH	Different ratios	Organophosphorous pesticides	Chiralcel OD, OJ
Perchlorate solution–MeCN	75 : 25 and others	β-Adrenergic blockers	Chiralcel OD-R
Hexane–2-PrOH–DEA	15 : 85 : 0.4 and others	β-Adrenergic blockers	Chiralcel OD and OD-R
MeCN–water	50 : 50	*o,p*-DDT and *o,p*-DDD	Chiralcel OJ-R
MeCN–2-PrOH	50 : 50	*o,p*-DDT and *o,p*-DDD	Chiralcel OD-R and Chiralcel OJ-R
MeCN–18 mM NH_4NO_3 (pH 7.0)	25 : 75, 55 : 45	Pipiridine derivatives	Chiralcel OD-R

(continued)

TABLE 2 Continued.

Solvents[a]	Ratios (v/v)	Racemates	CSPs
Water–MeCN–TEA	80 : 20 : 0.08	Clenbuterol, cimaterol, and mabuterol	Chiralcel OD-R and Chiralcel OJ-R
Water–MeCN	45 : 55	Tri- and tetrazole aromatase inhibitors	Chiralcel OD-R and OJ-R
Hexane–EtOH–MeOH–TFA	480 : 9.75 : 9.75 : 0.5	Methylphenidate	Chiralcel OD and OB
2-PrOH–MeCN	90 : 10, 50 : 50	Tri- and tetrazole aromatase inhibitors	Chiralpak OD-R and OJ-R
MeCN–water	50 : 50, 80 : 20, and 95 : 5	Tri- and tetrazole aromatase inhibitors	Chiralpak OD-R and OJ-R
EtOH	Pure	Mandelic amide	Chiralcel CTA
EtOH	Pure or 10%	Biaryl compounds	Chiralcel CTA
Hexane–EtOH	90 : 10	Biaryl compounds	Chiralcel OF and OG
EtOH or MeOH	Pure	Polycyclic aromatics	Chiralcel CA and OB
Hexane–EtOH	90 : 10	Sulfur compounds	Chiralcel OB and OC
Hexane–2-PrOH	90 : 10	Phosphorus compounds	Chiralcel OB, OK, and OC
Hexane–EtOH	90 : 10	Nitrogen and cyano compounds	Chiralcel OA, OB, OC, and OK
Hexane–EtOH	95 : 5	Amines	Chiralcel CA, OA, OC, OB, and OK
Hexane–2-PrOH	95 : 5-10	Carboxylic acids and derivatives	Chiralcel OC, OB, OF, and OG
EtOH	pure	Alphatic alcohols	Chiralcel CA, OB, OK, and OK
Hexane–EtOH	95 : 5, 90 : 10	Alphatic alcohols	Chiralcel CA, OB, OK, and OK
–	–	Triacylglycerol	Chiralcel OF

Amylose CSPs			
Hexane–EtOH	95 : 5, 90 : 10 and 88 : 12	β-Adrenergic blockers	Chiralpak AD
Hexane–2-PrOH	90 : 10	β-Adrenergic blockers	Chiralpak AD
Hexane–EtOH–MeOH–TFA	480 : 9.75 : 9.75 : 0.5	Methylphenidate	Chiralpak AD
Hexane–2-PrOH, hexane–EtOH	Different ratios	Vincamine	Chiralpak AD
–	–	2,4-Dioxo-5-acetamido-6-phenylhexanoic acid	Chiralpak AD
Hexane–2-PrOH–DEA	400 : 99 : 1	Imidazole antifungal agents	Chiralpak AD, AS, and AR
Hexane–2-PrOH	Different ratios	3-*tert*-Butylamino-1,2-propanediol	Chiralpak AS
MeCN–water	50 : 50	*o*, *p*-DDT and *o*, *p*-DDD	Chiralpak AD-R
MeCN–2-PrOH	50 : 50	*o*, *p*-DDT and *o*, *p*-DDD	Chiralpak AD-R
Water–MeCN–TEA	80 : 20 : 0.08	Clenbuterole, cimaterol, and mabuterol	Chiralpak AD-R
Water–MeCN	60 : 40	Flurbiprofen	Chiralpak AD-R
MeCN–water–TEA	50 : 50 : 0.03	Tetralone derivatives	Chiralpak AD-R
MeCN–water–AcOH	60 : 40 : 0.03	Tetralone derivatives	Chiralpak AD-R
2-PrOH–MeCN	90 : 10, 50 : 50	Tri- and tetrazole aromatase inhibitors	Chiralpak AD-R
MeCN–water	50 : 50, 80 : 20, and 95 : 5	Tri- and tetrazole aromatase inhibitors	Chiralpak AD-R
EtOH, 1-PrOH, and 2-PrOH separately	Pure	Nebivolol	Chiralpak AD and AD-R

[a] Abbreviations: AcOH, acetic acid; DEA, diethylamine; MeCN, acetonitrile; MeOH, methanol; MPH, methylphenidate; EtOH, ethanol; 1-PrOH, 1-propanol; 2-PrOH, 2-propanol; TEA, triethylamine; TFA, trifluoroacetic acid.
Source: Refs. 1, 6, 7, 9–116.

In one example of preparative chromatographic separation on a large scale [9], 20 g of the sample per run was injected onto a column (50 cm × 10 cm i.d.) filled with 20 μm Chiralcel OJ material. The eluent was hexane–ethanol (40 : 60, v/v). This column has been used for more than 2 years, and the column quality remains excellent, indicating long-term durability [9]. Recently, Cirrili et al. [117] described semipreparative-scale chiral resolution of imidazole derivatives by HPLC. To achieve the resolution of pure enantiomers, recycling [86] and peak shaving techniques are sometimes very important in preparative chromatography. The recycling techniques can be performed by two methods: by means of a closed-circuit system and by the alternating-two-column mode of operation. In general, the first method is preferred because a closed-circuit system is less complicated to construct and more economical than the construction of a unit that functions in the alternating-two-column mode. Moreover, recycling and peak shaving techniques are advantageous because they can be used for the separation of pure enantiomers in good yields, are suitable for the separation of partially resolved enantiomers, offer a higher production rate, consume less solvent, and can be run on smaller columns with less labor. Preparative enantioseparation by simulated moving bed (SMB) chromatography has been described by Schulte and Strube [118]. Their paper details system development, selectivity, saturation capacity, and mechanical and chemical stability, as well as methods for the optimization of pure enantiomeric separation.

2.6 OPTIMIZATION OF HPLC CONDITIONS

Chiral resolution on polysaccharide-based CSPs is sensitive, and therefore, the optimization of HPLC conditions on these phases is very important. The most important factors that control enantiomeric resolution are the composition, pH, and flow rate of the mobile phase and parameters, including temperature and solute structure. The optimization of these parameters on polysaccharide-based CSPs is discussed next.

2.6.1 Mobile Phase Composition

The selection of the mobile phase is the key aspect in chiral resolution. The mobile phase is selected according to the solubility and the structure of the drugs to be resolved. In the normal phase mode, the use of pure ethanol or 2-propanol is recommended. To decrease the polarity of the mobile phase and increase the retention times of the enantiomers, investigators use hexane, cyclohexane, pentane, or heptane as one of the main constituents of the mobile phase. However, other solvents (e.g., alcohols, acetonitrile) are also used in the mobile phase. Normally, if pure ethanol or 2-propanol is not well suited for the mobile phase, pure ethanol and hexane, 2-propanol, or ethanol in the ratio of 80 : 20 is used as

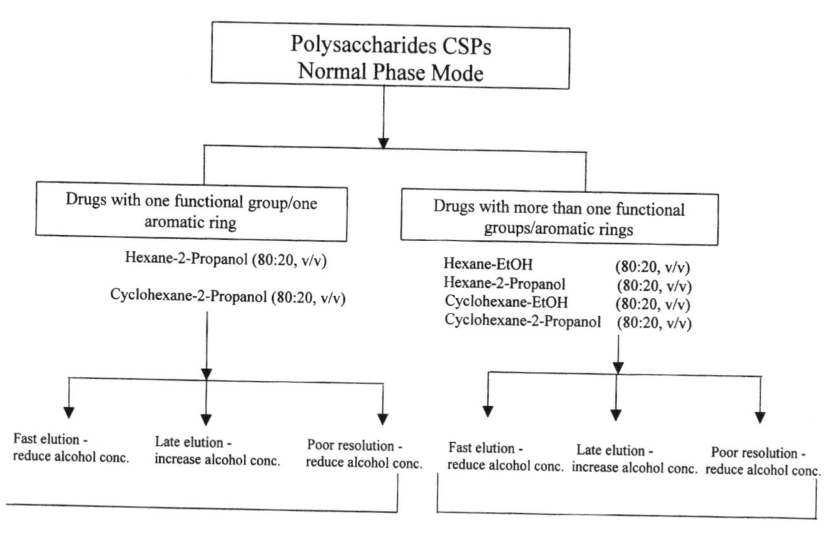

SCHEME 1 Protocol for the development and optimization of mobile phases on polysaccharide-based CSPs in the normal phase mode. This brief outline gives the procedure for developing a resolution on polysaccharide-based CSPs in the normal phase mode. However, other mobile phases may be used.

the mobile phase, and the mobile phase composition is changed on the basis of ensuing observations. Finally, chiral resolution is optimized by adding small amounts of amines or acids (0.1–1.0%). In 2001 Ye and Stringham [119,120] used basic and acidic organic modifiers to optimize the chiral resolution of certain racemates on amylose CSPs. The protocol of the selection and optimization of mobile phases for the enantiomeric resolution of drugs on polysaccharide-based CSPs in normal phase mode is presented in Scheme 1. The presence of small impurities may change the reproducibility. Moreover, the same solvent from two different suppliers may give rise to different results for the enantiomeric resolution of particular racemates under the identical conditions of chromatography.

The effect of the percentage of 2-propanol on the enantiomeric resolution of benzetimide is shown in Figure 13a. It may be concluded that retention factor values decreased with an increase in 2-propanol content. According to Wainer et al. [87], the diminution of retention factor values with increases in the content of polar modifiers indicates that competition for the binding sites on the CSP is a saturable process and that a maximum effect on retention factor will be reached at a certain polar modifier concentration. It is interesting to note that although only a

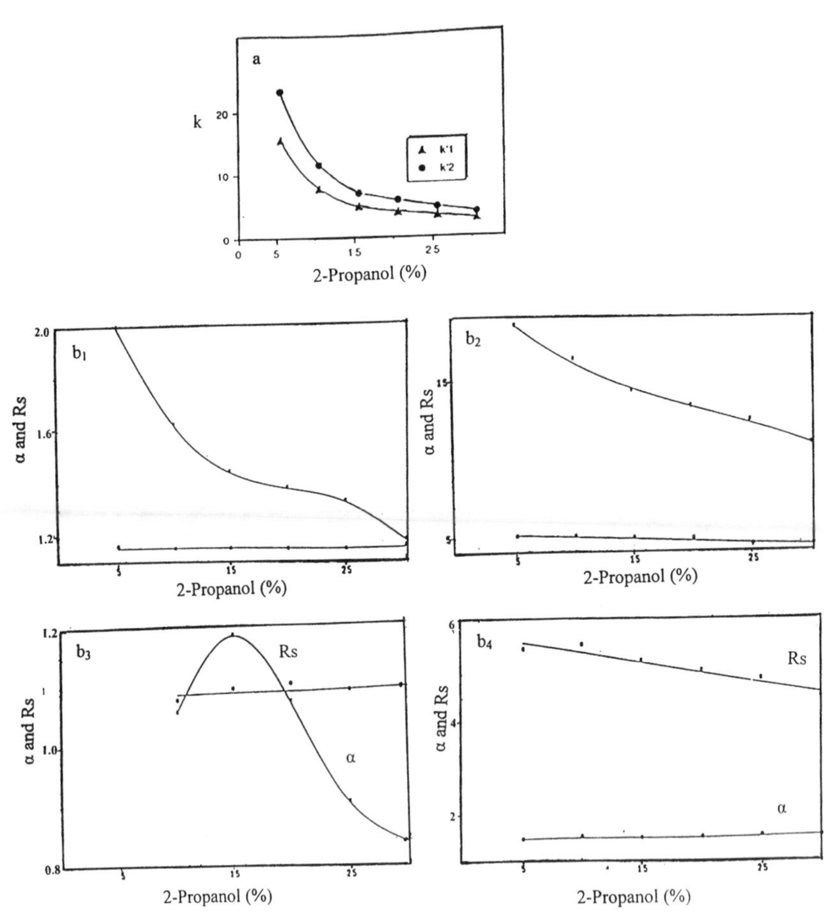

FIGURE 13 Effects of the concentration of 2-propanol on the enantiomeric resolution *Rs* of (a) benzetimide, (b_1) metomidate, (b_2) ambucetamide, (b_3) isoconazole, and (b_4) parconazole. (From Ref. 9.)

FIGURE 14 (A) Effects of the concentration of 2-propanol in the mobile phase hexane–2-propanol: △, second eluted isomer solute 1; ◇, first eluted isomer solute 1; ◆, second eluted isomers solute 2; ▲, first eluted isomer solute 2; ○, solute 3; □, second eluted isomer solute 4; V, first eluted isomer solute 4; ●, solute 5; ■, solute 6; and ▼, solute 7. Solutes: 1, 1-phenylethanol; 2, 1-phenylpropanol; 3, 1-phenyl propanol-2; 4, 2-phenyl-propanol; 5, benzylalcohol; 6, 3-phenyl propanol-1; 7, 2-phenyl propanol-2. (From Ref. 62.) (B) Effects of chain length of primary alcohol used as polar modifier on *k* and α for three different products: (b_1) Chiralcel OC, (b_2) Chiralcel OD, and (b_3) Chiralcel OJ.

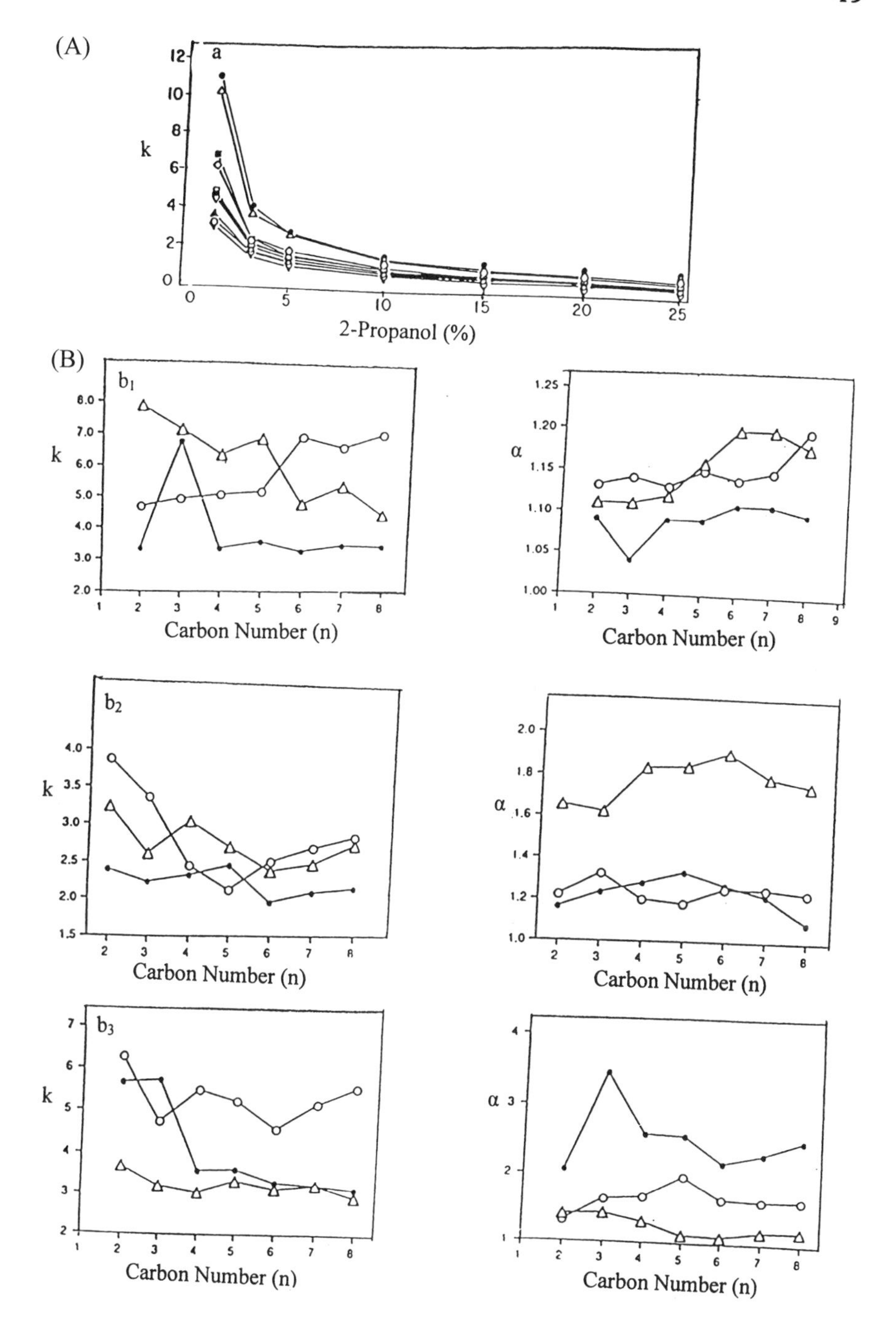
(A)
a
k
2-Propanol (%)
(B)
b1
k
Carbon Number (n)
α
Carbon Number (n)
b2
k
Carbon Number (n)
α
Carbon Number (n)
b3
k
Carbon Number (n)
α
Carbon Number (n)

small change in retention factor values occurred when the polar modifier content was varied from 5% to 25%, the effect on resolution can be completely different (Fig. 13b). The effect of isopropanol on the chiral resolution of certain drugs has been studied by Wainer et al. [62] and is given in Figure 14a. The effect of mobile phase composition was also studied by Bonato et al. [88] for the enantiomeric resolution of propafenone, 5-hydroxypropafenone and *N*-despropylpropafenone on Chiralpak AD column. The effect on the length of the carbon chain of the alcohol (modifier) in the mobile phase was studied and is depicted in Figure 14b. It may be concluded from this figure that the retention k and separation α factors usually decrease with an increase in the chain length of the polar modifier, although at times there are minor changes in or cyclic patterns of the values of k and α. Based on these observations, it may be concluded that modifier structure has an effect on stereoselectivity, but the steric fit between solute and CSP certainly plays an important role, as well. Since no clear trend could be observed delineating the effect of the polar modifiers on enantioselectivity, it is necessary to use a specific modifier for each specific application. In CTA, selection of buffers in the pH range of 5 to 10 is useful. However, eluents with a high water content or low pH value should be used sparingly because CTA can be hydrolyzed under these conditions [69].

Gaffney [89] reported the reverse order of elution of 2-phenoxypropanoic acid on Chiralcel OB CSP when different alcohols were used. In 2001 Aboul-Enein and Ali [63] observed the reverse order of elution of nebivolol enantiomers on a Chiralpak AD chiral stationary phase when ethanol and 2-propanol were used separately as the mobile phases. However, the best resolution was obtained when ethanol served as the mobile phase. The inversion of the elution may be due to the different conformation of the polysaccharide CSPs [63]. The pattern of conversion of order of elution using different ratios of ethanol and 2-propanol is shown in Figure 15.

The energies of interaction of ethanol and 2-propanol with amylose tris(3,5-dimethylphenylcarbamate) phases were calculated. The energies of interaction of ethanol, 2-propanol, and their mixtures in different ratios are calculated and presented in Table 3. The energy of interaction G and the energy factor (for ethanol and 2-propanol) F are calculated by the following equations [63].

$$G = RT \ln k \tag{1}$$

$$F_{et} \text{ or } F_{pr} = VE \tag{2}$$

where G, R, T, F_{et}, F_{pr}, V, and E are the energy of interaction, the gas constant, absolute temperature, energy factors for ethanol and 2-propanol, volume of ethanol or 2-propanol, and energy of interaction of ethanol or 2-propanol with amylose tris(3,5-dimethylphenylcarbamate), respectively. It has

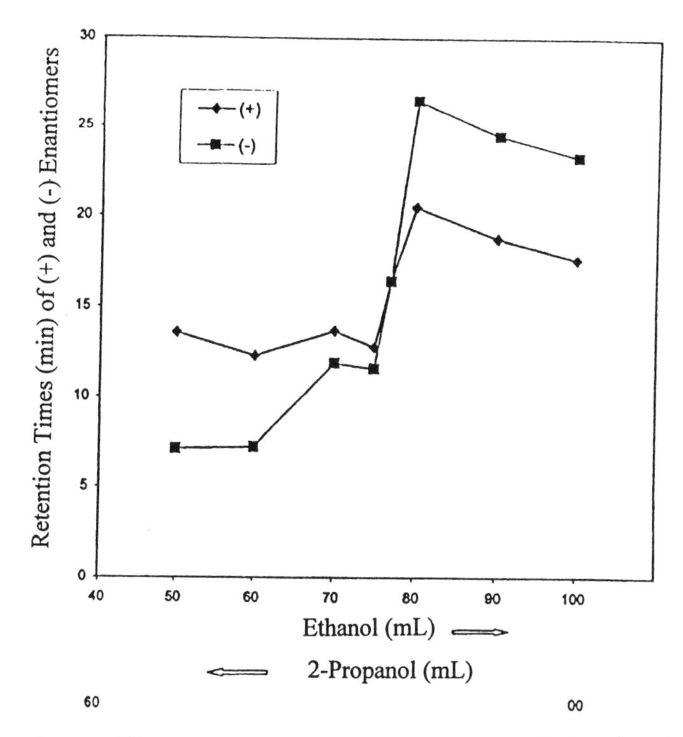

FIGURE 15 Relationship between the percentage of ethanol and 2-propanol and the retention times of (+)- and (–)-enantiomers of nebivolol. (From Ref. 63.)

been concluded that the reversal in elution order was due in part to the chiral grooves on the amylose CSP, which in turn were responsible for the different bond magnitudes between the CSP and the enantiomers, and that the type of alcohol used as the mobile phase influenced the conformation of the 3,5-dimethylphenylcarbamate moiety on the pyranose ring system of the amylose. Generally, the order of elution was reversed when the carbamate was changed to benzoate or the cellulose to amylose types of CSP [121,122]. The possibility of a change in the elution order may be an interesting tool in analytical and preparative resolution. In quality control applications, where small amounts of one enantiomer must be determined in the presence of a large excess of the other enantiomer, it is far easier to quantify a small peak in front of a large one than to do the reverse. In preparative chromatography, the first eluting peak is, in general, easy to isolate with good yields. This is certainly not always the case for the second elution product.

TABLE 3 Calculated Values of F_{et}, F_{pr}, and $F_{et} + F_{pr}$ on Chiralpak AD CSP

Volume (mL)		Values[a]		
EtOH	2-PrOH	F_{et}	F_{pr}	$F_{et} + F_{pr}$
00	100	0.00	71.00	71.00
50	50	−10.50	35.50	25.00
55	45	−11.55	31.95	20.40
60	40	−12.60	28.40	15.80
65	35	−13.65	24.85	11.20
70	30	−14.70	21.30	6.35
75	25	−15.75	17.75	2.00
77	23	−16.17	16.33	0.16
80	20	−16.80	14.20	−2.60
85	15	−17.85	10.65	−7.20
90	10	−18.90	7.10	−11.80
95	05	−19.95	3.55	−16.40
100	00	−21.00	0.00	−21.00

[a] $F = V \times E$, where F, V and E are factor, volume of alcohols, and E energy of interaction, respectively. F_{et} and F_{pr} are factors for ethanol and 2-propanol, respectively.
Source: Ref. 63.

Chiral resolution on polysaccharide-based CSPs in the reversed-phase mode is carried out by using aqueous mobile phases. Again the selection of the mobile phase depends on the solubility and other properties of the drugs to be analyzed. Choices of a mobile phase in the reversed-phase mode are very limited. Water is the main constituent of the mobile phases. The modifiers used are acetonitrile, methanol, and ethanol. Chiral resolution is optimized by adding a small percentage of amines or acids (0.1–1.0%). Some resolutions are pH dependent and require a constant pH for the mobile phase. Under such conditions, generally, the resolution is not reproducible when mobile phases such as water–acetonitrile or water–methanol are used, and therefore, buffers containing organic modifiers (acetonitrile, methanol, etc.) have been used as the mobile phase.

Resolution is optimized by adjusting the buffer pH and the amount of organic modifiers. The most commonly used buffers are perchlorate, acetate, and phosphate. The protocol of the selection and optimization of the mobile phase for the enantiomeric resolution of drugs on polysaccharide-based CSPs in reversed-phase mode is presented in Scheme 2. Table 4 correlates the effects of separation conditions for neutral, acidic, and basic drugs on polysaccharide-based CSPs. From Table 4, it may be concluded that a simple mixture of water and an organic modifier will produce chiral separation of a neutral molecule because there is no competing ionic interaction. For acidic analytes, it is essential to use an acidic mobile phase to suppress the dissociation of the analyte and to minimize ionic interactions. For basic compounds, a mobile phase having a pH greater than 7 is suitable because hydronium cation is less likely to form at this pH and ionic interactions do not occur. Recently, Ye et al. [123] reported that sulfonic acid is a better organic mobile phase modifier than trifluoroacetic acid for the chiral resolution of amino acids. In another study, the same authors [124] observed a memory effect of mobile phase additives in the chiral resolution of some amino acids on a Chiralpak AD column. The authors studied the effects of acidic and basic mobile phase additives in detail. Similarly, Perrin and coworkers [125] developed a strategy for rapid screening of the chiral resolution of β-adrenergic blockers, profens, benzodiazepines, and some other drugs on cellulose- and amylase-based CSPs. The authors reported a dramatic change in chiral resolution when diethylamine, triethylamine, ethanol, and 2-propanol were used separately as organic mobile phase modifiers.

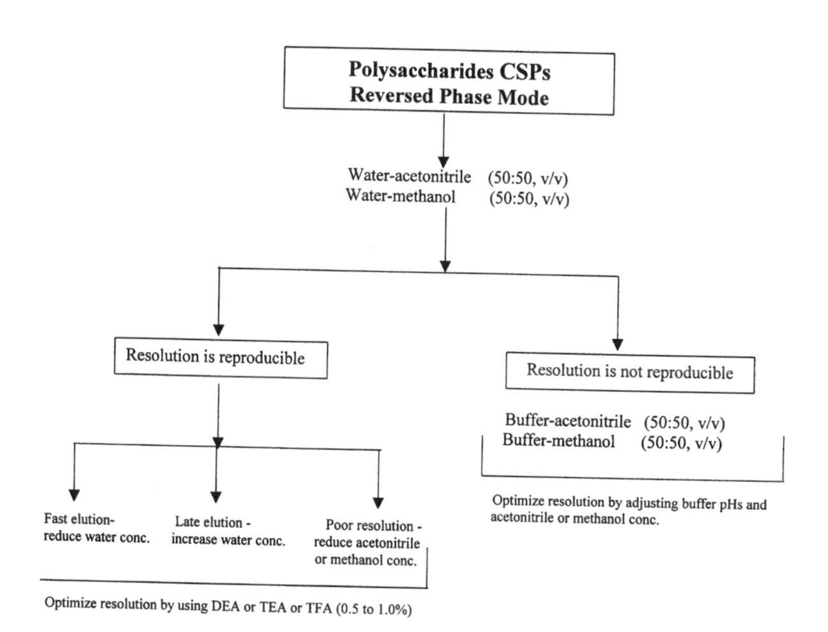

SCHEME 2 Protocol for the development and optimization of mobile phases on polysaccharide-based CSPs in the reversed-phase mode. This brief outline gives the procedure for developing a resolution on polysaccharide-based CSPs in the reversed-phase mode. However, other mobile phases may be used.

TABLE 4 Correlation of Separation Conditions for Neutral, Acidic, and Basic Compounds

	Systems[a]	
Compounds	Normal phase	Reversed phase[b]
Neutral	MP = IPA–hexane pH has no effect on the resolution	MP = water–ACN pH has no effect on the resolution
Acidic	MP = IPA–hexane–TFA pH near 2.0	MP = pH 2.0 perchlorate, acid–ACN
Basic	MP = IPA–hexane–DEA, IPA–hexane–TFA with pH near 2.0, ion-pair separation	MP = pH < 7 buffer–ACN Typical buffer is 0.5 M $NaClO_4$

[a] Abbreviations: MP, mobile phase; IPA, isopropanol; ACN, acetonitrile; TFA, trifluoroacetic acid; DEA, diethylamine.
[b] Columns normally are not run under basic conditions.

Resolution was effected by changing the polarity of the mobile phases. It is very interesting to note that the change in resolution with respect to mobile phase compositions varied from compound to compound. Resolution on polysaccharide-based CSPs in the reversed-phase mode was improved by adding cations and anions. For propranolol enantiomers on Chiralcel OD-R with sodium perchlorate salt–acetonitrile (60 : 40, v/v) as the mobile phase, in the presence of cations, the order of retention was $Na^+ > Li^+ > K^+ > NH_4^+ > N(C_2H_5)_4^+$, while in the presence of anions this order was $ClO_4^- > SCN^- > I^- > NO_3^- > Br^- > Cl^- > AcO^-$ [90].

Another study recorded the effect of the concentration of ammonium acetate on the chiral resolution of metomidate and etomidate on a Chiralpak AD-R column.. An increase in k values was observed with an increase in ammonium acetate concentration. On the other hand, the values of α decreased with an increase in ammonium acetate concentration [9]. The same trend was also observed on Chiralcel OD-R CSP for the chiral resolution of propranolol and trimepramine racemates [9].

Kazusaki et al. [91] studied the effect of acetonitrile concentration on the resolution of *RS*-2-(4-bromo-2-fluorobenzyl)-(1,2,3,4-tetrahydropyrrolo[1,2-*a*]-pyrazine-4-spiro-3′-pyrrolidine)-1,2′,3,5′-tetrone, and the results are given in Figure 16a,b. These authors reported that the resolution factor decreased with an increase in acetonitrile concentration. Aboul-Enein and Ali [92] have studied the effect of acetonitrile content on the chiral resolution of flubiprofen on Chiralpak AD-RH columns at different temperatures. Figure 16c indicates the

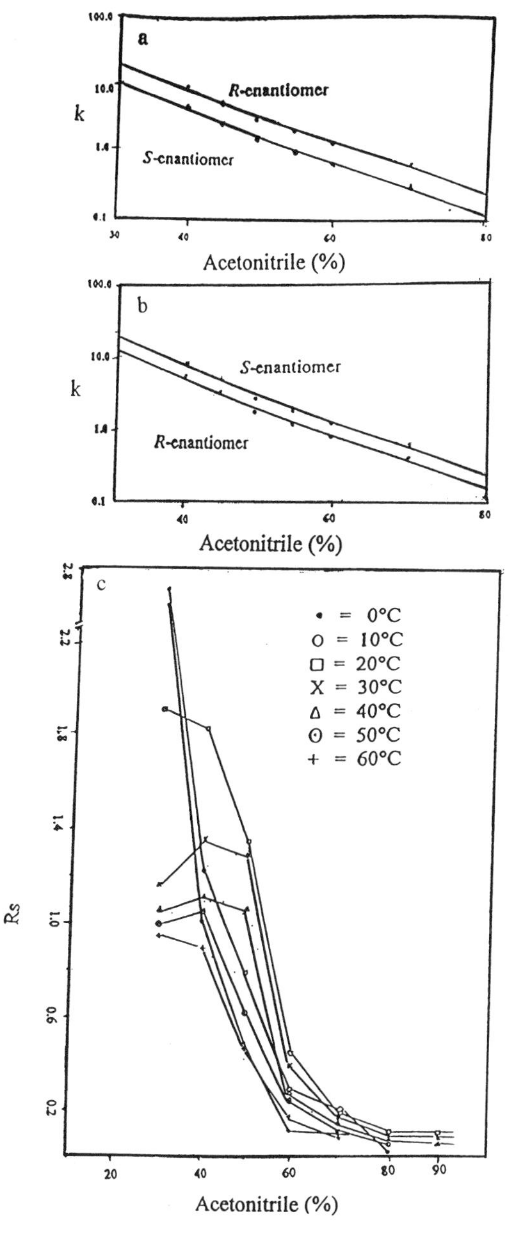

FIGURE 16 Effects of acetonitrile percentage on the enantiomic resolution of (a) *RS*-2-(4-bromo-2-fluorobenzyl)-(1,2,3,4-tetrahydropyrrolo[1,2-*a*]pyrazine-4-spiro-3′-pyrrolidine)-1,2′,3,5′-tetrone on Chiralpak AD-RH and (b) Chiralcel OD-RH, and (c) flubiprofen on Chiralpak AD-RH CSP at different temperatures. (From Refs. 91, 92.)

pattern of flubiprofen resolution with different percentages of acetonitrile. Again, a high concentration of acetonitrile resulted in poor resolution.

Many researchers have documented the effect of the mobile phase on the enantioselectivity of different racemates on polysaccharide-based CSPs. But no comprehensive study aimed at identifying an association between the structural features of the solute and the appropriate mobile phase conditions has yet been proposed. Piras et al. [126] have studied the characteristic features of about 2363 racemic molecules separated on a Chiralcel OD CSP. The mobile phases used for these racemates were compared with their structures available from CHIRBASE (www.chirbase). The data setup was submitted to data mining programs for molecular pattern recognition and mobile phase predictions for new cases. Some substructural solute characteristics were related to the efficient use of some specific mobile phases. For example, the application of acetonitrile–salt buffer at pH 6 to 7 was found to be convenient for reversed-phase separation of compounds bearing a tertiary amine functional group. Furthermore, cluster analysis allowed the arrangement of the mobile phase according to similarities found in the molecular patterns of the solutes.

2.6.2 pH of the Mobile Phase

Also controlling the chiral resolution of different racemic drugs on polysaccharide-based CSPs is pH value. Aboul-Enein and Ali [93,94] have observed that chiral resolution on polysaccharide-based CSPs is pH dependent in the normal phase mode. Only partial resolution of certain antifungal agents was achieved at lower pH, while resolution was improved by increasing the pH with triethylamine on amylose and cellulose chiral columns. Aboul-Enein and Ali [94] studied the chiral resolution of tetralone derivatives over a range of pH values on Chiralpak AD-RH columns [results shown later (Fig. 18)]. The mobile phase containing triethylamine was better than the mobile phase having acetic acid for the resolution of these derivatives. Therefore, the enantiomeric resolution of tetralone derivatives was pH dependent and was better at higher pH values. The poor resolution at lower pH might be due to the steric effect exerted by the protonated nitrogen of the pyridine ring of tetralone derivatives. The effect of pH on the retention factors of flubiprofen and alprenolol was investigated on polysaccharide-based CSPs in the reversed-phase mode [127] (Fig. 17). It can be observed that the resolution of flubiprofen and alprenolol was better at about 3.0 pH. Recently, Aboul-Enein and Ali [92] studied the effect of pH on the chiral resolution of flubiprofen on Chiralpak AD-RH columns, and results are shown in Figure 17c, which indicates that 3.5 was the best pH value for the resolution of flubiprofen. In reversed-phase mode, the chiral resolution on polysaccharide CSPs are pH dependent; therefore, buffers have been used to achieve the best resolution.

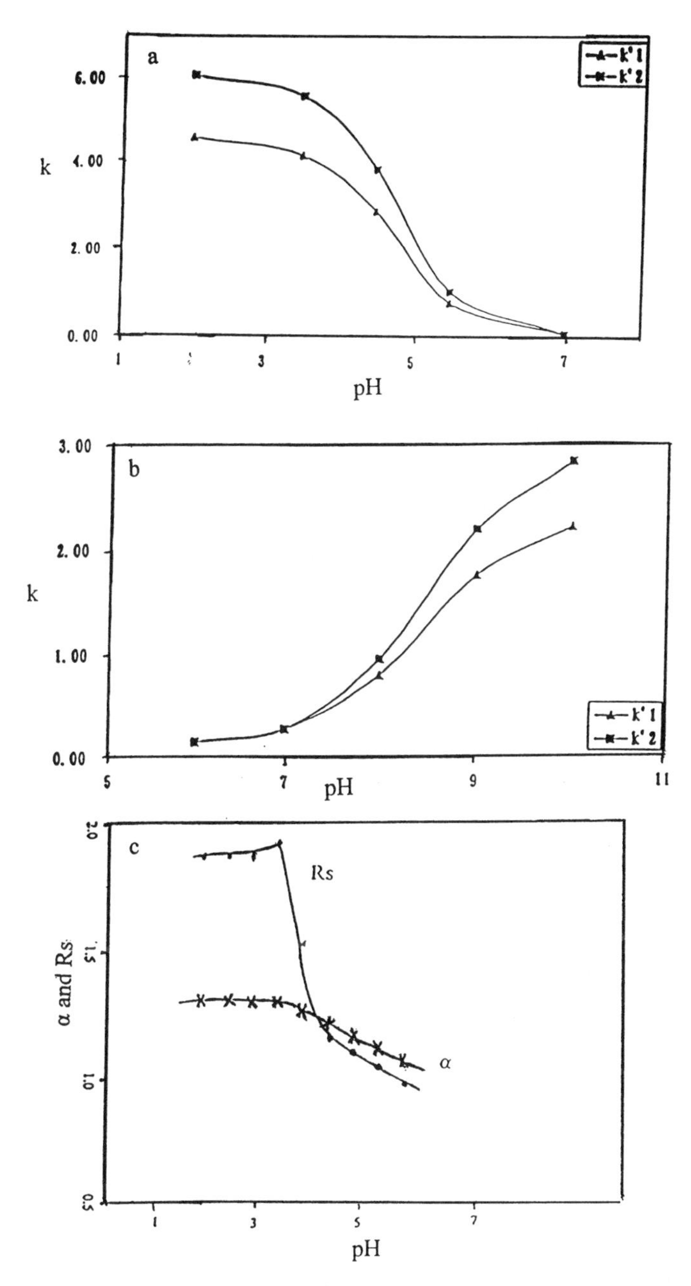

FIGURE 17 Effects of pH on the retention, separation, and resolution factors of (a) flubiprofen, (b) alprenolol, and (c) flubiprofen. [Figures (a) and (b) from Ref. 127; Fig. (c) from Ref. 92.]

2.6.3 Flow Rate

Chiral resolution can be controlled by regulating flow rate on polysaccharide-based CSPs. However, there are only a few studies dealing with the optimization of chiral resolution by adjusting flow rates. Aboul-Enein and Ali [93] have optimized the chiral resolution of some antifungal agents (Fig. 18) on Chiralpak AD, AS, and AR CSPs by adjusting flow rates. The flow rates were varied between 0.5 and 2.0 mL/min, but the best resolution was achieved at a flow rate of 0.5 mL/min. Table 5 shows values for separation factor α and resolution *Rs* of the antifungal agents at flow rates of 0.5 and 1.0 mL/min. There was no resolution for miconazole and sulconazole, while only partial resolution of econazole was obtained when the flow rate was 1 mL/min. Therefore, a 0.5 mL/min flow rate was found suitable for this study. It is interesting to note that there was no improvement in resolution when the flow rate was decreased below 0.5 mL/min. In another study, Aboul-Enein and Ali [94] optimized the chiral resolution of certain tetralone derivatives (Fig. 18) on a Chiralpak AD-RH CSP. The results of these findings are given in Table 6, where k_1 and k_2 are, respectively, negative and positive retention factors. Similarly, the flow rates were varied from 0.5 mL/min to 2.0 mL/min. It was observed that the 0.5 mL/min flow rate was suitable for the chiral resolution of most of the derivatives. All the tetralone derivatives studied resolved completely at 0.5 mL/min when water–acetonitrile–triethylamine (50 : 50 : 0.03, v/v/v) served as the mobile phase. However, partial resolution of tetralone derivative IV was observed at 0.5 mL/min when water–acetonitrile–acetic acid (60 : 40 : 0.03, v/v/v) was the mobile phase. This study shows that the optimization of chiral resolution can be achieved by adjusting the flow rate.

2.6.4 Temperature

Temperature also contributes to the chiral resolution of racemic drugs on polysaccharide-based CSPs. Working at elevated temperatures can often be beneficial, especially in preparative chromatographic separation on a larger scale or for analytical purposes, if a good separation of the enantiomers can be achieved only at very high values of k. Even then only a few studies dealing with the influence of temperature on chiral resolution are available. The retention and separation factors may be related to temperature by the following equations:

$$\ln k = -\frac{\Delta H^\circ}{RT} + \frac{\Delta S^\circ}{R} + \ln \Phi \qquad (3)$$

$$\ln \alpha = \ln\left(\frac{k_2}{k_1}\right) = -\frac{\delta H^\circ}{RT} + \frac{\delta S^\circ}{R} \qquad (4)$$

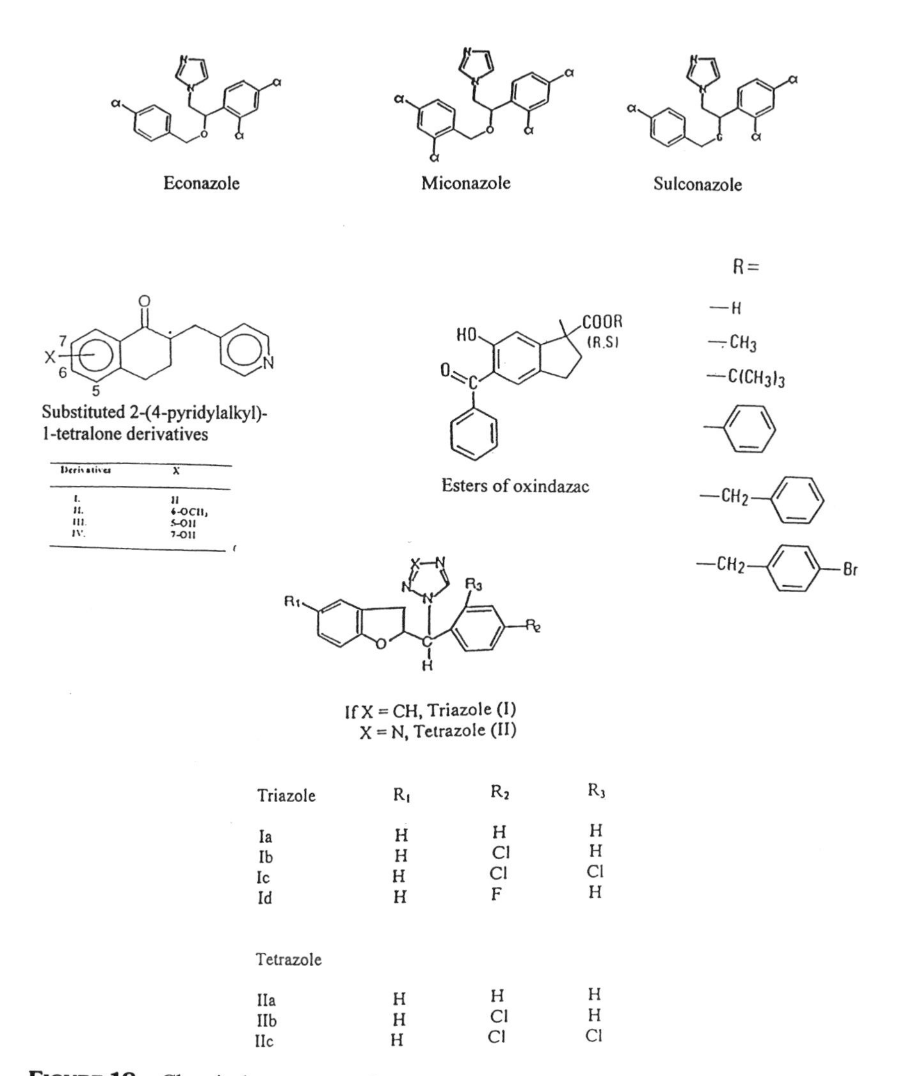

Triazole	R_1	R_2	R_3
Ia	H	H	H
Ib	H	Cl	H
Ic	H	Cl	Cl
Id	H	F	H
Tetrazole			
IIa	H	H	H
IIb	H	Cl	H
IIc	H	Cl	Cl

FIGURE 18 Chemical structures of econazole, miconazole, sulconazole, tetralone derivatives, and aromatase inhibitors.

where R is the ideal gas constant, T is the absolute temperature, and Φ is the phase ratio [128]; ΔH° and ΔS° represent the enthalpy and entropy terms for each enantiomer, δH° and δS° represent their differences, and k_2 and k_1 are the retention factors of the two resolved enantiomers. According to these equations, both retention factors and separation factors are controlled by an enthalpic

TABLE 5 Effects of Flow Rate and Substituents on Chiral Resolution of Imidazole Antifungal Agents on Amylose CSPs Using Hexane–2-Propanol–Diethylamine (400 : 99 : 1, v/v/v) as the Mobile Phase

	0.5 mL/min		1.0 mL/min	
	α (−)	*Rs* (+)	α (−)[a]	*Rs* (+)
Chiralpak AS				
Econazole	1.63	5.32	1.62	2.66
Miconazozle	1.56	4.69	1.54	2.89
Sulconazole	1.48	5.68	1.48	3.08
Chiralpak AD				
Econazole	1.05	1.42	1.05	0.37
Miconazole	1.06	1.26	1.06	0.32
Sulconazole	1.16	3.60	1.04	1.36
Chiralpak AR				
Econazole	1.07	0.45	1.07	0.40
Miconazole	1.05	0.32	NR	—
Sulconazole	—	—	NR	—

[a] NR, not resolved.
Source: Ref. 93.

contribution, which decreases with elevation of temperature, and a temperature-dependent entropic contribution. The selectivity is a compromise between differences in enantiomeric binding enthalpy and disruptive entropic effects [129].

The effect of temperature on the retention factor of nebivolol enantiomers on Chiralpak AD, given in Figure 19 [9], indicates poor resolution at higher temperature. Moreover, 45°C was found to be the best temperature for the resolution of nebivolol enantiomers, posting the best resolution of all 10 enantiomers, as indicated in Figure 20, which gives their peak shapes and resolution at 25, 35, and 45°C. Furthermore, when the resolution of benzotriazole derivatives on Chiralcel OJ is given as a function of temperature (Fig. 21a) [9], we again see a decrease in resolution at higher temperature. Moreover, the lowest resolution is observed when methanol was used. For this solvent, the resolution factor diminishes by a very small extent with increasing temperatures. The resolution factors are the highest for the hexane–ethanol mixture. The effect of temperature on the chiral resolution of *RS*-2-(4-bromo-2-fluorobenzyl)-(1,2,3,4-tetrahydropyrrolo[1,2-*a*]pyrazine-4-spiro-3′-pyrrolidine)-1,2′,3,5′-tetrone on a

TABLE 6 Effects of Flow Rate and Substituents on the Chiral Resolution of Tetralone Derivatives (Fig. 18)

Conditions, compounds	k_1	k_2	α	R_s
MP, water–acetonitrile–triethylamine (50 : 50 : 0.03, v/v/v); flow rate, 0.5 mL/min				
I	6.59	8.42	1.28	1.55
II	6.59	9.82	1.42	2.02
III	1.67	4.79	2.84	2.64
IV	1.53	2.98	1.95	1.10
MP, water–acetonitrile–triethylamine (50 : 50 : 0.03, v/v/v); flow rate, 1.0 mL/min				
I	6.83	8.71	1.28	1.10
II	6.97	9.87	1.42	1.50
III	1.49	4.04	2.71	1.61
IV	1.30	1.82	1.23	1.00
MP, water–acetonitrile–acetic acid (50 : 50 : 0.03, v/v/v); flow rate, 0.5 mL/min				
I	5.45	6.63	1.22	1.50
II	5.82	5.35	1.43	1.60
III	4.12	7.92	1.92	1.80
IV	3.23	3.76	1.16	0.40
MP, water–acetonitrile–acetic acid (50 : 50 : 0.03, v/v/v); flow rate, 0.5 mL/min				
I	5.54	6.74	1.22	0.80
II	6.78	8.48	1.26	1.00
III	4.21	7.42	1.76	1.10
IV	3.26	3.80	1.17	0.20

Source: Ref. 94.

Chiralpak AD-RH column was studied by Kazusaki et al. [95], and the results are shown in Figure 21b. The chiral resolution of flubiprofen on Chiralpak AD-RH columns was studied by Aboul-Enein and Ali [92]. They observed that the best resolution occurred at 25°C. Finally, the influence of temperature on enantioselectivity differs from one analyte to another. Thus researchers have an ideal tool for optimizing throughput per unit time in preparative chromatography or a possibility for improving the sensitivity of analytical methods.

Dingenen [9], who studied the effect of the mobile phase velocity on the height equivalent to a theoretical plate (HETP) at different temperatures for benzotriazole derivatives, obtained the results shown in Figure 22, which represents the HETP values found for methanol and the hexane–ethanol mixture. Both curves of Figure 22 clearly demonstrate that the kinetic circumstances are less favorable at low temperatures. A slow mass transfer between the two phases clearly determines the band-broadening process at temperatures below 20°C. This

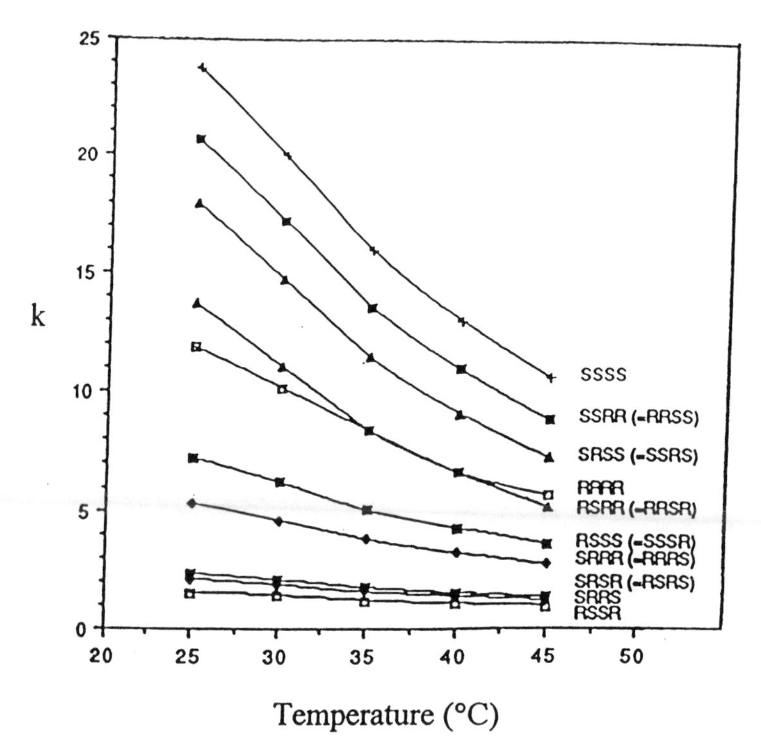

FIGURE 19 Effects of temperature on the retention factor of nebivolol enantiomers on Chiralpak AD column, with ethanol as the mobile phase. (From Ref. 9.)

effect is least pronounced when methanol is used as mobile phase. Thus the favorable effect of a temperature decrease on enantioselectivity will be neutralized, to a great extent, by strongly diminishing column efficiency. Finally, within the temperature range investigated, some of the racemates posted decreases of about 45% in retention factor values, while for some other racemates the k values remained unchanged. Briefly, since the thermodynamic and kinetic parameters govern the interactions between CSPs and enantiomers, the effect of temperature on enantiomeric resolution is very important.

2.6.5 Solute Structures

Chiral resolution on polysaccharide-based CSPs is due to the different types of bonding between racemates and CSP, as discussed later in this chapter. Therefore, different racemate structures provide bondings of different types, which in turn means that different patterns of chiral recognition will be observed. The effects of

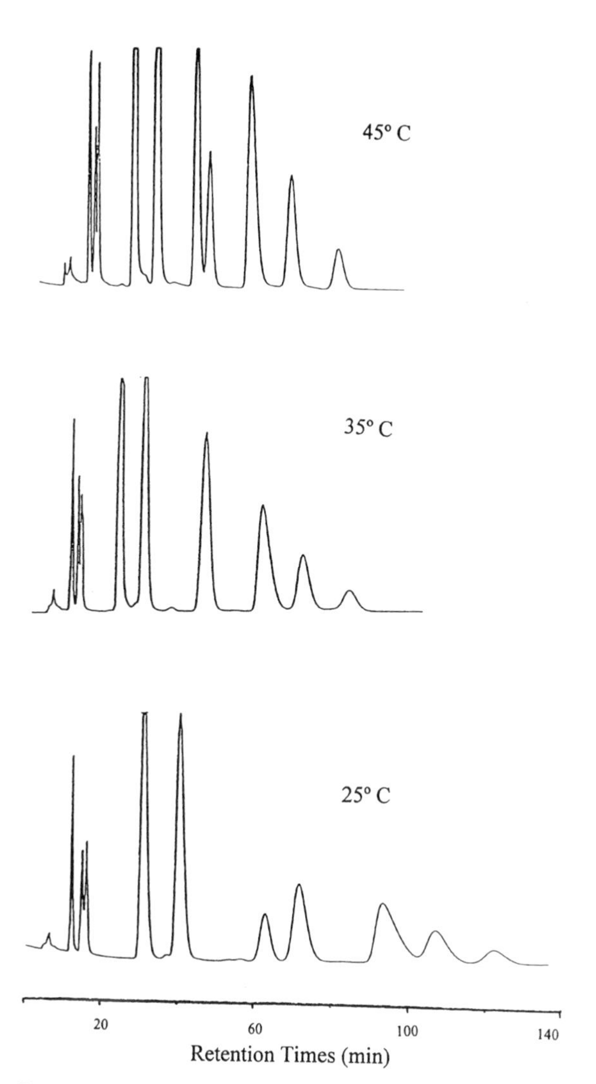

FIGURE 20 Chromatograms of nebivolol enantiomers on a Chiralpak AD column at 45°, 35°, and 25°C temperature. (From Ref. 9.)

different groups on the chiral resolution of oxindazac esters were studied by Francotte et al. [96]. From their results (Table 7), it is clear that the separation and resolution factors increased when hydrogen was replaced by a phenyl group or a phenyl group containing an electron-withdrawing group, which indicates, probably, the major role of π–π interactions and hydrogen bonding between the CSP

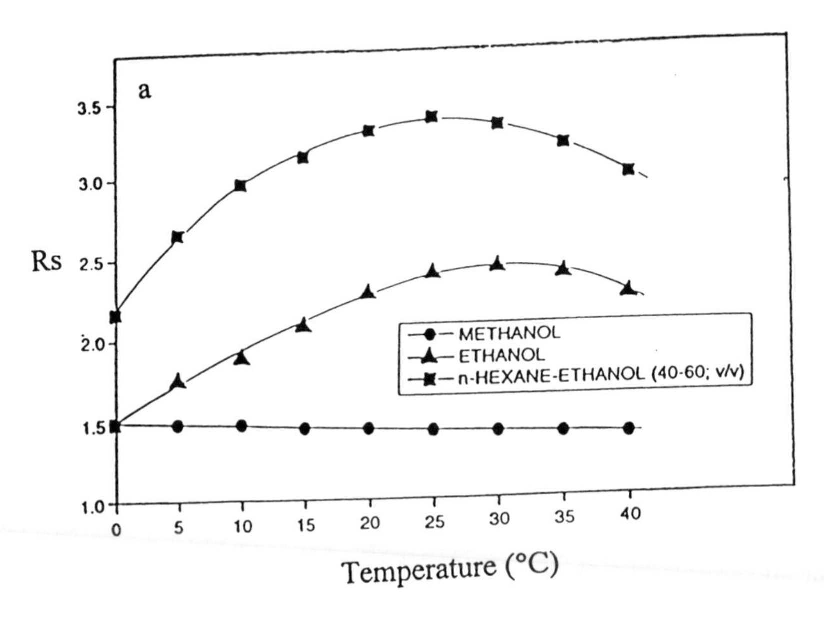

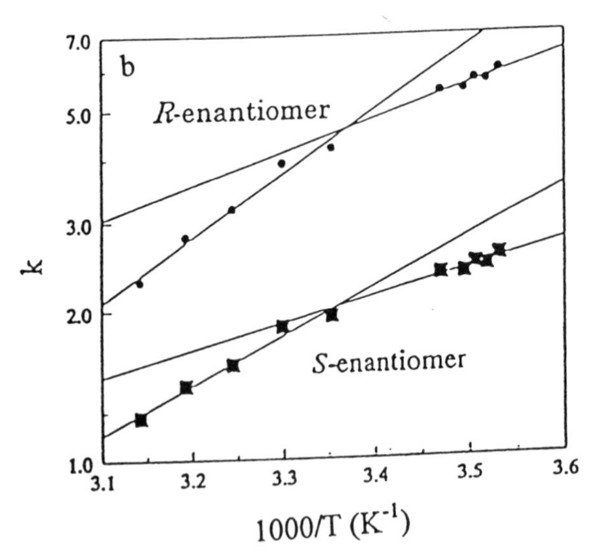

FIGURE 21 (a) Effects of temperature on the resolution of benzotriazole derivatives on a Chiralcel OJ column with methanol, ethanol, and hexane–ethanol (40 : 60, v/v) as the mobile phases separately and respectively. (b) Effects of temperature (van't Hoff plot) on the resolution of *RS*-2-(4-bromo-2-fluorobenzyl)-(1,2,3,4-tetrahydropyrrolo[1,2-*a*]pyrazine-4-spiro-3′-pyrrolidine)-1,2′,3,5′-tetrone on a Chiralpak AD-RH column; mobile phase was 0.01 M acetate buffer (pH 4.7)–acetonitrile (50 : 50, v/v). (From Ref. 95.)

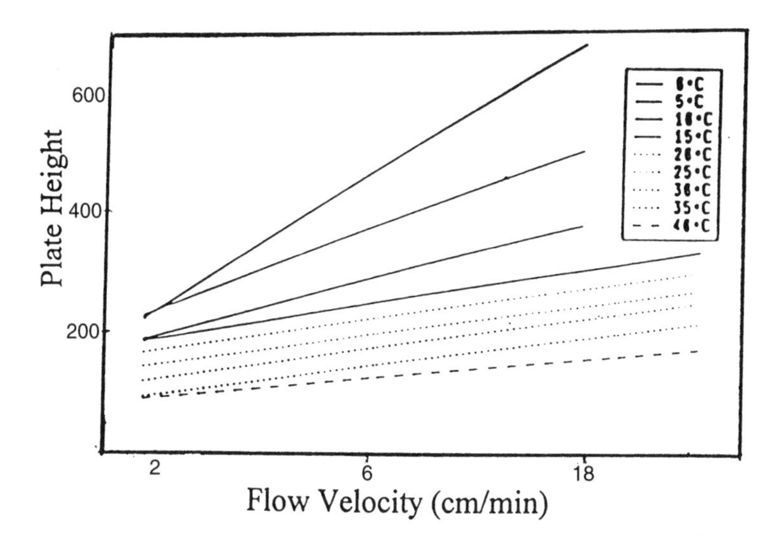

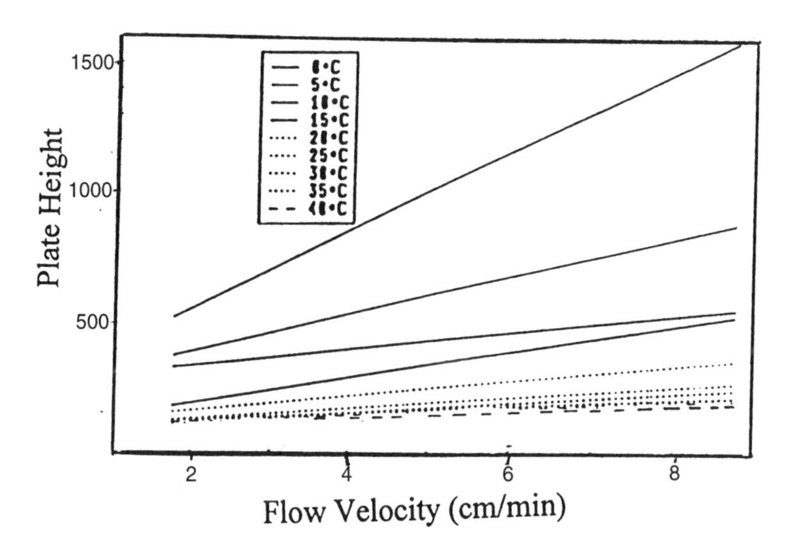

FIGURE 22 Dependence of height equivalent theoretical plate (HETP) on flow velocity at different temperatures on a Chiralcel OJ CSP. (From Ref. 9.)

and the enantiomers. Further, Francottte and Wolf [50] studied the effects of alkyl groups on the chiral resolution of some aromatic alcohols. They found that resolution decreased when the alkyl chain was increased in length. These authors

TABLE 7 Chromatographic Resolution of Esters of Oxindazac (Fig. 18) on a Triacetyl Column (60 × 1.3 cm) Using 95% Ethanol as the Mobile Phase

Compounds, R	Retention factors $k_1(-)$	$k_2(+)$	Separation factor α	Resolution R_s
Methyl	2.67	4.16	1.56	2.10
Isopropyl	1.10	1.72	1.55	1.50
Phenyl	4.13	8.27	2.00	3.00
Benzoyl	4.15	7.94	1.91	2.90
4-Bromobenzoyl	3.68	15.23	4.14	2.10

Source: Ref. 96.

concluded that the decrease in chiral resolution may be due to the steric effect. In the same study [50], they also investigated the effect of electron-donating (alkyl) and electron-withdrawing (halogen) groups on the chiral resolution of phenyl-cycloalkane and lactone derivatives. The substitution of halogens for hydrogen in phenylcycloalkane improved the resolution, while the resolution decreased when hydrogen was replaced by alkyl groups. The resolution in lactone derivatives was decreased by increasing the size of the alkyl chain. Francotte and Wolfe explained their results on the basis of different magnitudes of bonding.

Later, Wainer et al. [62,97] observed that the increase in the π basicity of the solute (substituting a phenyl group for hydrogen) resulted in improved chiral resolution of amides on cellulose tribenzoate (OB CSP) [97], while for certain aromatic alcohols resolution was decreased when the π basicity on a Chiralcel OB column was increased [62].

More recently, Aboul-Enein and Ali [94,98,99] studied the effects of substituents of certain molecules on their chiral resolution on polysaccharide-based CSPs. The racemates used for this purpose were aromatase inhibitors, tetralone derivatives, and antifungal agents (Fig. 18); for the results, refer again to Tables 5 and 6. The effects of substituents on the chiral resolution of aromatase inhibitors (triazole and tetrazole derivatives) on Chiralpak AD-RH, Chiralcel OD-RH and Chiralcel OJ-R columns are presented in Tables 8 to 10. It is interesting to note that more sulconazole is retained than econazole and miconazole. This may be because the strong coordination bonding between sulconazole and the stationary phases of sulfur allows stronger coordination bonding than is possible with oxygen. Miconazole is less well retained than econazole owing to the steric effect exerted on miconazole by the extra chlorine atom. Furthermore, micona-zole is less well retained than sulconazole owing to steric effects (due to the presence of the extra chlorine atom in miconazole) and the lack of coordination

TABLE 8 Effects of Substituents on the Chiral Resolution of Aromatase Inhibitors (Triazole and Tetrazole Derivatives, Fig. 18) on a Chiralpak AD-RH CSP

Compounds	Retention factors $k_1(-)$	$k_2(+)$	Separation factor α	Resolution R_s
MP, 2-PrOH–MeCN (90 : 10, v/v)				
Ia	1.03	2.04	1.98	0.90
Ib	1.16	2.42	2.08	1.12
Id	0.93	1.71	1.84	0.71
IIa	0.40	0.55	1.26	0.11
IIb	0.42	0.75	1.79	1.50
MP, 2-PrOH–MeCN (50 : 50, v/v)				
Ia	5.91	13.29	2.29	4.92
Ib	6.89	16.72	2.43	3.86
Ic	5.75	6.79	1.18	1.39
Id	6.18	12.61	2.04	3.65
IIa	3.09	3.72	1.20	1.28
MP, MeCN–H_2O (50 : 50, v/v)				
Ia	4.81	5.51	1.16	1.52
Ib	8.32	12.45	1.50	6.72
Ic	14.89	17.08	1.15	1.64
Id	6.98	8.05	1.15	2.64
IIa	8.82	9.84	1.12	1.24
MP, MeCN–H_2O (80 : 20, v/v)				
IIa	0.47	1.12	2.39	1.77
IIb	0.66	2.05	3.11	3.86
IIc	0.57	1.55	2.72	2.72
MP, MeCN–H_2O (95 : 5, v/v)				
IIb	0.12	0.68	5.63	1.98

Source: Ref. 98.

bonding in miconazole. In the case of tetralone derivatives, the order of elution, which was III > II > I > IV, may be attributable to the type of substitution on the aromatic ring of these compounds, which resulted in hydrogen bonds of different magnitudes. Additionally, substitution on the aromatic ring of tetralone derivatives also exerted steric effects of different magnitude. The chiral resolution behavior of aromatase inhibitors differed on Chiralpak AD-RH, Chiralcel OD-RH, and Chiralcel OJ-R columns because of the different groups attached to the aromatic rings of these derivatives. Therefore, different racemates exhibit

TABLE 9 Effects of Substituents on the Chiral Resolution of Aromatase Inhibitors (Triazole and Tetrazole Derivatives, Fig. 18) on a Chiracel OD-RH CSP

Compounds	Retention factors $k_1(-)$	$k_2(+)$	Separation factor α	Resolution R_s
MP, 2-PrOH–MeCN (90 : 10, v/v)				
IIb	1.04	2.12	2.03	0.67
IIc	1.06	2.26	2.13	1.95
MP, MeCN–H_2O (50 : 50, v/v)				
Ia	9.78	9.98	1.02	0.38
Ib	9.95	10.35	1.04	0.43
Ic	12.05	12.90	1.07	0.51
IIc	7.58	9.45	1.25	0.48
MP, MeCN–H_2O (80 : 20, v/v)				
Ia	0.57	0.63	1.11	0.26
Ib	0.59	0.70	1.19	0.31
Ic	0.82	1.48	1.80	2.73
Id	0.94	0.99	1.05	0.80
IIb	0.32	0.41	1.28	0.25
IIc	0.75	1.12	1.49	1.55

Source: Ref. 98.

TABLE 10 Effects of Substituents on the Chiral Resolution of Aromatase Inhibitors (Triazole and Tetrazole Derivatives, Fig. 18) on Chiracel OJ-R CSP

Compounds	Retention factors $k_1(-)$	$k_2(+)$	Separation factor α	Resolution R_s
MP, MeCN–H_2O (50 : 50, v/v)				
Ib	4.04	4.52	1.12	0.49
Ic	6.03	6.56	1.09	1.60
Id	4.76	5.00	1.05	0.31
IIb	3.22	3.69	1.15	0.52
IIc	9.65	10.40	1.08	1.55
MP, MeCH–H_2O (80 : 20, v/v)				
Ia	0.48	0.73	1.52	0.33
Ib	0.34	0.57	1.68	0.30
Ic	0.68	0.89	1.31	0.29
Id	0.55	0.80	1.46	0.33

Source: Ref. 98.

different chiral recognition on the same CSP because of various bondings between the racemates and the CSP. Therefore, a single CSP cannot be used for the chiral resolution of all racemic compounds. Ding et al. [130] studied the effects of various groups on the chiral resolution of indanediol and related racemates. They observed that the inclusion mechanism is the major contributor to enantioselectivity, with the size and shape of the racemates rather than electrostatic interactions involving the functional groups of the solute.

2.6.6 Other Parameters

In addition to the optimization of HPLC factors as just discussed, chiral resolution may be improved by adjusting other parameters. In 1996 Okamoto et al. [66] studied the effects of silica gel pore size, coating amount, and coating solvent on the chiral discrimination of some aromatic racemates. They found that CSPs with a silica gel having a large pore size and small surface area showed higher chiral recognition. CSPs coated with acetone solvent showed good chiral resolution capacity. Resolution was improved by preparing new derivatives of cellulose and amylase, which contain groups capable of good capacities for bonding with racemates [71]. Recently we observed very interesting results during the resolution of a methylphenidate racemate. The partial resolution of methylphenidate on Chiralcel OB was improved to complete resolution when phenol and benzoic acid were used separately as mobile phase additives [100]. In addition to the foregoing parameters, the optimization of chiral resolution on these CSPs can be achieved by varying other HPLC conditions such as particle size of CSP, column dimensions, racemic compound concentration, and choice of detector.

2.7 CHIRAL RECOGNITION MECHANISMS

For applications and development of chiral resolution methods, it is essential to have knowledge of the chiral resolution mechanism on polysaccharide-based CSPs. At the molecular level, this mechanism is still unclear because of difficulties associated with spectroscopic studies, as discussed earlier. Nevertheless, some experimental efforts have been made, and chiral resolution reportedly has been achieved through different types of bonding on the chiral grooves of polysaccharide-based CSPs.

Hesse and Hagel [41] and later Francotte et al. [64] proposed an inclusion mechanism by which enantiomers may be adsorbed in the chiral grooves of the CTA-I matrix. Other theoretical [110] and X-ray studies of the model compound, a fully acetylated *D*-glucopyronose-(*R*)-phenylethylamine inclusion complex [111], also support an inclusion mechanism. It has been established that the main chiral sites of bonding are polar carbonyl groups of esters, which can interact with racemic compounds through hydrogen bonding and dipole–dipole interactions for chiral discrimination [17].

Wainer and Alembic [97] proposed a similar mechanism based on the separation properties of a series of aromatic amides [17] and alcohols [62] on CTB phases; the mechanism of retention is an attractive binding–steric fit formulation involving hydrogen bonding and dipole–dipole interactions rather than inclusion. Wainer et al. [62] explained the mechanism by citing (1) the formation of a CSP complex through hydrogen-bonding interaction between the solute and the CSP, (2) the stabilization of this complex through insertion of the aromatic portion of the solute into a chiral groove of the CSP, and (3) chiral discrimination between enantiomeric solutes due to differences in their steric fit in the chiral groove.

The most important adsorbing sites on the phenylcarbamate derivatives are probably the polar carbamate groups as shown in Figure 23. These groups are capable of interacting with a racemic compound via hydrogen bonding with –NH– and >CO groups and dipole–dipole interaction on >C=O [43]. Therefore, the nature of the substituents on the phenyl ring affects the polarity of the phenyl group, which in turn leads to different chiral recognition capacities. Molecular dynamics calculations were carried out, and the structure of cellulose phenylcarbamate was predicted. Figure 24 shows the stable structure obtained by means of molecular mechanics calculations based on the structure of cellulose tris(phenylcarbamate) (CTPC) proposed on the basis of X-ray analysis (Fig. 25) [30,31]. CTPC has a left-handed threefold (3/2) helix, with glucose units regularly arranged along the helical axis. A chiral helical groove with polar carbamate residues exists along the main chain. The polar carbamate groups are favorably located inside, permitting the insertion of polar enantiomers into the grooves, where they can interact with the carbamate residues via hydrogen

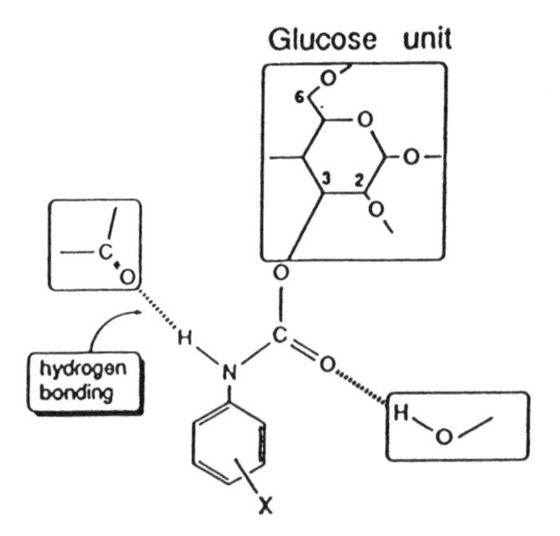

FIGURE 23 Schematic interactions between phenylcarbamate residue and racemates. (From Ref. 1.)

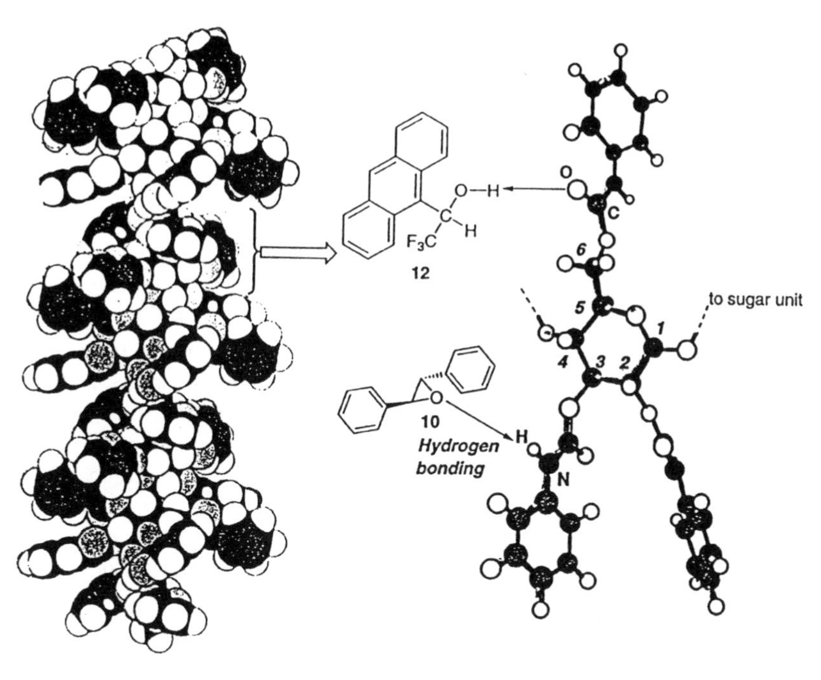

FIGURE 24 Optimized structure (left) and possible interaction sites (right) of a cellulose triphenylcarbamate derivative. (From Ref. 32.)

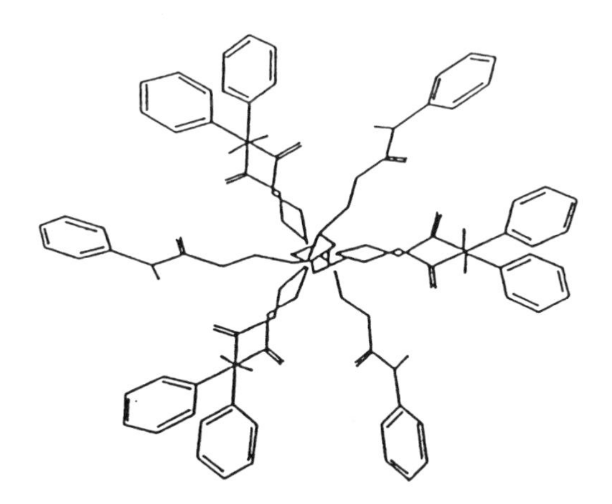

FIGURE 25 The structure of cellulose tris(phenylcarbamate). (From Ref. 1.)

bonding. This interaction results in efficient chiral discrimination. Besides these polar interactions, π–π interactions between the phenyl group of a CSP and aromatic group of a solute may play a role in chiral resolution because several nonpolar aromatic compounds have also been resolved (Table 2).

Lipkowitz [131,132] has extensively studied the mechanism of chiral recognition from a theoretical viewpoint, carrying out computational studies on a chiral discrimination mechanism in cellulose triphenylcarbamate. Yashima et al. [114] concluded that –NH– and >C=O groups are the most important bonding sites. More recently, the same group extended their work [29] and compared chiral recognition between cellulose tris(phenylcarbamate) (CTPC) and cellulose tris(3,5-dimethylphenylcarbamate) (CDMPC), with *trans*-stilbene oxide and benzoin as the racemates. The calculations of interaction energies between CTPC or CDMPC and *trans*-stilbene oxide or benzoin were performed by various force-field methods. Before the interaction energy was calculated, one of two methods was applied: (1) enantiomers were generated and tumbled around the –NH– proton and the >C=O oxygen of the carbamoyl group of CTPC and CDMPC, which are considered to be the most important adsorption sites, or (2) enantiomers were randomly generated by the Monte Carlo method on the surface of CTPC and CDMPC defined by the particular van der Waals radius, using the technique of "blowing up" the atomic radii [133].

For the chiral resolution of *trans*-stilbene oxide or benzoin, the results obtained from both methods were in agreement with the results obtained by chromatographic studies for both CTPC and CDMPC CSPs. However, significant differences of interaction energies between enantiomers appeared only when enantiomers were generated inside CTPC or CDMPC. The results indicate that the polar carbamate residues of cellulose derivatives may be the most important adsorbing sites for polar racemates and may play a crucial role in chiral recognition. The method of energy calculation is shown in Figure 26. These methods are useful for a qualitative understanding of the chiral recognition mechanism of cellulose phenylcarbamate, although the use of molecular dynamics calculations will be needed to simulate the dynamic behavior of the interactions occurring in chromatography.

Recently, Aboul-Enein and Ali [100] carried out certain studies on the chiral resolution of methylphenidate on derivatives of cellulose and amylose. They observed that π–π interactions are also the important binding forces for the chiral resolution of aromatic racemates. The best resolution of methylphenidate (MPH) on a Chiralcel OB column was achieved when phenol or benzoic acid, separately, was used as the mobile phase additive. Phenol (benzoic acid) forms the MPH–phenol (MPH–benzoic acid) pair in which the possibility of π–π interaction between one of these pairs and CSP is greater than the possibility of π–π interactions between MPH and CSP. Therefore, the resolution of MPH enantiomers was improved when phenol or benzoic acid served as the mobile

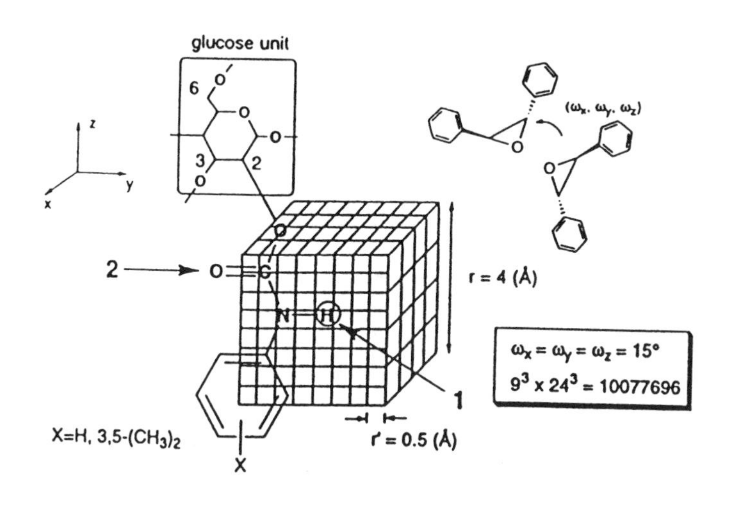

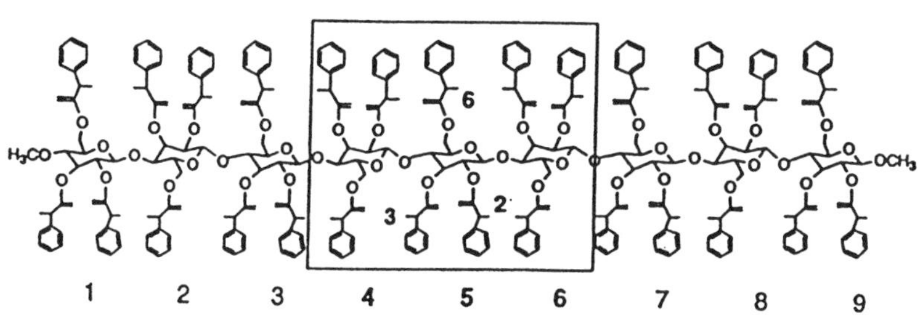

FIGURE 26 Method for calculating the interaction between cellulose tris(phenylcarbamate) derivatives or cellulose tris(3,5-dimethylphenylcarbamate) derivatives and enantiomers of *trans*-stilbene or benzoin. (From Ref. 29.)

phase additive, as evidenced by an enhancement of π–π interactions. In another study, Aboul-Enein and Ali [101] observed that coordination bonding also plays an important role in the chiral resolution of the racemates having a sulfur atom. Therefore, the chiral resolution of sulcanozole was better than the chiral resolution of econazole and miconazole on cellulose and amylose CSPs.

Finally, a look at the structures of polysaccharide-based derivatives (cellulose and amylose) (Fig. 1) clearly shows the presence of one or more chiral grooves on these CSPs. Electronegative atoms such as oxygen, nitrogen, and halogens of racemates form hydrogen bonds and there are dipole–dipole induced interactions with polysaccharide-based CSP. Besides, π–π interactions also occur between the phenyl rings of aromatic racemates and the CSP. During chiral resolution, the enantiomers fit stereogenically in different ways into the

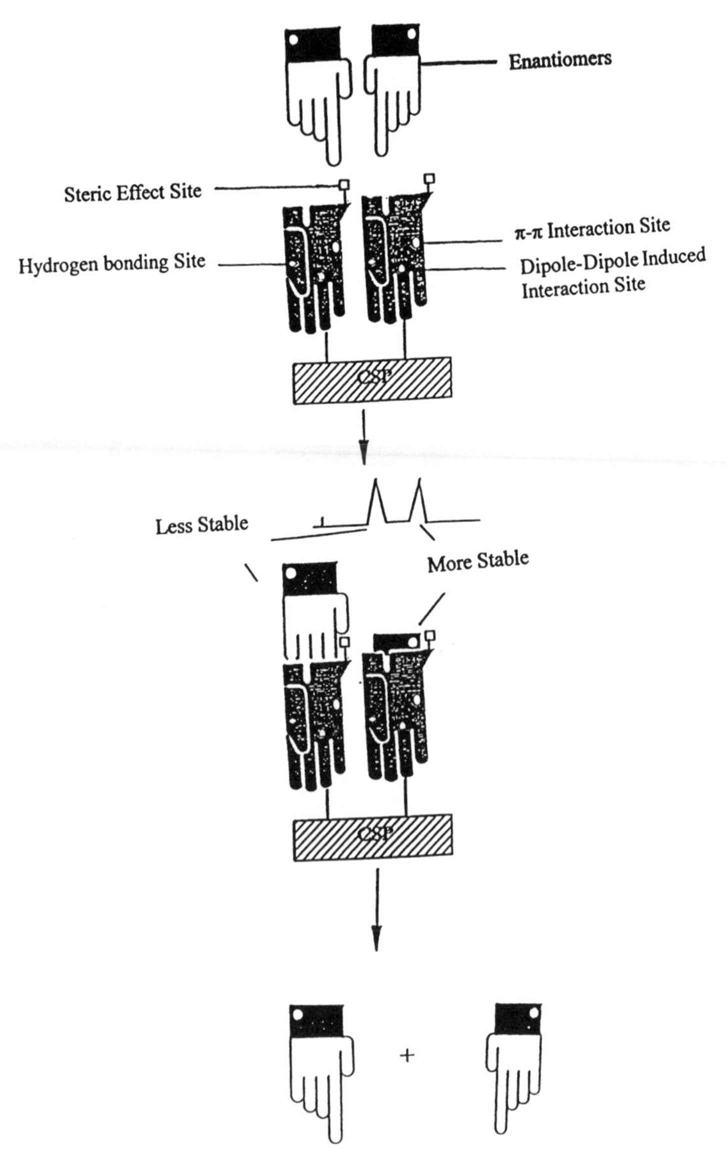

FIGURE 27 Graphical representation of the mechanism of chiral resolution on a polysaccharide-based CSP.

chiral grooves of the CSP, which is stabilized by bonds of various types (as discussed earlier) and magnitudes, and hence the enantiomers were resolved. In addition to these bonds, steric effects govern chiral resolution on polysaccharide-based CSPs. Besides, some other achiral weak bonds like van der Waal forces and ionic bonds may also contribute to the chiral resolution. To clarify the picture of chiral resolution, the graphical representation of chiral resolution on a polysaccharide-based CSP shown in Figure 27 indicates the different interactions of the two enantiomers.

2.8 APPLICATIONS IN SUB- AND SUPERCRITICAL FLUID CHROMATOGRAPHY

Polysaccharide-based CSPs have been used for chiral separations in subcritical fluid chromatography (subcritical-SFC) and supercritical fluid chromatography (SFC) since 1988 [134]. Two groups have reported the separation on cellulose tris(3,5-dimethylphenylcarbamate) of enantiomers of β-adrenergic blockers CSPs [135,136]. Bargmann-Leyder et al. [137] achieved the chiral resolution of enantiomers of β-adrenergic blockers and other drugs on Chiralcel OD and Chiralpak AD CSPs. Wang et al. [138] utilized Chiralcel OD for the chiral resolution of camazepam and its metabolites. Phinney et al. [139] demonstrated the separation of a series of benzodiazepines. An amino column was coupled in series with the Chiralcel OD to achieve the desired separation. Overbeke et al. [140] resolved racemic drugs such as benzoxaprofen, temazepam, and mephobarbital on Chiralcel OJ. The four optical isomers of calcium channel blockers were resolved on Chiralcel OJ by Siret et al. [141]. Amylose CSPs have been used for the resolution of nonsteroidal anti-inflammatory drugs (ibuprofen, flurbiprofen, and related drugs) [142,143].

Attempts have also been made to compile the results of chiral resolutions by SFC using polysaccharide-based CSPs, as summarized in Table 11. To show the nature of the SFC chromatograms, Figure 28 represents the SFC chiral resolution of ibuprofen on a Chiralpak AD CSP. Stringham et al. [144] resolved the enantiomers of four intermediates encountered in the process of developing synthetic antiviral drugs on a Chiralcel OD CSP. In another report, Blackwell [145] used various mobile phase additives to resolve the enantiomers of isoxazoline-based IIb/IIIb receptor antagonists on Chiralcel OD-H columns. Besides, polysaccharide-based CSPs have also been used for preparative SFC. Saito et al. [146] separated the enantiomers of *DL*-flavanone on a preparative scale using a Chiralcel OD column. Oka and coworkers resolved four optical isomers of the antidiabetic drug troglitazone on cellulose CSPs by means of preparative SFC [147]. Blum et al. [148] published a comparison of chiral resolution of a variety

TABLE 11 Chiral Resolution by Means of SFC Using Polysaccharide-Based CSPs

Racemates	SFC mode	CSPs	Ref.
Alkyl alkanols	Packed column	Chiralcel OB, Chiralcel OD	158
Benzodiazepines	Packed column	Chiralcel OD, Chiralcel OD-H, Chiralpak AD	139 155, 152
Benzothiazepines	Packed column	Chiralcel OC, Chiralcel OJ	156
Calcium channel blockers	Packed column	Chiralcel OD	158
β-Adrenergic blockers	Packed column, packed capillary column	Chiralcel OD, Chiralpak AD	137,139, 155
Imidazole derivatives	Packed column	Chiralcel OD, Chiralcel OJ, Chiralpak AD, Chiralpak AS	145, 158
N-protected amino acids and esters	Packed column	Chiralpak AD	158
Nonsteroidal anti-inflammatory drugs	Packed column	Chiralpak AD	143
Camazepam and metabolites	Packed column	Chiralcel OD	138
Fluoxetine	Packed column	Chiralpak AD	152
Barbiturates	Packed column	Chiralcel OJ	158
β-Adrenergic blockers	Packed column	Chiralcel OD	158

of drugs on Chiralcel CSPs in preparative SFC and HPLC, and better results were obtained on SFC.

The effect of various SFC parameters on chiral resolution were also studied. Modifiers can provide control over both retention and selectivity and, therefore, certain modifiers were used to optimize the separation in sub-FC and SFC. The effect of the enantioselectivity of carbon dioxide on acidic drugs (benzoxaprofen, temazepam, and mephobarbital), profen, and barbiturate derivatives was carried out on Chiralcel OJ, with acetonitrile or methanol as organic modifier [140]. Acetonitrile proved to be a good alternative to methanol, especially for the profen compounds that were not well resolved when methanol was used. Wilson [143] studied the effects of methanol, ethanol, and 2-propanol as organic modifiers on the chiral resolution of ibuprofen on Chiralpak AD CSPs. Methanol was found to be the best organic modifier.

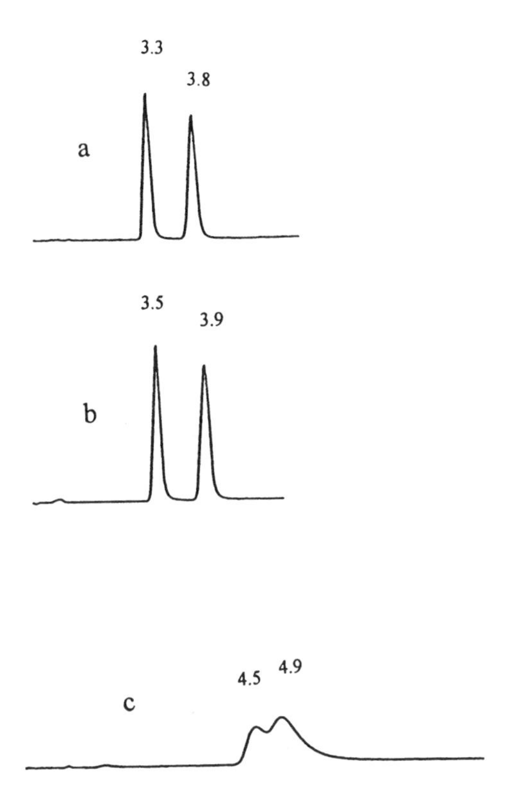

FIGURE 28 Chromatograms of the chiral resolution of ibuprofen by SFC on Chiralpak AD column, with (a) methanol, (b) ethanol, and (c) 2-propanol as the organic modifiers. (From Ref. 143.)

Temperature and pressure are rarely optimized in HPLC, but these parameters are very important in SFC, hence can alter retention, selectivity, and resolution. Toribio et al. [149] presented the chiral separation of ketoconazole and its precursors on Chiralpak AD and Chiralcel OD CSPs. The authors also reported that alcohol modifiers provided better separation than acetonitrile. Further, Wilson [143] studied the effects of composition, pressure, temperature, and flow rate of the mobile phase on the chiral resolution of ibuprofen on a Chiralpak AD CSP. It was observed that temperature affords the greatest change in resolution, followed by pressure and composition. An increase in methanol concentration, pressure, and temperature has resulted in poor chiral resolution. At first chiral resolution increased with an increase of flow rate (up to 1.5 mL/min) but then started to decrease. Contrary to this, Biermann et al. [135] described the rapid chiral resolution of β-adrenergic blockers at a flow rate of 4 mL/min. Therefore, chiral resolution in SFC can be optimized by controlling the flow rate.

The resolution of certain calcium channel blockers was optimized by temperature variations on cellulose tris(4-chlorophenylcarbamate) CSP coated on silica gel [150]. The van't Hoff plots for retention were curved, while those for selectivity were linear.

The chiral separation of *cis* enantiomers was improved with a decrease in temperature, whereas that of *trans* enantiomers was improved with an increase in temperature. The temperature dependence of enantioselectivities was studied to determine the thermodynamic parameters H°, S°, and T_{iso}. The thermodynamic parameters revealed that the separation of *trans* enantiomers was controlled by entropy in the range of temperatures examined, whereas enthalpy-controlled separation was observed for *cis* enantiomers. The separations of both *cis* and *trans* enantiomers, however, were controlled by enthalpy in normal phase HPLC [150].

In sub-FC, a detailed study of the influence of mobile phase additives on the chiral resolution of isoxazoline-based IIb/IIIb receptor antagonists was carried out by Blackwell [145] on Chiralcel OD-H CSPs. The different mobile phase additives used were acetic acid, trifluoroacetic acid, formic acid, water, triethylamine, triethanolamine, *n*-hexylamine, trimethyl phosphate, and tri-*n*-butyl phosphate. In general, *n*-hexylamine and tri-*n*-butyl phosphate mobile phase additives resulted in better resolution. The chiral separation of four 1,3-dioxolane derivatives on an amylose-based column has been described [151]. The effects of mobile phase composition, temperature, and pressure have been investigated. The nature of the modifier is the main parameter: it has the highest impact on chiral resolution and is more important than the polarity of the mobile phase. Therefore, the organic modifier that gave the best enantiomeric separation was different for each compound.

The pressure-related effects on chiral resolution were studied by Bargmann-Leyder et al. [137] on Chiralpak AD columns, while no pressure effect was observed on Chiralcel OD. These investigators also studied and compared the chiral resolution of β-adrenergic blockers on HPLC and SFC [137]. Although resolution was generally higher in SFC than in HPLC, differences in selectivity between the two techniques were observed, and these discrepancies seemed to be compound specific. Examination of a series of propanolol analogues provided additional insight into differences in the chiral recognition mechanisms operative in HPLC and SFC [137].

Packed column SFC is now rapidly replacing many HPLC methods for chiral resolution [152]. These CSPs are coated on the silica gel, like the CSPs for HPLC [150,153]. Besides, chiral resolutions on capillary chiral CSPs (Table 11) have also been reported. Finding the best combination of chiral stationary and mobile phases can be time-consuming. CHIRBASE (www.chirbase.u-

3mrs.fr/chirbase) [154], a database specializing in chiral chromatographic separation including sub-FC and SFC, can often aid in reducing the work associated with method development [155–158].

2.9 APPLICATIONS IN CAPILLARY ELECTROCHROMATOGRAPHY

In 2000 Wistuba and Schuring [159] reviewed the chiral resolution of pharmaceuticals by capillary electrochromatography (CEC) and discussed the different aspects of chiral resolution by this technique. Polysaccharide-based CSPs have been packed in capillaries, and the chiral resolution of various racemates was obtained [160–163]. The packed capillaries were prepared by means of the slurry packing method [161,162]. Chankvetadze et al. [161,163] has carried out the chiral separation of different racemates by means of capillary electrochromatography. Further, Chankvetadze et al. [163] observed that Chiralpak AD packed with 25% amylose tris(3,5-dimethylphenylcarbamate) coated on macroporous aminopropylsilanized silica gel exhibited higher resolving ability for thalidomide than the similar cellulose tris(3,5-dimethylphenylcarbamate) (Chiralcel OD), as well as cellulose tris(4-methylbenzoate) (Chiralcel OJ) CSPs. Enantioseparations of β-adrenergic blockers, benzodiazepines, and diuretics were performed in fused silica capillaries packed with silica gel that was modified by coating with cellulose tris(3,5-dimethylphenylcarbamate) [160].

In addition to other modes of liquid chromatography, chiral resolution by capillary electrochromatography was also optimized by adjusting the different chromatographic parameters. Francotte et al. [164] chirally resolved certain enantiomers by applying packed capillary electrochromatography. Fused-silica capillaries of different lengths with an inner diameter of 100 μm were packed with a cellulose derivative immobilized on macroporous silica gel. Parameters such as content of modifier in the mobile phase, concentration, and buffer pH were varied for a set of tests. In packed CEC the highest influence on resolution of the test racemates was found when the acetonitrile content was changed, while variation of the buffer concentration mostly affected the electroosmotic velocity. The performance of packed CEC and nano-LC was also compared. Packed CEC showed much better column efficiency and enantioselectivity under similar conditions of flow and electroosmotic velocity. Chankvetadze et al. [161] studied the effect of the amount of cellulose tris(3,5-dimethylphenylcarbamate) loaded on silica gel. The reduced coating of the chiral selector allowed more effective exploitation of the advantages of electrokinetically driven flow over pressure-driven propulsion of the analyte. Plate numbers were in the range of 180,000 to 215,000 m^{-1}, which are among the highest efficiencies observed in CEC. The effect of the amount of cellulose tris(3,5-dimethylphenylcarbamate) loaded on

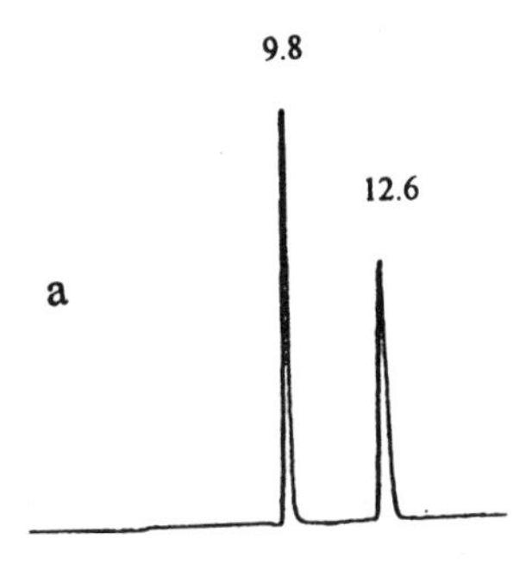

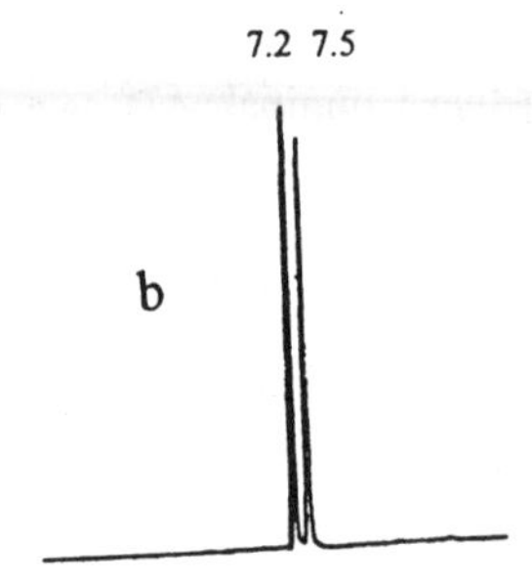

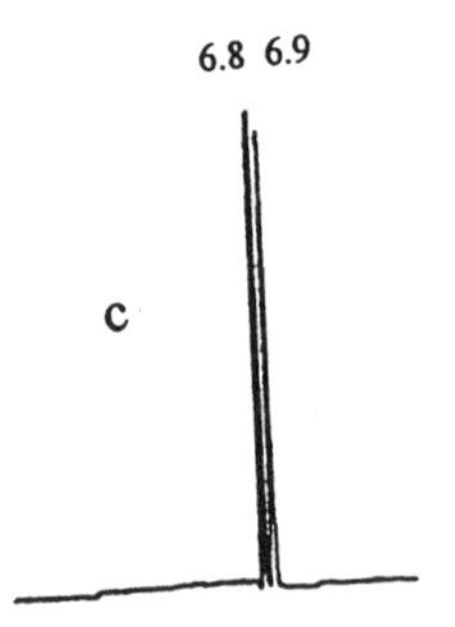

FIGURE 29 Chromatograms of the chiral resolution of 2-(benzylsulfinyl) benzamide in capillaries packed with tris(3,5-dichlorophenylcarbamate) at three different loadings: (a) 4.8 (w/w), (b) 1.0 (w/w), and (c) 0.5 percent (w/w). (From Ref. 161.)

silica gel on the chiral resolution of 2-(benzylsulfinyl) benzamide is shown in Figure 29 [161]. It was observed that 4.8% (w/w) resulted in the best resolution. Chankvetadze et al. [162] also investigated the effects of silica gel pore size and concentration of buffer on the chiral resolution of piprozolin. It was observed that the ionic strength of a buffer solution dramatically affected electroosmotic flow generation and intraparticle perfusive flow, especially for silica gel pore sizes below 12 nm. Optimization of packing material and separation conditions led to plate numbers above 400,000 m^{-1} and very good resolutions.

2.10 APPLICATIONS IN THIN-LAYER CHROMATOGRAPHY

In 1973 Hesse and Hagel [15] first described the chiral thin-layer chromatographic (TLC) separation of Tröger's base on cellulose acetate. Later, Faupel's systematic investigation of the chiral resolution capacities of this technique [165] resulted in the commercialization of the cellulose triacetate plates pioneered by Hesse and Hagel. These plates are stable and can be used in aqueous media (acid or base) and in nonaqueous mobile phases with the exception of glacial acetic acid and ketonic solvents. Other resolutions on cellulose plates include oxindanac benzylester and 2-phenylcyclohexanone [5]. Xuan and Lederer [166] reported the enantiomeric resolution of substituted tryptophans on microcrystalline cellulose. They observed that aqueous solvents and liquid–liquid systems yielded essentially the same separations, suggesting that adsorption can play a role in liquid–liquid (partition) systems in some instances. Yuasa et al. [167–171] used crystalline cellulose as the chiral stationary phase for TLC and investigated the separation of *DL*-tryptophan, *DL*-amino acids, and other derivatives of amino acids. Separation was performed in a variety of aqueous and nonaqueous solvents. In trying to explain the chiral recognition mechanism, the authors concluded that chiral resolution is attributable to the helical forms of the polysaccharide chains, which differ in aqueous and nonaqueous media. Suedee and Heard [172] reported the use of cellulose phenylcarbamate derivatives as the stationary phase for the resolution of enantiomers of β-adrenergic blockers. Lepri et al. [173] described the resolution of 21 racemates on microcrystalline cellulose triacetate plates eluted with aqueous/organic mixtures containing methanol or ethanol or 2-propanol. In 2000 Okamoto et al. [174] prepared TLC plates of cellulose derivatives, using the usual method as described in 1969 [175]. For the preparation of fluorescent chiral TLC plates, a fluorescent indicator is mixed with the cellulose derivatives in methanol and the resulting slurry is spread over the plate in a uniform layer. The plate is oven-dried for 30 min at 110°C [174]. TLC plates with cellulose (0.5%) coated with silica gel, immersed in Cu(II) acetate followed by *L*-arginine were used for the chiral resolution of amino acids and their derivatives [176]. Malinowska and Rozylo used chitin and chitosan as TLC plate material for the enantiomeric resolution of amino acids [177].

The cellulose derivatives used for chiral TLC are trisphenylcarbamate, 2,3-dichlorophenylcarbamate, 2,4-dichlorophenylcarbamate, 2,6-dichlorophenylcarbamate, 3,4-dichlorophenylcarbamate, 3,5-dichlorophenylcarbamate, 2,3-dimethylphenylcarbamate, and 3,5-dimethylphenylcarbamate. Aboul-Enein et al. [178] have reviewed the chiral resolution of racemates on polysaccharide chiral TLC plates. They discussed the role of the substituents of polysaccharide derivatives on chiral resolution. The effects of the substituents of cellulose derivatives and the mechanisms of chiral resolution on these plates are similar to what is found for HPLC CSPs.

2.11 CONCLUSION

Today polysaccharide-based derivatives are considered to be one of the best and most widely used chiral stationary phases because of their greater efficiencies and wide range of applications. Although many racemic compounds can be resolved on these CSPs, these CSPs are not yet fully developed because they are not capable of resolving all, or almost all, racemic compounds. A disadvantage of polysaccharide-based CSPs remains, however, the limited choice of solvents for the coated form. On the other hand, the bonded forms generally have poor chiral resolution power. The polysaccharide-based CSPs that are most widely used and usually achieve 80% successful resolution are Chiralcel OD, Chiralcel OD-R, Chiralcel OJ, Chiralcel OJ-R, Chiralpak AD, and Chiralpak AD-R CSPs, while the use of the other polysaccharide-based CSPs is limited. Research is still under way to improve the chiral resolution capacities of these phases in preparing new derivatives of cellulose and amylose. We expect that these CSPs will gain more and more interest in the field of chiral resolution in the near future. HPLC with polysaccharide-based CSPs has been used very frequently for chiral resolution, but there are few applications of chiral resolution by sub- and supercritical fluid chromatography modes, which are emerging as important modes of liquid chromatography for chiral resolution because of their versatility, sensitivity, and high speed of resolution. The further development of these two modalities of liquid chromatography certainly will lead to their use as routine methods of chiral resolution in the near future.

REFERENCES

1. Okamoto, Y, Yashima, E, Chiral recognition by optically active polymers, in "Molecular Design of Polymeric Materials" (Hatada K, Kitayama T, Vogl O, Eds.), Marcel Dekker, New York, pp. 731–746 (1997).
2. Okamoto Y, CHEMTECH 176 (1987).

3. Okamoto Y, Aburatani R, Polym News 14: 295 (1989).
4. Kotake M, Sakan T, Nakamura N, Senoh S, J Am Chem Soc 73: 2973 (1951).
5. Günther K, Enantiomer separations, in "Handbook of Thin Layer Chromatography" (Sherma J, Fried B, Eds.), Marcel Dekker, New York, p. 541 (1991).
6. Yashima E, Okamoto Y, Chiral recognition mechanism of polysaccharides chiral stationary phase in "The Impact of Stereochemistry on Drugs Development and Use" (Aboul-Enein, HY, Wainer IW, Eds.), John Wiley & Sons, New York, p. 345 (1997).
7. Shibata T, Mori K, Okamoto Y, Polysaccharide phases, in "Chiral Separations by HPLC" (Krstulovic AM, Ed.) Ellis Horwood, New York, p. 336 (1989).
8. Okamoto Y, Kawashima M, Yamamoto K, Hatada, K, Chem Lett 739 (1984).
9. Dingenen J, Polysaccharide phases in enantioseparations, in "A Practical Approach to Chiral Separations by Liquid Chromatography" (Subramanian G, Ed.), VCH Verlag, Weinheim, Germany, p. 115 (1994).
10. Allenmark, S. (Ed.), "Chromatographic Enantioseparation: Methods and Applications," 2nd ed., Ellis Horwood, New York (1991).
11. Beesley TE, Scott RPW (Eds.), "Chiral Chromatography," John Wiley & Sons, New York (1998).
12. Stivala SS, Crescenti V, Dea ICM (Eds.), "Industrial Polysaccharides: The Impact of Biotechnology and Advanced Methodologies: Proceedings," Ellis Horwood, New York (1987).
13. Harding SE, Tombs MP, Harding S, "Introduction to Polysaccharide Biotechnology," Ellis Horwood, New York (1997).
14. Luttringhaus A, Peters K, Angew Chem 78: 603 (1966).
15. Hesse G, Hagel R, Chromatographia 6: 277 (1973).
16. Ichida A, Shibata T, Okamoto I, Yuki Y, Namikoshi H, Toda Y, Chromatographia 19: 280 (1984).
17. Okamoto Y, Aburatani R, Hatada K, J Chromatogr 389: 95 (1987).
18. Okamoto Y, Kawashima M, Hatada K, J Chromatogr 363: 173 (1986).
19. Chankvetadze B, Yashima E, Okamoto Y, Chem Lett 671 (1992).
20. Chankvetadze B, Yashima E, Okamoto Y, J Chromatogr A 670: 39 (1994).
21. Okamoto Y, Aburatani R, Fukumoto T, Hatada K, Chem Lett 1857, (1987).
22. Okamoto Y, Aburatani R, Hatada K, Bull Chem Soc Jpn 63: 955 (1990).
23. Okamoto Y, Oshashi T, Kaida Y, Yashima E, Chirality 5: 616 (1993).
24. Chankvetadze B, Yashima E, Okamoto Y, J Chromatogr A 694: 101 (1995).
25. Shibata T, Sei T, Nishimura H, Deguchi K, Chromatographia 24: 552 (1987).
26. Buchanan CM, Hyatt JA, Lowman DW, J Am Chem Soc 111: 7312 (1989).
27. Oguni K, Matsumoto A, Isokawa A, Polym 11: 1257 (1994).
28. Yashima E, Yamamoto C, Okamoto Y, J Am Chem Soc 118: 4036 (1996).
29. Yamamoto C, Yashima E, Okamoto Y, Bull Chem Soc Jpn 72: 1815 (1999).
30. Steinmeir H, Zugenmaier P, Carbohydr Res 164: 97 (1987).
31. Vogt U, Zugenmaier P, Ber Bunsenges Phys Chem 89: 1217 (1985).
32. Yashima E, J Chromatogr A 906: 105 (2001).
33. Matsumoto K, Yamamoto C, Yashima E, Okamoto Y, Anal Commun 35: 63 (1998).

34. Matsumoto K, Yamamoto C, Yashima E, Okamoto Y, Rapid Commun Mass Spectrom 13: 2011 (1999).
35. Sawada M, Mass Spectrom Rev 16: 73 (1997).
36. Yashima E, Fukaya H, Sahavattanapong P, Okamoto Y, Enantiomer 1: 193 (1996).
37. Lightner DA, Reisinger M, Landen GL, J Biol Chem 261: 6034 (1986).
38. Lightner DA, Gawronski JK, Gawronska K, J Am Chem Soc 107: 2456 (1985).
39. Kano K, Yoshiyasu K, Hashimoto S, J Chem Soc Chem Commun 801 (1988).
40. Murakami Y, Hayashida O, Nagai Y, J Am Chem Soc 116: 2611 (1994).
41. Hesse G, Hagel R, Liebigs Ann Chem 966 (1976).
42. Shibata T, Okamoto Y, Ishii K, J Liq Chromatogr 9: 313 (1986).
43. Okamoto Y, Kaida Y, J Chromatogr A 666: 403 (1994).
44. Yashima E, Kasashima E, Okamoto Y, Chirality 9: 63 (1997).
45. Yashima E, Yamamoto C, Okamoto Y, Polym J 27: 856 (1995).
46. Rimbock K-H, Kastner F, Mannschreck A, J Chromatogr 351: 346 (1986).
47. Francotte E, Wolf RM, Chirality 3: 43 (1990).
48. Francotte E, Baisch G, European Patent EP 0316270A2 (1988).
49. Juvancz Z, Grolimund K, Francotte E, Chirality 4: 459 (1992).
50. Francotte E, Wolf RM, J Chromatogr 595: 63 (1992).
51. Castells CB, Carr PW, J Chromatogr A 904: 17 (2000).
52. Franco P, Senso A, Oliveros L, Minguillon C, J Chromatogr A 906: 155 (2001).
53. Okamoto Y, Aburatani R, Miura S, Hatada K, J Liq Chromatogr 10: 1613 (1987).
54. Yashima E, Fukuya H, Okamoto Y, J Chromatogr A 677: 11 (1994).
55. Minguillon C, Franco P, Oliveros L, Lopez P, J Chromatogr A 728: 407 (1996).
56. Enomoto N, Furukawa S, Ogasawara Y, Akano H, Kuwamura E, Yashima E, Okamoto Y, Anal Chem 68: 2789 (1996).
57. Enomot N, Furukawa S, Ogasawara, Y, Yashima E, Okamoto Y, 66th Annual Meeting of the Chemical Society of Japan, 2B233: 140 (1994).
58. Felix G, J Chromatogr A 906: 171 (2001).
59. Kubota T, Kusano T, Yamamoto C, Yashima E, Okamoto Y, 13th International Symposium on Chirality 118: 35 (2001).
60. Kimata K, Tsuboi R, Hosoya K, Tanaka N, Anal Methods Instrum 1: 23 (1993).
61. Hampe TRE, Schluter J, Brandt KH, Nagel J, Lamparter E, Blaschke G, J Chromatogr 634: 205 (1993).
62. Wainer IW, Stiffin RM, Shibata T, J Chromatogr 411: 139 (1987).
63. Aboul-Enein HY, Ali I, J Biochem Biophys Methods 48: 175 (2001).
64. Francotte E, Wolf RM, Lohmann D, Mueller R, J Chromatogr 347: 25 (1985).
65. Francotte E, Wolf RM, Chirality 2: 16 (1990).
66. Yashima E, Sahavattanapong P, Okamoto Y, Chirality 8: 446 (1996).
67. Mannschreck A, Koller H, Wernick R, Kontake 1: 40 (1995).
68. Rizzi A, J Chromatogr 478: 87 (1989).
69. Isaksson R, Erlandsson P, Hansson L, Holmberg A, Berner S, J Chromatogr 498: 257 (1990).
70. Francotte E, J Chromatogr A 666: 565 (1994).
71. Chankvetadze B, Yamamoto C, Okamoto Y, Comb Chem High Throughput Screen 3: 497 (2000).

72. Okamoto Y, Aburatani R, Hatano K, Hatada K, J Liq Chromatogr 11: 2147 (1988).
73. Witte DT, Bruggeman FJ, Franke JP, Copinga S, Jansen JM, de Zeeuw RA, Chirality 5: 545 (1993).
74. Matlin SA, Tiritan ME, Crawford AJ, Cass QB, Boyd DR, Chirality 6: 135 (1994).
75. Toussaint B, Duchateau ALL, ven der Waal S, Albert A, Hubert P, Crommen J, J Chromatogr A 890: 239 (2000).
76. Caccamese S, Principato G, J Chromatogr A 893:47 (2000).
77. Fadnavis NW, Babu RL, Sheelu G, Deshpande A, J Chromatogr A 890: 189 (2000).
78. Belloli E, Foulon C, Yous S, Vaccher MP, Bonte JP, Vaccher C, J Chromatogr A 907: 101 (2001).
79. Feng J, May SW, J Chromatogr A 905: 103 (2001).
80. Iwasaki Y, Yasui M, Ishikawa T, Irimescu R, Hata K, Yamane T, J Chromatogr A 905: 111 (2001).
81. Cirilli R, Costi R, Santo RD, Ferretti R, Torre FL, Angiolella L, Micocci M, J Chromatogr A 942: 107 (2002).
82. Aboul-Enein HY, Biomed Chromatogr 12: 116 (1998).
83. Aboul-Enein HY, J Chromatogr A 906: 185 (2001).
84. Maier NM, Franco P, Linder W, J Chromatogr A 906: 3 (2001).
85. Blaschke G, J Liq Chromatogr 9: 341 (1986).
86. Schlogl K, Widhalm M, Monatsh Chem 115: 1113 (1984).
87. Wainer IW, Alembic MC, Smith E, J Chromatogr 388: 65 (1987).
88. Bonato PS, Pires de Abreu LR, de Gaitani CM, Lanchote VL, Bertucci C, Biomed Chromatogr 14: 227 (2000).
89. Gaffney MH, Chromatographia 27: 15 (1989).
90. Ishikawa A, Shibata T, J Liq Chromatogr 16: 859 (1993).
91. Kazusaki M, Kawabata H, Matsukura H, J Liq Chromatogr Relat Technol 23: 2819 (2000).
92. Aboul-Enein HY, Ali I, Seventh International Symposium on Hyphenated Techniques in Chromatography and Hyphenated Chromatographic Analyzers (7-HTC), Feb. 6–8, 2002, Brugge, Belgium, p. 193 (2002).
93. Aboul-Enein HY, Ali I, Fresenius J Anal Chem 370: 951 (2001).
94. Aboul-Enein HY, Ali I, J Sep Sci 24: 831 (2001).
95. Kazusaki M, Kawabata H, Matsukura H, J Liq Chromatogr Relat Technol 23: 2937 (2000).
96. Francotte E, Stierlin H, Faigle JW, J Chromatogr 346: 321 (1985).
97. Wainer IW, Alembic M, J Chromatogr 358: 85 (1986).
98. Aboul-Enein HY, Ali I, Gubitz G, Simons C, Nicholls PJ, Chirality 12: 727 (2000).
99. Aboul-Enein HY, Ali I, Schmid MG, Jetcheva V, Gecse O, Gubitz G, Anal Lett 34: 1107 (2001).
100. Aboul-Enein HY, Ali I, Chirality 14: 47 (2002).
101. Aboul-Enein HY, Ali I, J Pharm Biomed Anal 27: 441 (2002).
102. Chimonczyk Z, Ksycinska H, Mazgajska M, Aboul-Enein HY, Chirality 11: 790 (1999).
103. Aboul-Enein HY, Bakr SA, J Liq Chromatogr Relat Technol 21: 1137 (1998).

104. Van Overbeke A, Sandra P, Medvedovici A, Baeyens W, Aboul-Enein HY, Chirality 9: 126 (1997).
105. Aboul-Enein HY, Ali I, Pharmazie, 56: 214 (2001).
106. Aboul-Enein HY, Ali I, Arch Pharm 334: 258 (2001).
107. Aboul-Enein HY, Ali I, Chromatographia 54: 200 (2001).
108. Schmid MG, Gecse O, Szabo Z, Kilár F, Gübitz G, Ali I, Aboul-Enein HY, J Liq Chromatogr Relat Technol 24: 2493 (2001).
109. Huhnerfuss H, Chemosphere 40: 913 (2000).
110. Ellington JJ, Evans JJ, Prickett KB, Champion WL Jr, J Chromatogr A 928: 145 (2001).
111. Wolf RM, Francotte E, Lohmann D, J Chem Soc Perkin Trans II 893 (1988).
112. Francotte E, Rihs G, Chirality 1: 80 (1989).
113. Ronden NG, Nyquist RA, Gillie JK, Nicholson LW, Goralski CT, Fourth International Symposium, Montreal, Canada, Abstract 162, p. 90 (1993).
114. Yashima E, Yamada M, Kaida Y, Okamoto Y, J Chromatogr A 694: 347 (1995).
115. Tachibana K, 13th International Symposium on Chirality, Abstract 138, p. 41 (2001).
116. Geiser F, Champion WL Jr., Pickett KB, 13th International Symposium on Chirality, abstract 157, p. 48 (2001).
117. Cirrili R, Costi R, Di Santo R, Ferretti R, L Torre F, Angiolella L, Micocci M, J Chromatogr A, 942: 107 (2002).
118. Schulte M, Strube J, J Chromatogr A 906: 399 (2001).
119. Ye YK, Stringham R, J Chromatogr A 927: 47 (2001).
120. Ye YK, Stringham R, J Chromatogr A 927: 53 (2001).
121. Wang T, Chen YW, Vailaya A, J Chromatogr A 902: 345 (2000).
122. Yamamoto C, Okamoto Y, J Chromatogr A 922: 127 (2001).
123. Ye YK, Lord B, Yin L, Stringham RW, J Chromatogr A, 945: 147 (2002).
124. Ye YK, Lord B, Stringham RW, J Chromatogr A, 945: 139 (2002).
125. Perrin C, Vu VA, Matthijs N, Maftouh M, Massart DL, Heyden YV, J Chromatogr A, 947: 69 (2002).
126. Piras P, Roussel C, Pierrot-Sanders J, J Chromatogr A 906: 443 (2001).
127. Tachibana K, Ohnishi A, J Chromatogr A 906: 127 (2001).
128. O'Brien T, Crocker L, Thompson R, Thompson K, Toma PH, Conlon DA, Moeder C, Bicker GR, Grinberg N, Anal Chem 69: 1999 (1997).
129. Stringham RW, Blackwell JA, Anal Chem 69: 1414 (1997).
130. Ding H, Grinberg N, Thompson R, Ellison D, J Liq Chromatogr Relat Technol 23: 2641 (2000).
131. Lipkowitz KB, Modeling enantioseparation in chiral chromatography, in "A Practical Approach to Chiral Separations by Liquid Chromatography," (Subramanian G, Ed.), VCH Verlag, Weinheim, Germany, p. 19 (1994).
132. Lipkowitz KB, J Chromatogr A 694: 15 (1995).
133. Theodorou DN, Suter UW, Macromolecules 18: 1467 (1985).
134. Macaudiere C, Rosset M, Tambute R, J Chromatogr 450: 255 (1988).
135. Biermanns P, Miller C, Lyon V, Wilson W, LC-GC 11: 744 (1993).
136. Lee CR, Porziemsky JP, Aubert MC, Krstulovic AM, J Chromatogr 539: 55 (1991).

137. Bargmann-Leyder N, Tambute A, Caude M, Chirality 7: 311 (1995).
138. Wang MZ, Klee MS, Yang KS, J Chromatogr B 665: 139 (1995).
139. Phinney KW, Sanders LC, Wise SA, Anal Chem 70: 2331 (1998).
140. Overbeke AV, Sandra P, Medvedovici A, Baeyens W, Aboul-Enein HY, Chirality 9: 126 (1997).
141. Siret L, Macaudiere P, Bargmann-Leyder N, Tambute A, Caude M, Guogeon E, Chirality 6: 440 (1994).
142. Kot A, Sandra P, Venema A, J Chromatogr Sci 32: 439 (1994).
143. Wilson WH, Chirality 6: 216 (1994).
144. Stringham RW, Lynam KG, Grasso CC, Anal Chem 66: 1949 (1994).
145. Blackwell JA, Chirality 10: 338 (1998).
146. Saito M, Yamauchi Y, Higashidate S, Okamoto I, in "35th Anniversary Research Group on Liquid Chromatography" (Hatano H, Hanai T, Eds.), World Scientific, Singapore, p. 863 (1995).
147. Oka K, Shoda S, Kato K, Watanabe T, Nakazawa H, Hiroyki T, Kawasaki T, Ikeda M, Jpn Chromatogr 20: 310 (1999).
148. Blum AM, Lynam KG, Nicolas EC, Chirality 6: 302 (1994).
149. Toribio L, Bernal JL, Nozal MJ, Jimenze JJ, Nieto EM, J Chromatogr A 921: 305 (2001).
150. Yaku K, Aoe K, Nishimura N, Morishita F, J Chromatogr A 848: 337 (1999).
151. Bernal JL, Toribio L, Nozal MJ, Nieto EM, Jimneze JJ, J Chromatogr A 87: 127 (2000).
152. Phinney KW, Anal Chem 72: 204A (2000).
153. Nitta T, Yakushijin Y, Kametani T, Katayama T, Bull Chem Soc Jpn 63: 1365 (1990).
154. Koppenhoefer B, Graf R, Holzschuh H, Nothdurf A, Trittin U, Piras P, J Chromatogr A 826: 217 (1998).
155. Medvedovici A, Sandra P, Toribio L, David F, J Chromatogr A 785: 159 (1997).
156. Yaku K, Aoe K, Nishimura N, Sato T, Morishita F, J Chromatogr 785: 185 (1997).
157. Villeneu MS, Anderegg RJ, J Chromatogr A 826: 217 (1998).
158. Terfloth G, J Chromatogr A 906: 301 (2001).
159. Wistuba D, Schuring V, J Chromatogr A 875: 255 (2000).
160. Krause K, Girod M, Chankvetadze B, Blaschke G, J Chromatogr A, 837: 51 (1999).
161. Girod M, Chankvetadze B, Okamoto Y, Blaschke G, J Sep Sci 24: 27 (2001).
162. Chankvetadze B, Kartozia I, Okamoto Y, Blaschke G, J Sep Sci 24: 635 (2001).
163. Meyring M, Chankvetadze B, Blaschke J Chromatogr A 876: 157 (2000).
164. Mayer S, Briend X, Francotte E, J Chromatogr A 875: 331 (2000).
165. Faupel M, Fourth International Symposium on Instrumental TLC, Selvino/Bergamo, Italy (1987).
166. Kieu Xuan HT, Lederer M, J Chromatogr 635: 346 (1993).
167. Yuasa S, Itoh M, Shimada AK, J Chromatogr Sci 22: 228 (1984).
168. Yuasa S, Shimada A, Isoyama M, Fukuhara T, Itoh M, Chromatographia 21: 79 (1986).
169. Yuasa S, Shimada A, Kameyama K, Yasui M, Adzuma K, J Chromatogr Sci 18: 311 (1990).

170. Yuasa S, J Biol Phys 20: 229 (1994).
171. Yuasa S, Fukuhara T, Isoyama M, Tanaka M, Shimada A, Biomed Chromatogr 11: 276 (1997).
172. Suedee R, Heard CM, Chirality 9: 139 (1996).
173. Lepri L, Bubba MD, Masi F, J Planar Chromatogr Mod TLC 10: 108 (1997).
174. Kubota T, Yamamoto C, Okamoto Y, J Am Chem Soc 122: 4056 (2000).
175. Stahl E, "Thin Layer Chromatography," 2nd ed., Springer-Verlag, Berlin (1969).
176. Deng QY, Zhang Z, Zhy YF, Su JY, Fenxi Huaxue 25: 197 (1997).
177. Malinowska I, Rozylo JK, Biomed Chromatogr 11: 272 (1997).
178. Aboul-Enein HY, El-Awady MI, Heard CM, Nicholls PJ, Biomed Chromatogr 13: 531 (1999).

3

Cyclodextrin-Based Chiral Stationary Phases

Cyclodextrins (CDs) are cyclic, nonreducing oligosaccharides obtained from starch. Villiers [1] described CDs in 1891 and called them cellulosine. Later Schardinger [2] identified three different naturally occurring forms and referred them as Schardinger's sugars, α-, β-, and γ-CDs. These are also called cyclohexamylose(α-CD), cycloheptamylose(β-CD), cycloctamylose(γ-CD), cyclo-glucans, glucopyranose, and Schardinger dextrins. From 1911 to 1935, Pringsheim, in Germany, carried out an extensive research and reported that these sugars form stable aqueous complexes with many compounds [3]. By the mid-1970s, all three naturally occurring CDs had been structurally and chemically characterized, and many more complexes had been studied [4]. Therefore, the ability of CDs to form complexes with a wide variety of molecules has been documented [5–10]. The details of CD complex formation and their binding constants were determined and found to be controlled by several different factors: hydrophobic interactions, hydrogen bondings, and van der Waals interactions. Therefore, CDs and their derivatives have been widely used in separation science since the early 1980s [4,11].

The evolution of CDs as chiral selectors in the liquid chromatographic separation of enantiomers has been a subject of interest for the last two decades. The presence of the chiral hollow basket, or cavity, makes these molecules suitable for the chiral resolution of a wide range of racemic compounds. At present, the use of CDs as chiral selectors for enantiomeric resolution by liquid chromatography is very common. Indeed, CDs have been used in the form of both chiral stationary phases (CSPs) and chiral mobile phase additives (CMPs). This chapter describes the techniques of chiral resolution by liquid chromatography using cyclodextrins as the stationary phases. The use of CDs as chiral selectors in the form of mobile phase additives is covered in Chapter 10.

3.1 STRUCTURES AND PROPERTIES

Cyclodextrins are cyclic oligosaccharides comprising six to twelve D-(+)-glucopyranose units in α-(1,4) linkage with the chair conformation. The three types of CD have different numbers of glucopyranose units: α-, β-, and γ-CDs contain six, seven, and eight glucopyranose units, respectively. Secondary 2- and 3-hydroxyl groups line the mouth of the CD cavity, while primary 6-hydroxyl groups are found on the opposite end of the molecule. This arrangement makes the CD cavity hydrophilic, which results in the formation of inclusion complexes with a variety of molecules. However, the interior of the cavity is slightly hydrophobic.

Many derivatives of CDs have been synthesized with various derivatizing groups, including acetyl, hydroxypropyl, naphthyl, ethylcarbamoyl, 3,5-dimethylphenyl carbamoyl, tosyl, trimethyl, dimethyl, phenyl carbamoyl, propyl carbamoyl, and aminomethylbenzyl. These derivatives modify the CD hydroxyl groups without destroying the CD cavity. Derivatized CDs exhibit behavior significantly different from native CDs. The derivatization results in partial blocking of the mouth of the CD cavity and eliminates some or all of the hydrogen bonds, leading to significant differences in enantioselectivity. The internal diameter of these CDs varies because of the different numbers of glucopyranose units present on the α, β, and γ forms. The most stable three-dimensional molecular configuration for these nonreducing cyclic oligosaccharides is a toroid, its upper (larger) and lower (smaller) openings having secondary and primary hydroxyl groups, respectively. The interior of the toroid is hydrophobic as a result of the electron-rich environment provided in large part by the glycosidic oxygen atoms. CDs are highly stable in acidic and basic conditions. The chemical structures of CDs and their derivatives are shown in Figure 1.

As noted earlier, CDs are capable of forming inclusion complexes with a variety of molecules. The stoichiometry (guest:CD) is 1:1, 1:2, or 2:1 depending on the structure of the molecule [12]. Complexation in CDs is molecule specific, and generally ternary complexes, stable in water solutions and containing lower alcohols, amines, and nitriles, are favored. However, α-CD includes single phenyl and naphthyl groups (small molecules), β-CD accepts naphthyl and heavily substituted phenyl groups, and γ-CD attracts bulky steroid-type molecules. The stability constants of complex formation vary from 10^{-2} to $10^{-5}\,M^{-1}$ with the complex formation speed $T_{1/2} \sim 0.001$–1.0 ms. These

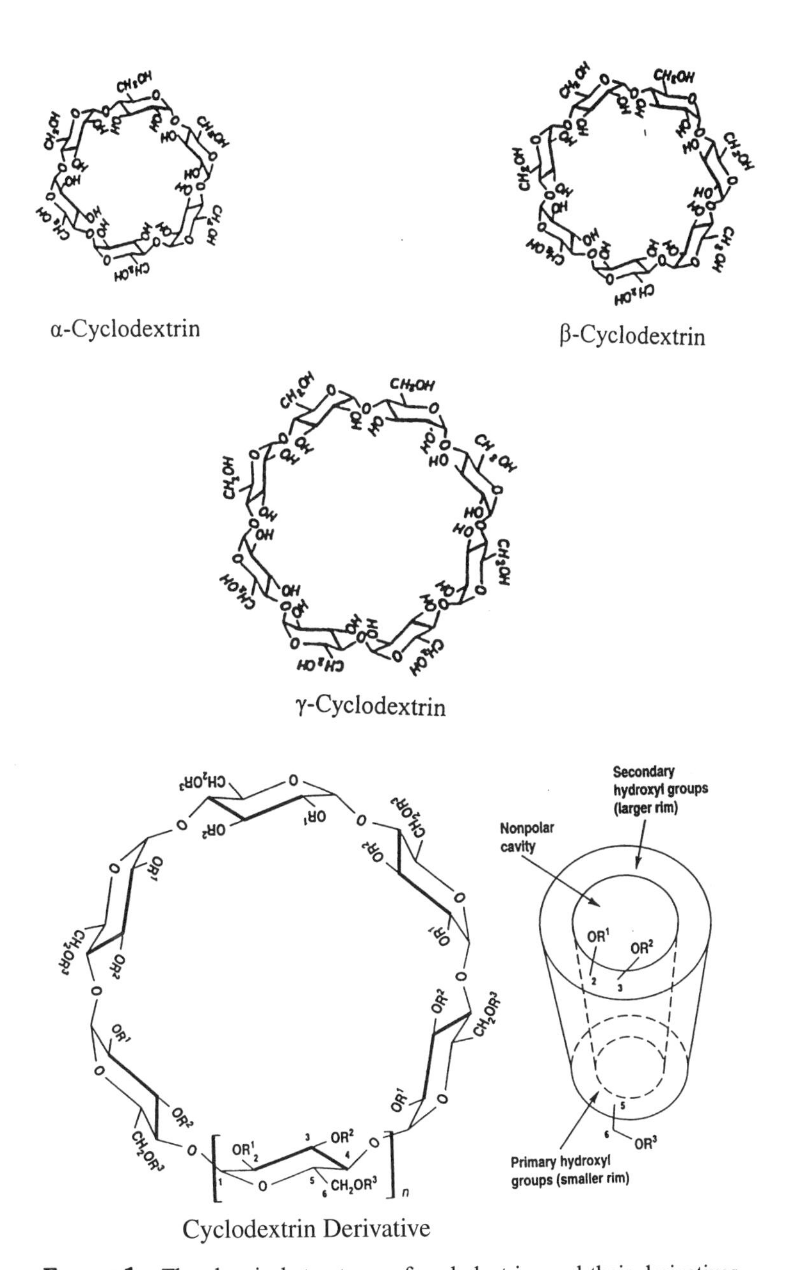

FIGURE 1 The chemical structures of cyclodextrins and their derivatives.

TABLE 1 The Properties of CDs

Properties	α-CD	β-CD	γ-CD
Number of glucopyranose units	6	7	8
Molecular weight	972	1135	1297
pK_a	12.33	12.20	12.08
Inner diameter, nm	0.45–0.57	0.62–0.78	0.79–0.95
Outer diameter, nm	1.37	1.53	1.69
Height, nm	79	79	79
Cavity volume, nm^3	0.174	0.262	0.472
Cavity volume			
per milliliter	104	157	256
per gram	0.10	0.14	0.20
Solubility in water, g/mL	14.5	1.85	23.2
Chiral centers	30	35	40
$[\alpha]_D^{25}$	+150.5	+162.0	+177.4

complex formation properties are advantageous in the chiral resolution of enantiomers because two enantiomers form inclusion diastereoisomeric complexes. The other properties of CDs are given in Table 1.

3.2 PREPARATION AND COMMERCIALIZATION

CDs are produced from starch by the action of *Bacillus macerans amylase* or the enzyme cyclodextrin transglycosylate (CTG) [13–16]. The latter enzyme can be used to produce CDs of specific sizes by controlling the reaction conditions. Initially, CDs were linked to solid supports. In 1978, Harada et al. [17] polymerized and cross-linked a cyclodextrin with a gel support and the CSP that developed was tested for chiral resolution of mandelic acid and its derivatives. Later, various workers bonded all the three CDs with different solid supports [18–21]. Of course, the CDs bonded to gel supports were used for the chiral resolution of different racemates, but their poor mechanical strength and low efficiency were distinct drawbacks. An improvement in these CSPs was introduced through covalent bonds between CDs and silica gel via propylamine, ethylenediamine, and other related linkages [22–27]. Again inherent problems prevented these developed phases from achieving the status of the best CSPs. The drawbacks related to these CD CSPs were tedious synthesis, involving the formation of nitroxide in the reaction mixture, as well as poor loading capacities, hydrolytic instability, and the effect of amines on enantioseparation. These drawbacks were overcome, however, by preparing CSPs that were more hydrolytically stable [28–43]. These CSPs were utilized in the reversed-phase mode, as well, and commercialized. Most of the chiral resolution work was carried out on these CSPs. The development of derivatized CD-based CSPs constituted a major development in the field [44–52]. The biggest advantage of these CSPs was versatility, for they could be used in normal, reversed, and new polar organically modified phases.

The three CDs and their derivatives can be bonded to silica gel by using silane reagents. The most commonly used silane reagents are 3-glycidoxypropyl trimethoxysilane, 3-glycidoxypropyldimethylchlorosilane, and 3-glycidoxypropyltriethoxysilane. The preparation of CD-based CSPs is a two-step process. The silane reagent is attached to silica gel (spherical; 5 μm particle diameter) by refluxing in dry toluene at 95°C for 3 h. The mixture is then cooled, filtered,

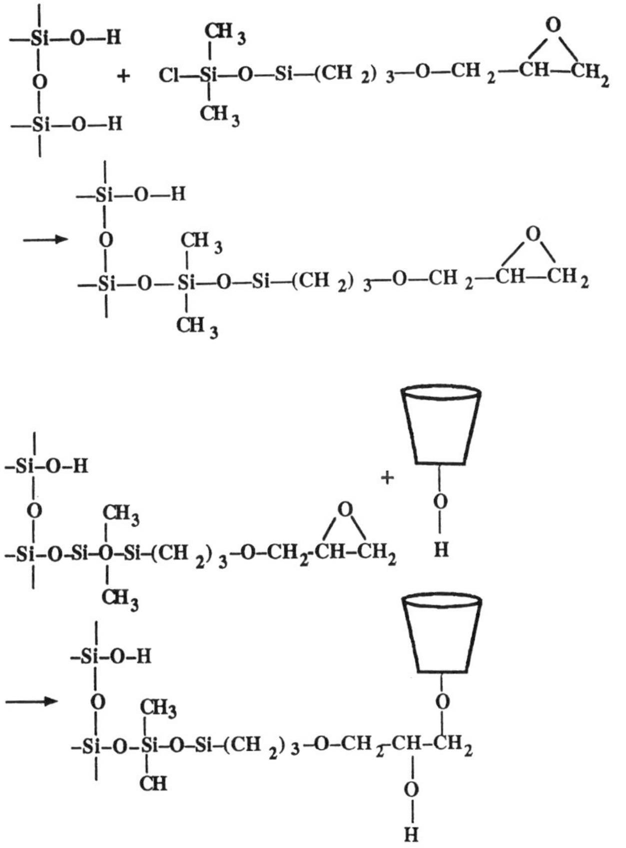

FIGURE 2 Chemical immobilization pathways for cyclodextrins on silica gel (see text).

washed several times with toluene and methanol, and then dried at 60°C over phosphorus pentoxide under vacuum. In the second step, the treated silica gel is allowed to react with CD in pyridine at 80°C [29,53]. The chemical pathways for these reactions are shown in Figure 2. The synthetic procedure varies slightly depending on the use of silane reagent and the CD type. From the theoretical point of view, linkage can occur through either the primary or the secondary hydroxyl groups, but in practice primary hydroxyl groups are more reactive, since they are less sterically hindered. If more than one chloro or methoxy groups are present on the silane reagent, more than one bonding link may occur between the silane reagent and the silica gel. CD has many free hydroxyl groups and, therefore, bonding with silica gel may occur through more than one silane bond. The high stability of CD bonded phases supports the binding of CDs through more than one bond, but no experimental evidence of this has been presented till today. The structures of some bonded CD chiral stationary phases are shown in Figure 3. In general, CD columns are only moderately robust but easily regenerated. At present, CD-based CSPs are among the most inexpensive tools for chiral resolution purposes. Table 2 lists the commercialized CD-based columns offered by different companies.

TABLE 2 Various Cyclodextrin-Based Commercial CSPs

Commercialized CSPs	Companies
Cyclobond I, II, and III (CDs)	Advanced Separation Technologies, Whippany, NJ, U.S.A.
Cyclobond AC, RN, SN (derivatives of CDs)	
ApHpera ACD and BCD (polymer-bonded CD derivatives)	
Nucleodex β-OH and Nucleodex β-PM	Macherey-Nagel, Duren, Germany
ORpak CD-HQ (CDs bonded to polymer gel)	
ORpak CDB-453 HQ ORpak CDBS-453, ORpak CDA-453 HQ, and ORpak CDC-453 HQ (CDs bonded to polymer-based gel)	Showa Denko, Kanagawa, Japan
Keystone β-OH and Keystone β-PM	Thermo Hypersil, Bellefonte, PA, U.S.A.
YMC Chiral cyclodextrin BR, YMC Chiral NEA (R), and YMC Chiral NEA (S)	YMC, Kyoto, Japan
β-Cyclose-6-OH	Chiral Separations, La Frenge, France

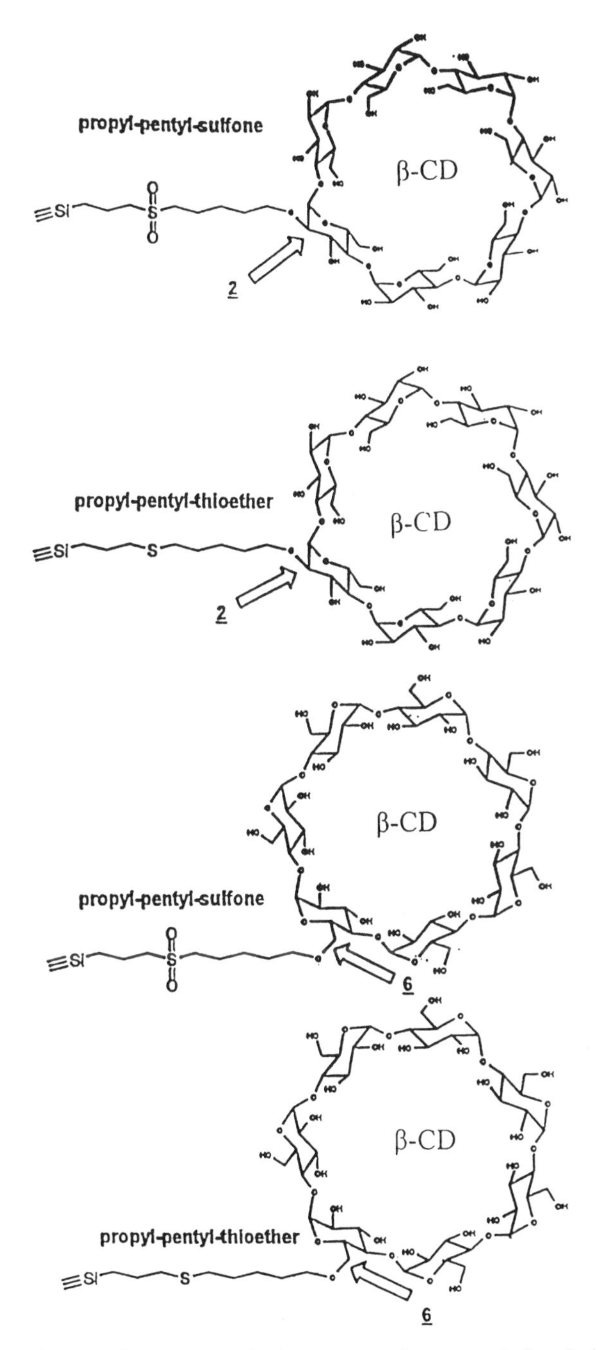

FIGURE 3 The chemical structures of some cyclodextrin-based CSPs.

3.3 APPLICATIONS

Cyclodextrin-based CSPs are among the most popular materials used for the chiral resolution of racemic compounds. These CSPs have a wide range of applications because they can be used successfully in all three mobile phase modes: normal, reversed, and polar organic. There are numerous examples of chiral separations on CDs and CSPs based on their derivatives. Some of the important chiral separations are discussed herein.

3.3.1 Analytical Separations

In 1978, Harada et al. [17] used polymerized CD with gel support for the chiral resolution of mandelic acid and its derivatives. Later Zsadon et al. [18–21] used cyclodextrin-based CSPs for the chiral resolution of indole alkaloids, with aqueous buffers as the mobile phases. Today CD-based CSPs have a good reputation. In separate studies, Fujimura [22] and Kawaguchi [23] and their colleagues resolved the enantiomers of aromatic compounds in the reversed-phase mode. Armstrong et al. [29,30,33,34,41,44–46,48,54–63] carried out extensive and remarkable work on the chiral resolution of various racemic compounds using CD-based CSPs.

Armstrong et al. [54] resolved the enantiomers of some amino acids and their derivatives on α-CD-based CSPs using 1% aqueous triethylammonium acetate (pH 5.1). The same authors also tested a β-CD CSP for the chiral resolution of amino acids [55]. In addition, they evaluated a γ-CD phase for the enantiomeric resolution of some dansyl amino acids and other drugs. The mobile phase was 38% methanol with 1% triethylammonium acetate [58]. In another study, the same authors reported the chiral resolution of 25 pairs of amino acids in less than 30 min [63]. The enantiomers of some β-adrenergic blockers were resolved on a β-CD stationary phase, with 1% aqueous triethylammonium acetate, containing methanol, as the mobile phase [9,48].

The enantiomers of baclofen, mandelic acid, tropic acid, chlorhexylphenylacetic acid, and chlorhexylphenylglyconic acid were resolved on β-CD phases, with a complex of potassium phosphate buffer (0.1 M) and acetonitrile (65 : 35, v/v) as the mobile phase [26]. The chiral resolution of rotenoids and their hydroxy analogues was carried out on β-CD, with different mixtures of water–acetonitrile and water–methanol, separately, as the mobile phases [64]. The enantiomers of epinephrine and norepinephrine were resolved on a β-CD phase by using a variety of mobile phases [65]. The chiral resolution of tetrahydroisoquinoline alkaloids was achieved on a β-CD phase by using mobile phases comprising citric acid monohydrate and phosphoric acid [66]. Furthermore, Seeman and Secor [57] resolved the enantiomers of nicotine analogues on a β-CD-bonded phase using 1% aqueous triethylammonium acetate (pH 7.1) containing acetone as the mobile phase modifier. A column bonded with β-CD carbamate was used for the chiral resolution of chlorpheniramine, with sodium phosphate and phosphoric acid containing acetonitrile as the mobile phase [67]. Chromatograms of the chiral resolution of amino acids and nicotine alkaloids and derivatives are shown in Figure 4.

Major advances have been made in recent years in the development and optimization of chiral resolutions of derivatized CD-based CSPs [68]. It has been reported that CSPs based on CD derivatives were more enantioselective than CSPs obtained from native CDs [68]. An acetyl β-CD column exhibited enhanced separation for scopolamine in comparison to the native β-CD CSP in the reversed-phase mode [69]. The enantiomeric resolution of some drugs was compared on the native β-CD and on CSPs based on (*S*)- and (*R*)-2-hydroxypropyl β-CD, and the best resolution was reported on the derivatized CSPs [44]. Five types of natural and chemically modified β- or γ-CD stationary phases were developed and used for the chiral resolution of dansyl amino acids. The best resolution of dansyl amino acids was provided by γ-CD CSPs [70].

In another study [71], native and chemically modified β-CD-bonded stationary phases of six types were evaluated for the chiral resolution of some

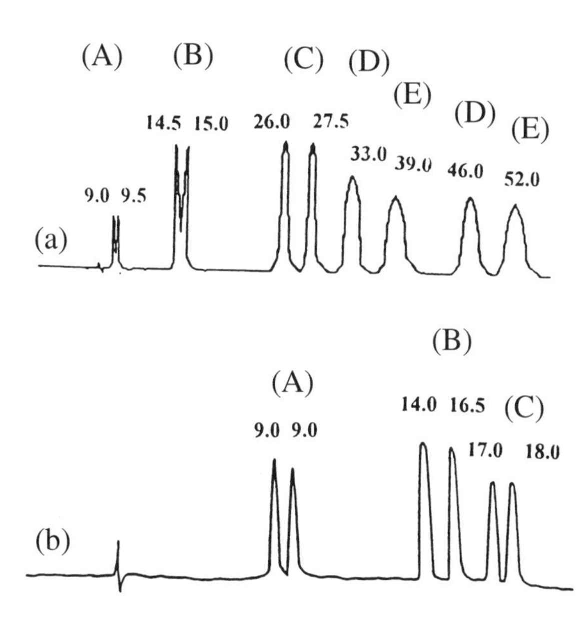

FIGURE 4 Chromatograms of (a) tyrosine (A), phenylalanine (B), tryptophan (C), 6-fluorotryptophan (D), and *o*-methyltyrosine (E) on an α-CD phase; and (b) *N*′-(2,2,2-trifluoroethyl)nornicotine (A), *N*′-benzylnornicotine (B), and *N*′-(2-naphthylmethyl)nornicotine (C) on a β-CD column. (From Refs. 54,57.)

derivatized amino acids in the reversed-phase mode. The chiral separations of derivatized amino acids were compared on (*R*)- and (*S*)-naphthylethylcarbamate derivatives of β- and γ-CD-based CSPs. The best separations were reported on β-CD-based CSPs. Han and Armstrong [4] used (*R*)- and (*S*)-naphthylethylcarbamate β-CD-based phases for the chiral resolution of pesticides (dyfonate, ruelene, ancylmidol, and coumachlor) and some pharmaceuticals, including trocanide, indapamide, and tolperisone. These CSPs were highly effective multimodal phases, (i.e., suitable for use in the normal, reversed, and organic mobile phases).

A novel heptasubstituted β-cyclodextrin bearing the methoxyethylamine group linked to the upper cyclodextrin rim was successfully used as the chiral selector for the enantiomeric separation of nonsteroidal anti-inflammatory drugs (NSAIDs) and phenoxypropionic acid herbicides (PPAHs) [72]. Results indicated that heptakis(6-methoxyethylamine-6-deoxy)-β-cyclodextrin (β-CD-OMe) performed exceptionally well in the enantiomeric resolution of two NSAIDs, indoprofen and fenoprofen. In addition, baseline enantiomeric separation of a mixture of six pairs of PPAHs was achieved within 30 min. Compared with other cationic β-cyclodextrins reported in the literature, the β-CD-OMe showed improved selectivity for both classes of the aforementioned anionic racemates [72]. Further, Ryu et al. [73] carried out the chiral separation of 2,4-dinitrophenyl (DNP) amino acids (Phe and Trp) on native β-CD and heptakis(3-*O*-methyl)-β-CD; the use of various substituents at different positions allowed the investigators to compare the enantioselectivity of these CSPs. Both native β-CD and heptakis(3-*O*-methyl)-β-CD showed good enantioselectivity for the DNP amino acids investigated. Heptakis(3-*O*-methyl)-β-CD, the cavity of which is more electron rich than that of native β-CD, showed in general much better enantioselectivity for the amino acid derivatives studied.

Generally, CD-based chiral stationary phases have been used in the reversed-phase mode. Earlier, it was assumed that in the normal phase mode, the more nonpolar component of the mobile phase would occupy the CD cavity, thereby blocking inclusion complexation between the chiral analyte and CD [4,11]. But with the development of CD derivatives, it has become possible to use the normal phase mode too [45,74]. Among the various CSPs based on CD derivatives, one based on a naphthylethyl carbamoylated derivative has shown excellent enantioselectivity in the normal phase mode [46,59]. Armstrong et al. [45] synthesized several β-CD derivatives and had them tested in the normal phase mode to resolve the enantiomers of a variety of drugs; hexane–2-propanol (90 : 10, v/v) served as the mobile phase. The authors discussed the similarities and differences of the enantioselectivities on the native and derivatized CD phases.

In another study, Armstrong et al. [59] resolved the enantiomers of profens on (*R*)- and (*S*)-naphthylethylcarbamate β-CD phases in the normal phase mode. Dichloro-, dimethyl-, and chloromethylphenylcarbamate derivatives of CDs were prepared and used for the chiral resolution of Tröger's base, flavanone, benzoin, and other drugs [49]. The mobile phase used was hexane–2-propanol (90 : 10, v/v). A phenylated β-CD CSP was developed and evaluated for the chiral resolution of β-adrenergic blockers, benzodiazepine anxiolytics, arylpropionic acid, and some herbicides. The column showed good chiral recognition capability when heptane, with 2-propanol or chloroform, was used as the mobile phase [75]. Chromatograms of the normal phase mode chiral separation of *trans*-2,3-cyclopropanedicarboxylic acid anilide, benzoin, fluoxetine, and norfluoxetine are shown in Figure 5.

The mobile phase conditions traditionally used for the chiral resolution on CD phases containing high concentrations of buffers adversely affected column stability and enantioselectivity. Therefore, in 1992, Armstrong et al. [60] developed a new mobile phase, comprising 90 to 99% polar organic solvent (i.e., methanol) for the chiral resolution of β-adrenergic blocking agents on β-CD based CSPs. The authors also used acetic acid and triethylamine as mobile phase additives. The same group [76] evaluated new polar organic mobile phases for the chiral resolution on CD-based CSPs.

Furthermore, Armstrong et al. [48] used the polar organic phase mode to resolve the enantiomers of β-adrenergic blockers on CSPs based on β-CD and its 2,3-methylated derivatives. The mobile phase used had methanol as the major component, with acetic acid and triethylamine (1%). In 1996, two reports were published on the polar organic phase mode chiral resolution of some drugs by means of CD-based CSPs [62,77]. The racemic compounds resolved were amino acids [62] and propranolol [75]. Zukowski [78] later resolved the enantiomers of Jacobson's catalyst on a hydroxypropyl β-CD phase with methanol–triethylamine (99.95 : 0.05, v/v) as the mobile phase. Even more recently, Péter et al. [79] separated the enantiomers of *diendo*- and *diexo*-3-aminobicyclo[2.2.1]heptane-2-methanol and *diendo*- and *diexo*-3-aminobicyclo[2.2.1]heptane-2-methanol derivatives on a naphthylethylcarbamate-derivatized β-CD cyclodextrin (Cyclobond I SN) stationary phase, which was used in the polar organic mode. This approach allowed the simultaneous separation of stereoisomers of alcohol and ester analogues of the bicyclic 1,3-amino alcohols. The chromatograms of the chiral resolution of β-adrenergic blockers on CSPs based on β- and γ-CDs are shown in Figure 6.

Direct analysis of the enantiomers in biological fluids is very important because it reduces both analysis time and sample preparation time. Indeed, when there is risk of the quick racemization of the enantiomer, direct analysis is essential. It has been observed that CD-based CSPs employ mobile phases that are generally compatible with biological samples, hence can be used for the direct analysis of the enantiomers in biological fluids [67,80]. Stalcup et al. [58] employed coupled column chromatography to isolate scopolamine from a plant extract and found that the extent of racemization depends on the isolation

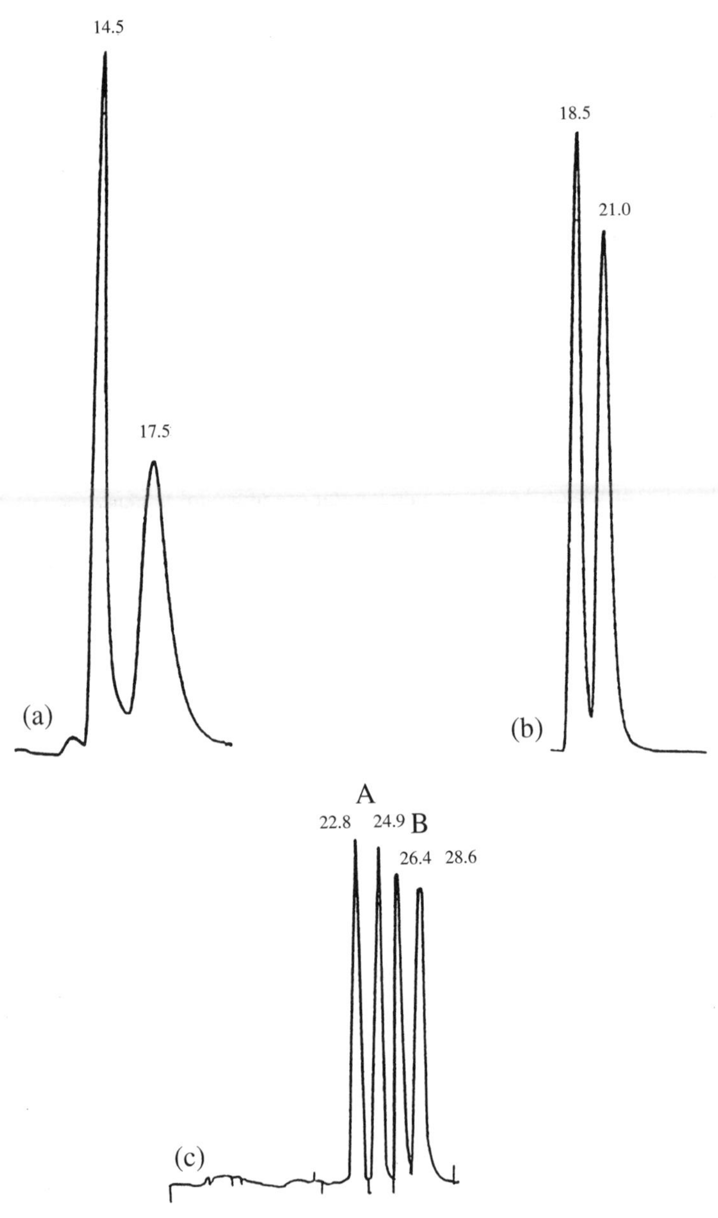

FIGURE 5 Chromatograms of (a) *trans*-2,3-cyclopropanedicarboxylic anilide, (b) benzoin, and (c) norfluoxetine (A) and fluoxetine (B) on β-CD phases using the normal mobile phase mode. (From Refs. 49,108.)

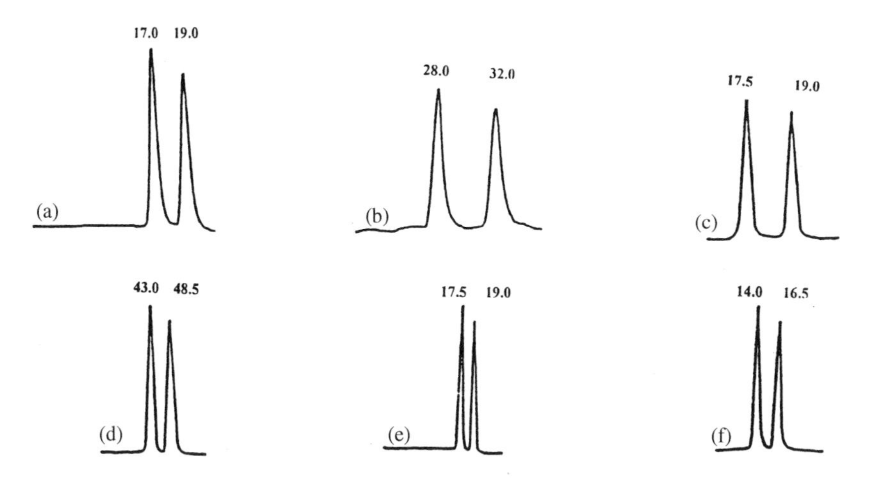

FIGURE 6 Chromatograms of (a) propranolol, (b) metoprolol, (c) timolol, (d) atenolol, (e) cateolol, and (f) alprenolol on β-CD columns, using the organic polar phase mode. (From Ref. 60.)

protocol. Furthermore, Stalcup and Williams [80] developed a special CD-based CSP with suitable spacer and used it successfully in the chiral resolution of hexabarbital in blood samples. Armstrong et al. [81] analyzed unnatural (D-) isomers of various amino acids in human urine samples. The chiral separation of lorazepam was achieved on a β-cyclodextrin derivative immobilized with silica gel under reversed-phase conditions. The method described allows the quantitation of the stereoisomers of lorazepam in human plasma following the administration of a therapeutic dose of the racemic drug [82]. The chiral resolution of various racemic compounds on native and derivatives of cyclodextrin-based CSPs is summarized in Table 3 [4,9,26,27,48,54–60,62–66,70–73,75,85–108].

3.3.2 Preparative Separations

CD-based CSPs can be used for chiral resolution at the preparative scale because their loading capacities are quite good. West and Cardellina [83] reported a semipreparative chiral separation of polyhydroxylated sterols on a β-CD phase. Later, Shaw et al. [84] presented a preparative chiral resolution of a substituted thienopyran on a β-CD phase. Advanced Separation Technologies, Inc., of Whippany, New Jersey, offers CD-based columns having diameters of 10, 20, and 50 mm on which 20 mg, 200 mg, and 5 g samples, respectively, can be loaded, in a single run [85]. The highest sample-loading capacity is in the polar organic mode, followed by the normal phase mode. A wide variation appears with the

TABLE 3 Enantiomeric Resolution of Some Racemic Compounds on Cyclodextrin-Based CSPs

Racemic Compounds	CSPs[a]	Refs.
Reversed-phase mode		
Alkaloids	β-CD and derivatives	9,56,57,66,87,88
Hydantoins	β-CD and derivatives	56,65,89–91
Antihistamines	β-CD	90
β-Adrenergic blockers	β-CD	9,48,75
Amino acids and derivatives	β-CD and derivatives	55,58,62,63,70,71,73,89,92
	α-CD	54
Peptides	β-CD	92
Crown ethers and analogues	β-CD and derivatives	93
Metallocenes	β-CD	94
Rotenoids	β-CD	64
Carboxylic acids and esters	β-CD and derivatives	9,25,26,27,56,89
Profens	β-CD	9,59,72,95
Propranolol	β-CD	96
Aminoalkyl phosphonic acid	β-CD	97
Dimethylaminobutyrophenone	β-CD	98
Antifungal agents	β-CD	99
Tetracyclic eudistomins	β-CD	100
Deltahedral carbonate	β-CD	101
9-Fluorenylmethyl chloroformate	γ-CD	102
NDP–amino acids	β-CD	103
Nomifensine hydrogen maleate	β-CD	104
Alkaloids and metabolites	β-CD	105
Warfarin	β-CD derivative	106
Terfenamide	β-CD	107
Pesticides	β-CD	4
Normal phase mode		
Tetracyclic eudistomins	β-CD	100
Ciprofibrate	Cyclobond I SN	85
Amino acids and derivatives	Cyclobond I SN	85
	Cyclobond I RN	
Terfenamide	β-CD	107
Fluoxetine and its metabolites	β-CD	108
Carotenes	β-CD	57
Herbicides	β-CD	75
New polar organic phase		
Amino acids and derivatives	β-CD	62,85
Ciprofibrate	Cyclobond I	85
Warfarin	Cyclobond I SN	85
β-Adrenergic blockers	β-CD	48,60,75,85
Profens	Cyclobond I RN	85
Methadone	Cyclobond I	85

[a] Cyclobond I, β-CD; Cyclobond I SN, *S*-naphthylethylcarbamated β-CD; Cyclobond I RN, *R*-naphthylethylcarbamated β-CD.

reversed-phase mode, which depends on the structures of the racemic compounds to be resolved. To reduce solvent consumption and improve purity and yield, recycling is the best technique. The enantiomeric resolution of warfarin was carried out on the preparative scale by separating enantiomers of about 500 mg on a Cyclobond 1 2000 column (25 cm × 2.5 cm), with methanol–acetic acid–triethylamine (100 : 0.3 : 0.2, v/v/v) as the mobile phase, at a flow rate of 12 mL/min [86]. AmeriChrom Global Technologies, of Beltsville, Maryland, markets VersaPrep CSP, which is also good for the preparative chiral separation of different racemic compounds. This CSP gave excellent results when used with recyclic technology in peak-shaving chromatography [85].

3.4 OPTIMIZATION OF HPLC CONDITIONS

The formation of CD inclusion complexes is favored in aqueous media, and hence most of the early chiral resolutions were carried out in the reversed-phase mode. However with the development of various derivatives of CDs, normal and polar organic mobile phases could also be used. The selectivities of these modes increase in the following order: normal > polar organic > reversed phase. This is due to certain advantages associated with the normal and polar organic phase modes. Now that all three mobile phase modes can be used, optimization on these phases can be achieved by varying a number of parameters such as composition of the mobile phase, use of organic modifiers, mobile phase pH, flow rate, and temperature. The details of the optimization of the chiral resolution on these CSPs are presented next.

3.4.1 Composition of the Mobile Phase

Most of the initial chiral resolutions on CD-based CSPs were carried out with aqueous mobile phases. Buffers of different concentrations and pH values were developed and used for this purpose. Triethylammonium acetate (TEAA), phosphate, citrate, and acetate are among the most commonly used buffers [4,11,26,68]. Also phosphate buffers such as sodium, potassium, and ammonium phosphate buffers are commonly used. The stability constant of the complexes decreases as organic solvents are added, hence organic modifiers are used to optimize the chiral resolution. The most commonly used organic solvents are methanol and acetonitrile. Acetonitrile is a stronger organic modifier than methanol. Some other organic modifiers, such as ethanol, 2-propanol, 1-propanol, *n*-butanol, tetrahydrofuran, triethylamine, and dimethylformamide were also used for the optimization of chiral resolution on CD-based CSPs [4,11,109,110]. Since the effect of the type and concentration of these organic modifiers varies from one analyte to another, it is very difficult to predict a successful strategy for their use as organic modifiers.

In 1987 Abidi [64] used an aqueous solution of methanol for the chiral resolution of rotenoids. The author reported that the best resolution occurred with 50% methanol as the mobile phase. The chiral resolution of hydantoin racemates was optimized by using 10 and 20% methanol as the mobile phase [88]. Armstrong et al. [54] achieved the chiral resolution of amino acids by using 60% methanol as the mobile phase. Similarly, the chiral resolution of nicotine analogues was achieved by using 40% methanol as the mobile phase [55]. Aboul-Enein et al. [104] optimized the chiral resolution of nomifensine hydrogen maleate by adding methanol in the mobile phase. The authors reported 15% as the optimum concentration of methanol for chiral resolution. According to Fujimura et al. [70], concentrations of 10 to 30% were better for the chiral resolution of dansyl amino acids. In another study, Stammel et al. [66] reported a decrease in the retention factors of the enantiomers of tetrahydroquinoline upon increasing the concentration of methanol. A very interesting study on the effects of organic modifiers on the chiral resolution of some drugs was presented by Piperaki et al. [109]. Of the organic modifiers tested (acetonitrile, methanol, ethanol, tetrahydrofuran, 1-propanol, and 2-propanol), methanol proved to be the best. Furthermore, Armstrong et al. [48] used a variety of solvent systems for the chiral resolution of different drugs. The effects of these organic modifiers on the chiral resolution of zopiclone are shown in Figure 7.

Armstrong et al. compared the chiral resolution of *N*-benzylnornicotine with two different organic modifiers: methanol and acetonitrile [57]. The maximum retention factor value ($k = 6.0$) was obtained with methanol as the organic modifier. On the other hand, the maximum retention factor value was 5.0 with acetonitrile as the mobile phase organic solvent. These findings indicate that methanol is a more suitable organic modifier than acetonitrile under the chromatographic conditions reported. When Abidi [64] compared the chiral resolution of rotenoids using methanol and acetonitrile as the organic modifiers, better resolution was achieved with methanol. Therefore, the role of organic modifiers in chiral resolution is not fixed and varies from one analyte to another. Figure 8 shows the effects of methanol and acetonitrile concentrations on the chiral resolution of *N*-benzylnornicotine [57] and nomifensine hydrogen maleate [104] racemates. The protocol for the development and use of the mobile phase in the reversed-phase mode is presented in Scheme 1.

Sometimes, the use of high-concentration buffers in the reversed-phase mode decreases column life and efficiency. Therefore, the use of an alternative mobile phase (i.e., normal phase) is an advantage in chiral resolution with these buffers. The most commonly used solvents in the normal phase mode are hexane, cyclohexane, and heptane. However, dichloromethane, acetone, propanol, ethyl-acetate, ethanol, and chloroform also have been used as mobile phase solvents. Hargitai and Okamoto [110] used hexane–2-propanol (in different ratios) as the mobile phase in the chiral resolution of several drugs. These authors also studied

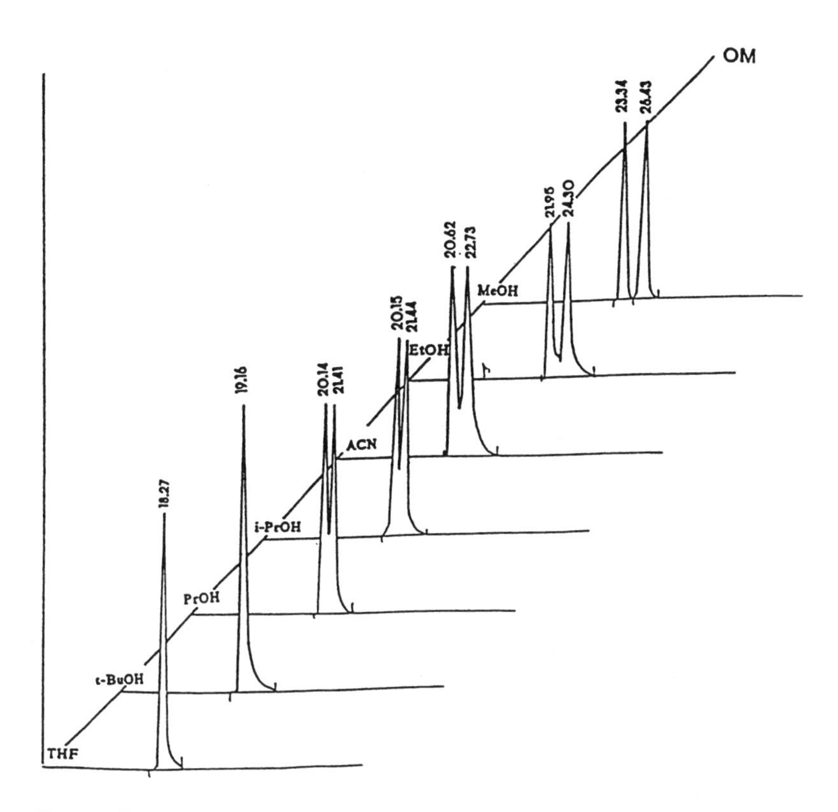

FIGURE 7 The effects of selected organic solvents on the enantioselectivity of zopiclone on a β-CD-based CSP. (From Ref. 109.)

chiral resolution in the reversed-phase mode. The highest selectivities were usually observed under normal phase chromatographic conditions. Hilton et al. [111] compared the chiral resolution of profens and amino acids, using methanol and acetonitrile as the major component of the mobile phase. The authors reported that the CSPs acted like cellulosic phases when used with 100% methanol or acetonitrile, which resulted in an improved chiral resolution. Chankvetadze et al. [49] resolved the enantiomers of benzoin using hexane–2-propanol (90 : 10, v/v) as the mobile phase. The chiral resolution of herbicides and β-adrenergic blockers [75] was achieved in the normal phase mode. The columns showed good chiral recognition ability for most of the solutes tested when heptane containing either 2-propanol or chloroform was one of the components. The protocol for the development and use of the normal phase mobile mode on these CSPs is presented in Scheme 2.

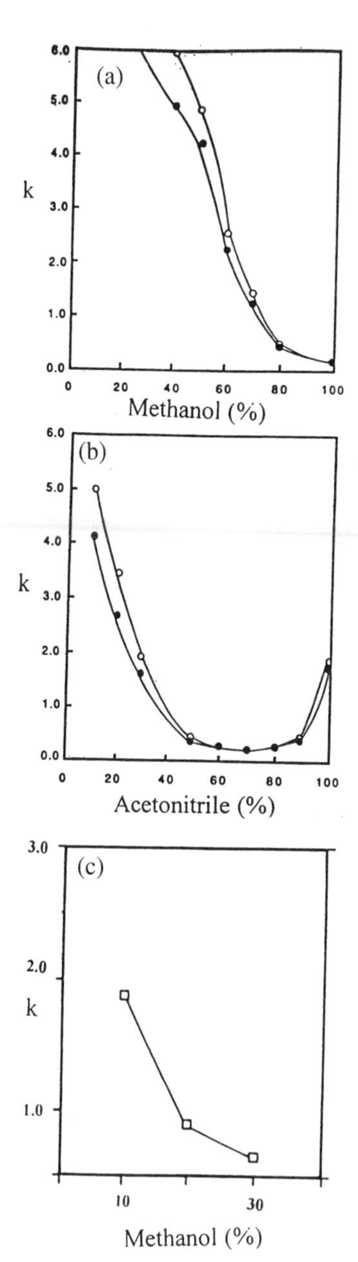

FIGURE 8 The effects of (a) methanol on the chiral resolution of *N′*-benzylnornicotine, (b) acetonitrile on the chiral resolution of *N′*-benzylnornicotine, and (c) methanol on the chiral resolution of nomifensine hydrogen maleate on β-CD-based CSPs. (From Refs. 57,104.)

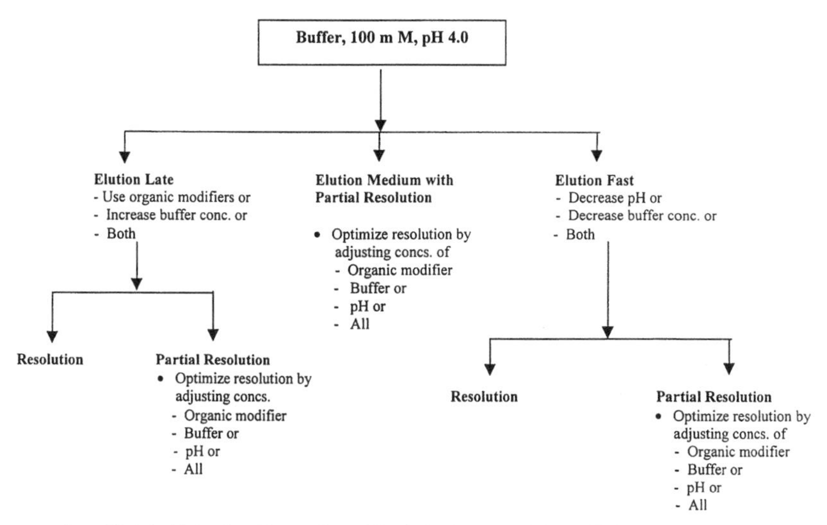

Note: This is the brief outline of the procedure to follow in developing a resolution on CDs based CSPs.

SCHEME 1 Protocol for the development and optimization of mobile phases on CD-based CSPs in the reversed-phase mode.

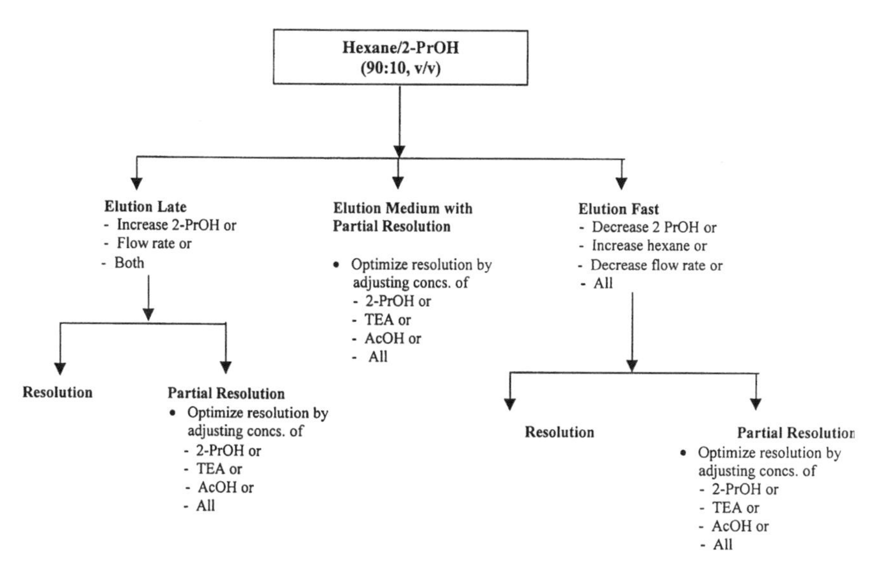

TEA and AcOH: Triethylamine and acetic acid respectively.

Note: This is the brief outline of the procedure to follow in developing a resolution on CDs based CSPs.

SCHEME 2 Protocol for the development and optimization of mobile phases on CD-based CSPs in the normal phase mode.

In 1996 it was reported [85] that it is possible to override inclusion complexation in favor of interacting directly with the secondary hydroxyl groups across the larger opening of the cyclodextrin toroid or the pendant carbamate, acetate, or hydroxypropyl functional groups. To accomplish this, the polar organic mode, comprising a polar organic solvent such as methanol or acetonitrile, was developed, and it produced very efficient separations not previously reported on these phases. Warfarin, β-adrenergic blockers, and other analytes have been separated by using the polar organic phase mode. This mode is also suitable for molecules that do not contain an aromatic group. Moreover, chiral separations achieved by using the polar organic phase mode are more efficient, more reproducible, and more effective at the preparative scale.

The ratio of acid to base is crucial, and the separation is optimized by monitoring the acid and base concentrations: if the analyte is eluting too fast, the concentration of acid or base is reduced; if the analyte is too well retained, the acid or base concentration is increased. The parameters for concentration are between 1 and 0.001%. In 1992 Armstrong et al. [60] used the polar organic phase mode to resolve the enantiomers of β-adrenergic blockers. The authors urged that other molecules could be resolved successfully under this mobile phase mode. A few years later, Matchett et al. [77] utilized this approach for the chiral resolution of propranolol and some of its analogues. Ekborg-Ott and Armstrong [62] used acetonitrile as the major component in the CSP resolution of amino acids. The separation was optimized by adjusting the concentrations of triethylamine and acetic acid. In another study, Zukowski [78] adopted the same polar organic mobile phase for the chiral resolution of Jacobson's catalyst. The author used 1000 mL of acetonitrile with 0.5 mL of triethylamine and 2.5 mL of acetic acid as the mobile phase; as recommended earlier [111], DNB and dansyl amino acids were resolved by using methanol or acetonitrile as the major component of the mobile phase. Optimization was achieved by adjusting the concentrations of triethylamine, acetic acid, and ethanol [111]. The effect of the concentration of methanol on the chiral resolution of some β-adrenergic blockers is shown in Figure 9 [77]. It may be concluded from Figure 9 that the retention factors decreased as a function of the increase in methanol concentration. Yet Han [68] has reported that in general, retention increases, without significantly affecting the enantioselectivity, when methanol concentration is increased. The protocol for the development and use of the polar organic mobile phase is given in Scheme 3.

3.4.2 pH of the Mobile Phase

Many of the early chiral resolutions were carried out in the reversed-phase mode. Therefore, the pH of the mobile phase, which affects the degree of complexation of analytes with CD, is very important. In 1985 Feitsma et al. [26] studied the

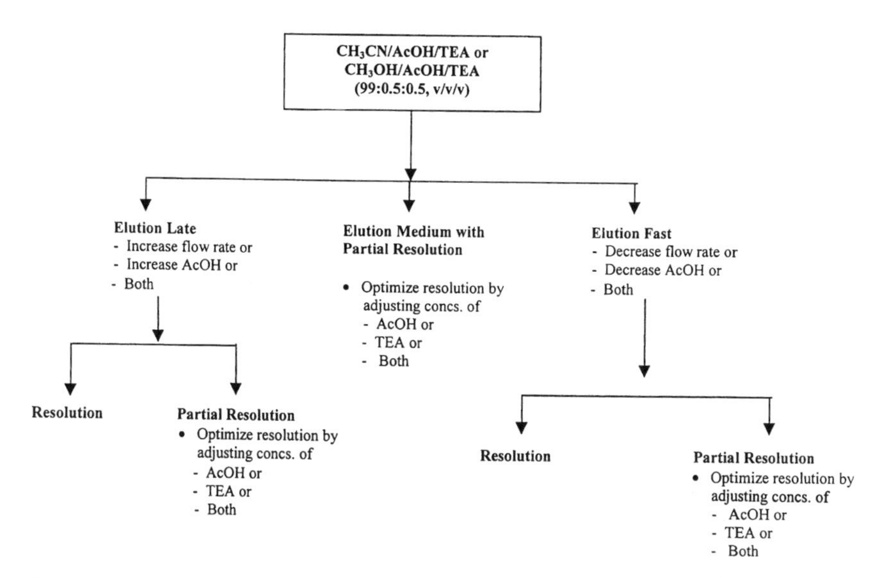

SCHEME 3 Protocol for the development and optimization of mobile phases on CD-based CSPs in the polar organic phase mode. TEA, triethylamine; AcOH, acetic acid.

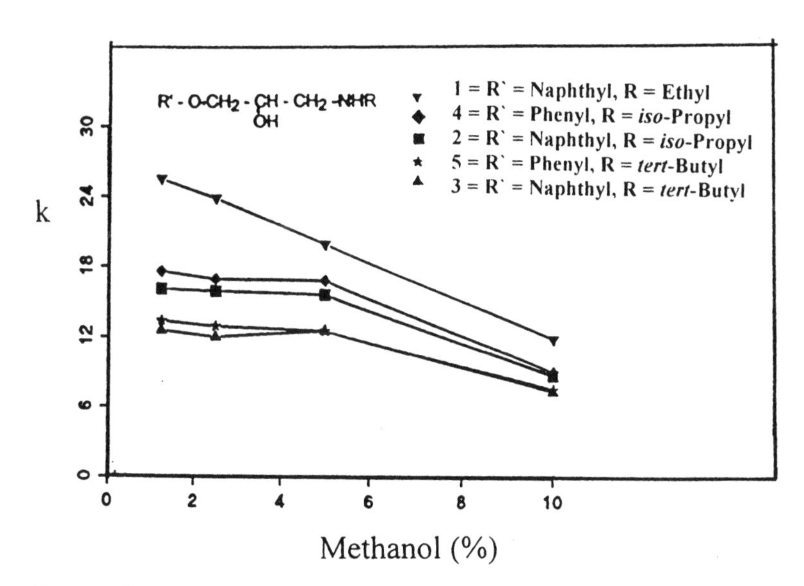

FIGURE 9 The effects of methanol on the chiral resolution of some β-adrenergic blockers on β-CD column, in the polar organic mobile phase mode. (From Ref. 77.)

chiral resolution of aromatic acids. They reported that mandelic acid could be separated at pH 4.2 but not at pH 6.5. Furthermore, it was observed that cyclohexylphenylglycolic acid and cyclohexylphenylacetic acid were highly retained at pH 4.2 and 6.5, respectively. Aboul-Enein et al. [104], who optimized the chiral resolution of nomifensine hydrogen maleate at pH 3.5, reported that an increase in pH resulted in an increase in retention factor. Armstrong et al. [105] observed three types of pH effect on the chiral resolution of rotenoids: (1) an increase in retention with pH, (2) an increase in retention followed by a decrease, and (3) an increase in retention that becomes constant. Therefore, chiral resolution varied from one rotenoid to another.

In another study, Piperaki and Parissi-Poulou [108] carried out the chiral resolution of fluoxetine and norfluoxetine at pH 3.5 to 7.0. The best resolution of both racemates was reported at pH 5.0 to 5.5. Lemer et al. [112] studied the chiral resolution of ephedrine, methamphetamine, and selegiline at pH 3.5 to 5.5, and their results are given in Table 4. The lower retention factor values at lower pH may be due to drug protonation, since the compounds studied were basic. It is interesting to observe from Table 4 that the separation factor values decreased when pH increased, while the reverse was true for the resolution factor values. Haynes et al. [72] reported a major effect of pH on the chiral resolution of anti-inflammatory drugs and phenoxypropionic acid herbicides. Recently, Chen et al. [113] studied the effect of pH on the chiral resolution of β-adrenergic blockers and amines. The authors reported the best resolution of these compounds between 4.65 and 6.30 pH range. The effects of pH on the chiral resolution of nomifensine hydrogen maleate, fluoxetine, and norfluxetine are shown in Figure 10. Optimization in the normal and polar organic phase modes was also achieved when acids and bases were used. Therefore, pH also controls optimization in the normal and polar organic phases. To the best of our knowledge, however, no report has been published on this issue.

TABLE 4 Effects of pH on the Chiral Resolution of Ephedrine, Methamphetamine, and Selegiline

	Ephedrine			Methamphetamine			Selegiline		
pH	*k*	α	*Rs*	*k*	α	*Rs*	*k*	α	*Rs*
3.5	0.59	0.73	0.53	0.80	1.00	1.00	0.87	0.83	0.69
4.5	0.65	0.71	0.58	0.84	0.90	0.95	0.92	0.79	0.91
5.5	0.77	0.69	0.66	0.92	0.78	0.90	1.00	0.67	0.84

Source: Ref. 112.

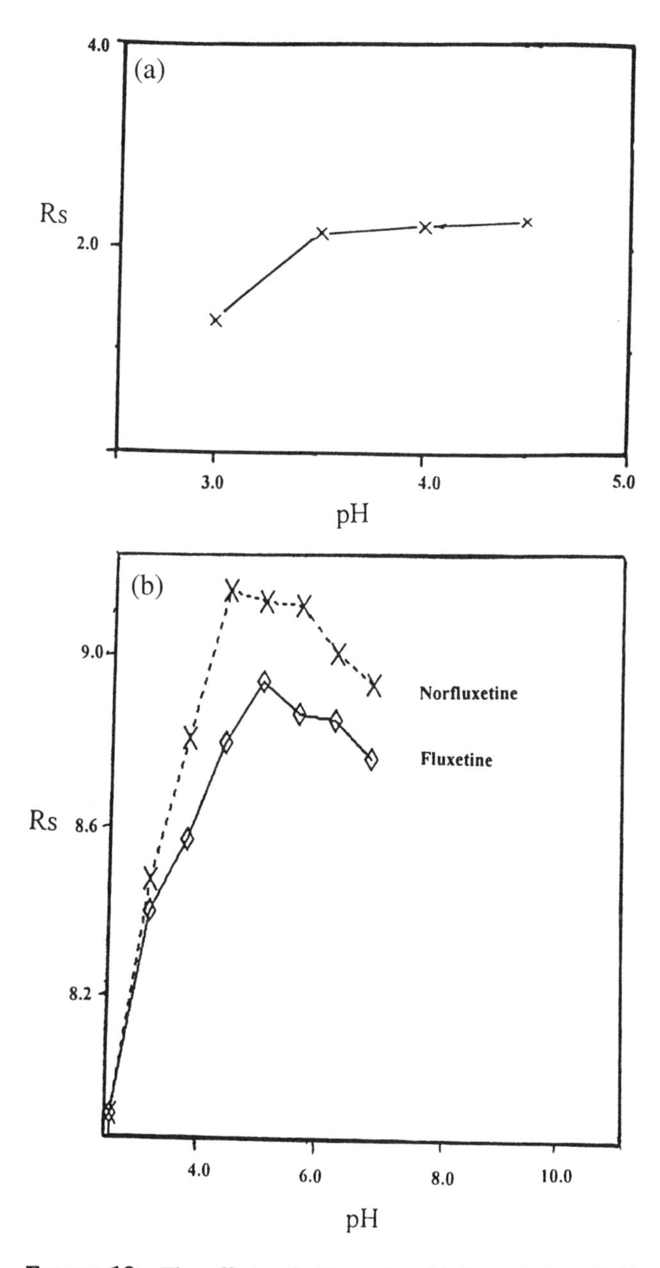

FIGURE 10 The effects of pH on the chiral resolution of (a) nomifensine hydrogen maleate and (b) fluoxetine and norfluoxetine on β-CD columns. (From Refs. 104 and 108.)

3.4.3 Ionic Strength of the Mobile Phase

Buffer concentration is a very important aspect of chiral resolution on CD-based phases in the reversed-phase mode. Addition of salts to the reversed mobile phase has improved chiral resolution [68]. In some cases, as the buffer concentration increased, the retention times decreased and the peaks sharpened. These results prompted some studies to evaluate the effect of buffer concentration on chiral resolution in these phases. Fujimura et al. [70], who studied the effects of phosphate buffer concentration on the chiral resolution of dansyl amino acids, reported that the best separations were obtained at 0.2 M buffer concentration. The retention factors were increased when the ionic strength of the buffer was raised. Lemer et al. [112] varied from 50 nmol/L to 750 nmol/L the concentration of triethylamine in a triethylamine–aqueous sulfuric acid mobile phase for the chiral resolution of ephedrine, selegiline, and methamphetamine racemates. The authors reported that the highest concentration was the best. In another study, Fukushima et al. [65] investigated the effects of ammonium phosphate concentration on the chiral resolution of ephedrine and norephedrine. It was observed that retention factors increased when the concentration of the buffer was increased. Haynes et al. [72] optimized the chiral resolution of anti-inflammatory drugs and phenoxypropionic acid herbicides by adjusting the concentration of the buffer. The effect of buffer concentration on the chiral resolution of ephedrine, norephedrine, and some amino acids is presented in Figure 11.

The effects of certain ions on the chiral resolution on CD-based CSPs also were evaluated. Fukushima et al. [65] used sodium, ammonium, and triethylamine cations in the mobile phase and assessed the effects of their presence on the chiral resolution of some drugs. The enantioselectivity was in the following order: triethylamine > ammonium > sodium. Lemer et al. [112] also studied the effects of acetate, bicarbonate, sulfate, chlorate, nitrate, and phosphate anions on the chiral resolution of ephedrine, methamphetamine, and selegiline racemates. The authors reported that the retention of all the three drugs increased in the following order: chlorate < acetate < nitrate < bicarbonate < phosphate < sulfate. They attributed the increase in retention factor values to the increasing hydration of the anions studied. However, the authors reported the best resolution in the presence of sulfate anion. Similarly, Aboul-Enein et al. [104] tried to improve the chiral resolution of nomifensine hydrogen maleate by adding ammonium chloride to a mobile phase consisting of TEAA (at 3.5 pH) and methanol (80 : 20, v/v). But the authors reported no effect on chiral resolution due to the presence of ammonium chloride in the mobile phase. The effects of the presence of anions on the chiral resolution of ephedrine, methamphetamine, and selegiline are shown in Figure 12.

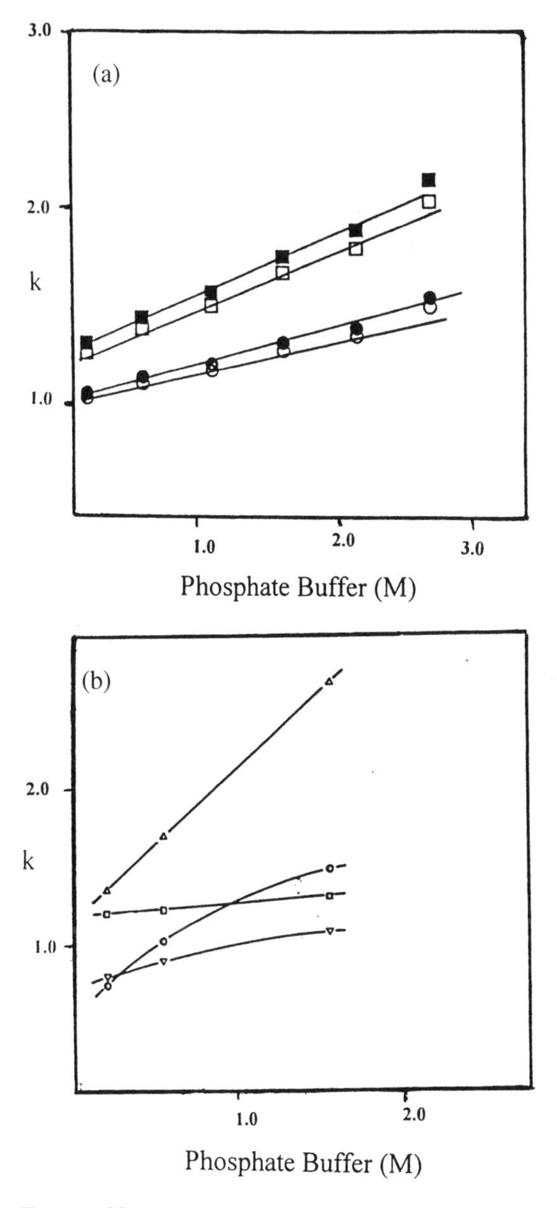

FIGURE 11 The effects of phosphate buffer concentration (ionic strength) on the chiral resolution of (a) *d*-ephedrine (■), *l*-ephedrine (□), *d*-norephedrine (●), and *l*-norephedrine (○) and (b) the dansyl amino acids aspartic acid (△), glutamic acid (○), serine (□), and phenylalanine (▽) on β-CD columns. (From Refs. 65,70.)

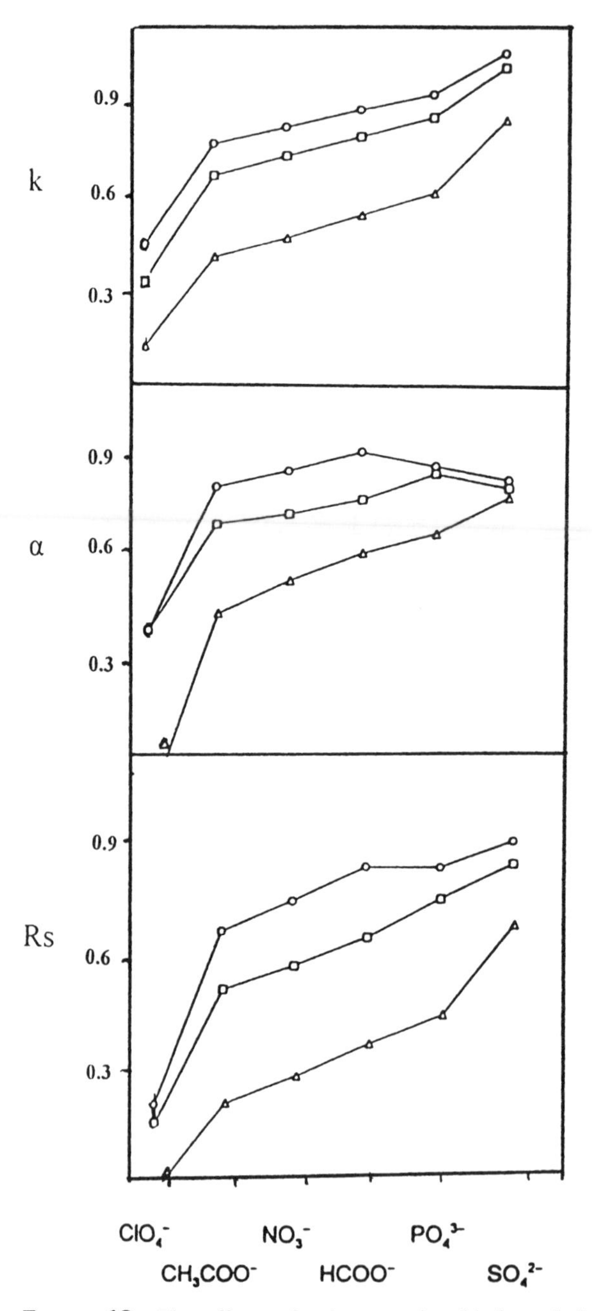

FIGURE 12 The effects of anions on the chiral resolution of ephedrine (△), methamphetamine (□), and selegiline (○) on ChiraDex columns. (From Ref. 112.)

3.4.4 Flow Rate of the Mobile Phase

Enantiomeric separations on mobile phases depend on the placement of the enantiomers in the cavities of the CDs. Therefore, upon displacing the enantiomers from the CD cavities, flow rate may be used as one of the optimizing parameters. This allows investigators to optimize chiral resolution in all the three mobile phase modes by means of flow rate. Piperaki and Parissi-Poulou [108] varied the flow rate from 0.2 mL/min to 1.0 mL/min and observed the chiral resolution of fluoxetine and norfluoxetine. The effects of flow rate on the chiral resolution of ephedrine, methamphetamine, selegiline, fluoxetine, and norfluoxetine are shown in Figure 13, which indicates that retention times are affected tremendously by decreases in flow rate. The authors suggest that by decreasing the flow rate, one affords the molecules a greater opportunity to interact with CDs, thus increasing the chiral discrimination ability of the cyclodextrins. Similarly, Lemer et al. [112], who varied flow rate from 0.2 mL/min to 1 mL/min in the chiral resolution of ephedrine, methamphetamine, and selegiline, reported that chiral resolution was decreased at higher flow rates for all three drugs. The best resolution was reported at a flow rate of 0.2 mL/min. Thus, 0.2 to 1.0 mL/min seems to be the best range of working flow rates on these CSPs. The effects of flow rate on the chiral resolution of ephedrine, methamphetamine, selegiline, fluoxetine, and norfluoxetine are shown in Figure 13.

3.4.5 Temperature

The kinetics of the formation of inclusion complexes between enantiomers and CDs is affected by changes in temperature and, therefore, temperature may be used to optimize chiral resolution on these phases. The degree to which temperature affects resolution is dependent on analyte and CD phase, respectively. Generally, lower temperatures enhance the weaker bonding forces between enantiomers and CD phase, and the net result is an improvement in chiral resolution. In one study, Feitsma et al. [26] were plagued by the tailing of peaks in the chiral resolution of mandelic acid, tropic acid, cyclohexylphenylglycolic acid, and cyclohexylphenylacetic acid racemates at room temperature. This problem was removed by increasing the experimental temperature. However, higher temperature resulted in shorter retention times owing to decreased stability of the inclusion complexes at higher temperature. Aboul-Enein et al. [104], who studied the effects of temperature by resolving the enantiomers of nomifensine hydrogen maleate found that temperature did not exert a prominent effect. The best resolution of nomifensine hydrogen maleate, however, was achieved at 8°C.

Lipkowitz and Stoehr [114] studied the binding of the enantiomers of methyl mandelate under buffered and unbuffered conditions at different temperatures. The authors reported that the stereodifferentiating binding energies for the

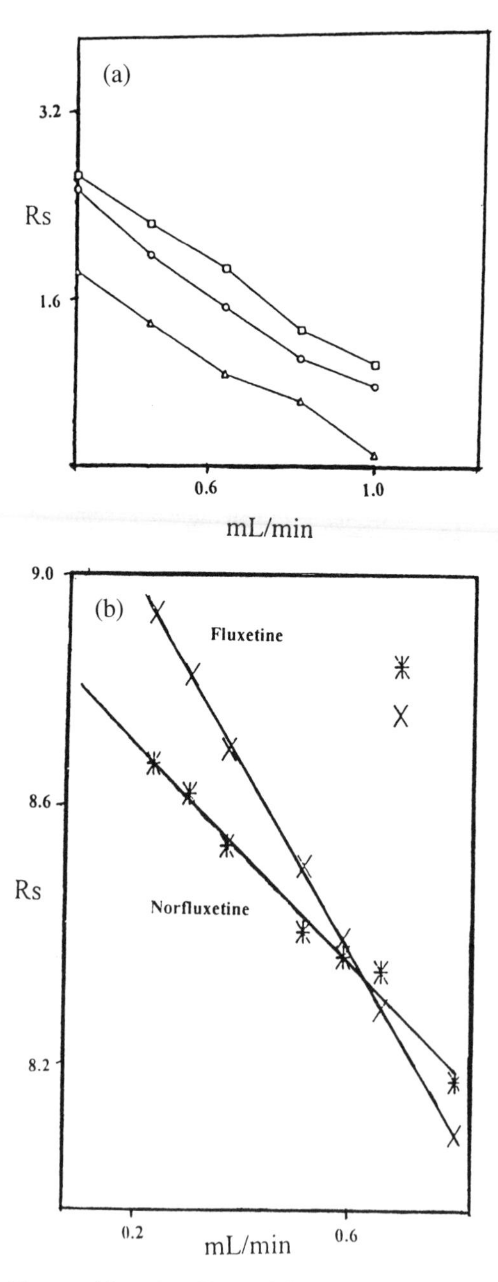

FIGURE 13 The effects of flow rate (mL/min) on the chiral resolution of (a) ephedrine (△), methamphetamine (□), and selegiline (○) on a ChiraDex column and (b) fluoxetine and norfluoxetine on β-CD based CSPs. (From Refs. 108,112.)

enantiomers were higher at lower temperatures. Lemer et al. [112], who studied the effects of temperature, in the range 5 to 30°C, on the chiral resolution of ephedrine, methamphetamine, and selegiline drugs, reported an increase in the retention, separation, and resolution factors at lower temperatures. This behavior was explained on the basis of more stable inclusion complexes at lower temperatures. The effects of temperature on the chiral resolution of ephedrine, methamphetamine, selegiline, and nomifensine hydrogen maleate are shown in Figure 14.

3.4.6 Solute Structures

The chiral resolution on CD-based CSPs depends on the formation of inclusion complexes in the cavities and, therefore, the structures and sizes of analytes are very important for the chiral resolution of racemates on these phases. Amino acids often are considered to be the best class of racemic compounds to use in structural studies. In 1987, Han and Armstrong [55] studied the chiral resolution of amino acids on β-CD-based CSPs. It was observed that different retention, separation, and resolution factor values were obtained for different amino acids under identical chromatographic conditions, which indicated that the structures and sizes of amino acids govern their chiral resolution. The same observations may be found in the work of Fujimura et al. [70].

Maguire [89] carried out an interesting study of the chiral resolution of hydantoins on a β-CD phase. The results (Table 5) indicate that retention, separation, and resolution factor values increased when bulky alkyl groups were introduced into the molecules. Similarly, Han et al. [56] studied the effects of hydantoin structures on their chiral resolution. The authors reported good chiral resolution when these compounds had two or more rings. Recently, Tazerouti et al. [75] used the normal phase mode (two mobile phases) to achieve chiral resolution of some related herbicides on the β-CD-based phase (Table 6). In general, retention, separation, and resolution factor values increased when bulky alkyl groups were introduced into the molecules. Furthermore, the retention, separation, and resolution factor values dramatically increased upon addition of a phenyl group. Thus selection of the structures and sizes of analytes is very important for successful chiral resolution on a specific CD-based CSP. It may be concluded that structures and sizes exert their effects on chiral resolution as a result of certain interactions and forces between the CSP and the analytes, which will be discussed later (Sec. 3.5).

3.4.7 Structures of the CDs and Their Derivatives

All three natural cyclodextrins have different selectivities and enantiorecognition capacities owing to differences in cavity diameter and length. Generally, α-, β-, and γ-CDs are suitable for the chiral resolution of small, medium, and large

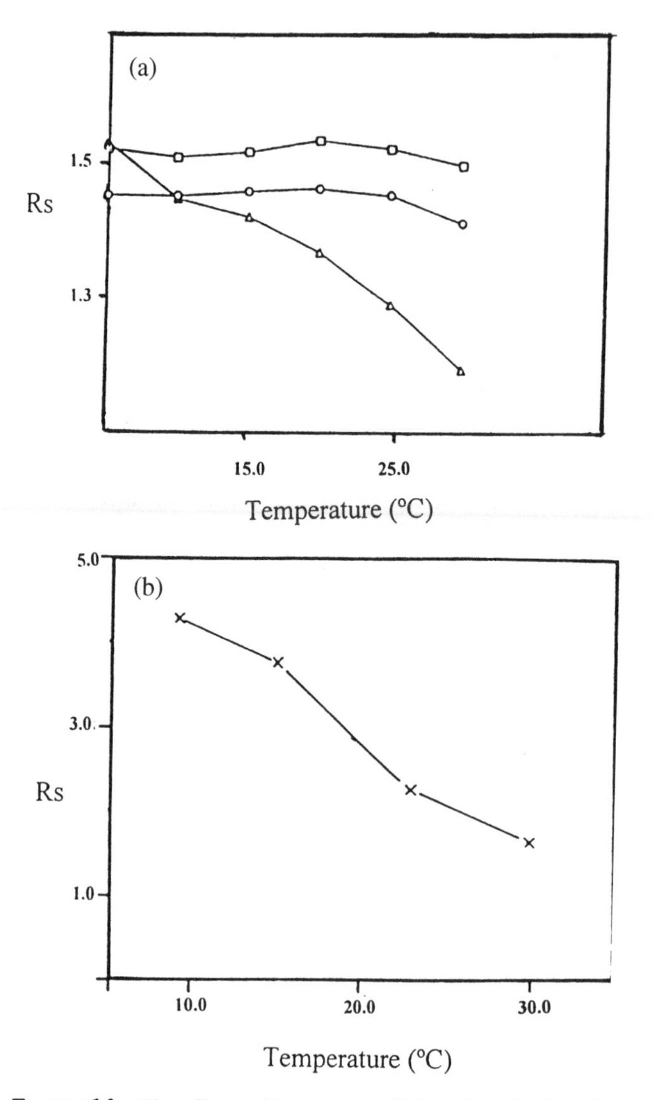

FIGURE 14 The effects of temperature (°C) on the chiral resolution of (a) ephedrine (△), methamphetamine (□), and selegiline (○) on a ChiraDex column and (b) nomifensine hydrogen maleate on β-CD-based CSP. [From Refs. 112 (a), 104 (b).]

molecules respectively. Moreover, the chiral recognition capacities of the different CD derivatives vary greatly. Thus published results of studies of chiral resolution on different CD-based CSPs have been compared and discussed.

TABLE 5 Effects of the Substituents on the Chiral Resolution of 5-Alkyl-5-phenylhydantoin Enantiomers on a β-CD Phase

Substituents	k	α	Rs
H	0.67	1.09	0.20
CH_3	1.17	1.11	0.50
C_2H_5	2.30	1.21	2.00
n-C_3H_7	3.30	1.26	2.20
iso-C_3H_7	4.96	1.27	2.50
n-C_4H_9	4.40	1.14	1.10

Source: Ref. 88.

Fujimura et al. [70] prepared five types of natural and chemically modified β- and γ-CD-based CSPs and evaluated their performance in the chiral resolution of amino acids. Chiral resolution data for some amino acids on β-CD, 6-*O*-phenylcarbamoylated β-CD, and 6-*O*-propylcarbamoylated β-CD are presented in Table 7. A perusal of Table 7 indicates that β-CD is the best CSP. In an interesting study, Hargitai and Okamoto [110] compared the chiral recognition capacities of 3,5-dimethylphenylcarbamoylated β-CD with those of Cyclobond I SN and Cyclobond I DMP (carbamoylated β-CD) in the chiral resolution of a variety of drugs. The authors reported the best resolution on 3,5-dimethylphenylcarbamoylated β-CD. Chankvetadze et al. [49] prepared dichloro-, dimethyl-, and chloromethylphenylcarbamate CD derivatives (Fig. 15) and tested them in the chiral resolution of drugs including flavanone and cyclopropanedicarboxylic acid dianilide in the normal phase mode (Fig. 16). Furthermore, the authors studied the effects of substitution at the narrow and wide ends of the CDs. The results of this work are given in Tables 8 to 10.

TABLE 6 Effects of the Substituents on the Chiral Resolution of *N*-(3,5-Dinitrobenzoyl)-2-aminoalkyl Propanoate Enantiomers on a β-CD Phase

Substituents	Heptane–2-propanol (v/v)	k	α	Rs	Heptane–chloroform (v/v)	k	α	Rs
CH_3	80 : 20	1.89	1.09	0.99	25 : 75	2.38	1.00	—
$CH(CH_3)_2$	80 : 20	1.40	1.11	1.04	75 : 25	15.02	1.07	1.13
$CH_2C_6H_5$	90 : 10	4.02	1.62	5.65	50 : 50	2.93	1.10	1.15

Source: Ref. 75.

TABLE 7 Effects of the Structures of β-CD Phases[a] on the Chiral Resolution of Dansyl Amino Acids

	CSP I		CSP II		CSP III	
Dansyl amino acids	*k*	*Rs*	*k*	*Rs*	*k*	*Rs*
Glutamic acid	2.79	1.19	1.55	0.99	0.82	0.63
Aspartic acid	3.03	2.18	1.68	1.58	0.84	0.89
Serine	4.24	1.05	2.42	1.14	1.40	0.80
Threonine	4.49	2.17	2.72	1.78	1.53	1.21
Norvaline	7.26	0.62	3.51	0.67	2.04	0.00
Valine	7.56	1.07	3.73	1.00	2.17	0.85
Norleucine	10.90	0.55	4.70	0.00	2.59	0.00
Leucine	17.10	0.88	4.95	0.00	2.89	0.00
Phenylalanine	22.90	0.85	7.46	0.85	4.33	0.70
Tryptophan	13.30	0.00	8.09	0.00	4.28	0.00

[a] CSP I, CSP II, and CSP III are β-CD, (6-*O*-phenylcarbamoylated)-β-CD, and (6-*O*-propylcarbamoylated)-β-CD, respectively.
Source: Ref. 70.

TABLE 8 Chiral Resolutions of Some Drugs on 4-Chloro-3-methylphenylcarbamate Derivatives of α-, β-, and γ-CDs[a]

	CSP 1		CSP 2		CSP 3	
Solutes	*k*	α	*k*	α	*k*	α
Tröger's base (I)	1.30	1.00	1.80	1.07	1.80	1.00
trans-2,3-Diphenyloxirane (II)	1.20	1.06	0.87	1.10	0.73	1.00
1-(9-Anthryl)-2,2,2-trifluoroethanol (III)	1.23	1.00	1.33	1.00	1.80	1.10
1,2,2,2-Tetraphenylethanol (IV)	0.97	1.34	1.18	1.00	1.40	1.06
2,2′-Dihydroxy-6,6′-dimethylbiphenyl (V)	2.10	1.00	2.00	1.00	2.50	1.00
2-Phenylcyclohexanone (VI)	1.23	1.00	1.83	1.00	1.70	1.06
Flavanone (VII)	1.10	1.00	3.20	1.60	1.83	1.56
Benzoin (VIII)	2.70	1.00	3.75	1.03	1.53	1.00
trans-Cyclopropanedicarboxylic acid anilide (IX)	1.30	1.15	2.17	1.65	1.93	1.41

[a] The structures of the CSPs and the solutes are given in Figures 15 and 16, respectively.
Source: Ref. 49.

TABLE 9 Chiral Resolutions of Some Drugs on Phenylcarbamate Derivatives of β-CDs with Variations in the Nature and Position of the Substituents on the Phenyl Ring

	Solutes[b]																	
	I		II		III		IV		V		VI		VII		VIII		IX	
CSPs[a]	k	α	k	α	k	α	k	α	k	α	k	α	k	α	k	α	k	α
2	1.80	1.07	0.87	1.10	1.33	1.00	1.18	1.00	2.00	1.00	1.83	1.00	3.20	1.60	3.75	1.03	2.17	1.65
4	1.32	1.06	0.48	1.10	2.37	1.00	1.17	1.00	3.00	1.00	1.07	1.00	1.70	1.00	3.75	1.03	2.17	1.65
5	1.40	1.00	0.63	1.00	0.90	1.00	0.60	1.00	1.73	1.00	1.13	1.00	0.90	1.00	2.10	1.00	1.07	1.00
6	0.90	1.00	0.60	1.00	2.10	1.00	1.17	1.00	2.13	1.03	1.36	1.00	1.03	1.03	1.05	2.23	1.07	2.23
7	1.67	1.00	0.68	1.00	1.20	1.00	1.10	1.00	1.80	1.00	1.50	1.00	1.30	1.04	2.20	1.00	0.95	1.00
8	1.18	1.11	0.67	1.00	1.50	1.00	0.95	1.00	1.87	1.00	1.40	1.00	1.84	1.84	2.50	1.05	1.33	1.35
9	1.07	1.00	0.67	1.00	1.83	1.00	6.67	1.00	1.27	1.00	1.18	1.00	1.28	1.05	2.07	1.11	0.67	1.00
10	1.23	1.00	1.53	1.00	2.77	1.00	1.20	1.00	2.58	1.05	1.58	1.00	1.70	1.09	2.70	1.00	1.90	1.00
11	1.40	1.00	0.67	1.00	1.66	1.00	1.10	1.20	2.23	1.00	2.03	1.05	1.80	1.11	2.83	1.00	1.87	1.07
12	1.17	1.00	0.67	1.00	2.53	1.00	1.00	1.00	2.67	1.00	1.23	1.00	0.98	1.15	2.20	1.00	1.70	1.00
13	1.23	1.00	0.55	1.00	1.17	1.00	0.77	1.00	1.17	1.00	1.30	1.00	0.90	1.11	2.05	1.14	1.58	1.09
14	1.10	1.00	0.60	1.00	3.00	1.00	1.37	1.00	2.57	1.00	1.07	1.00	1.20	1.00	2.70	1.00	2.50	1.08

[a] All CSPs are β-CD derivatives, the structures are given in Figure 15.
[b] Solute structures are given in Figure 16.
Source: Ref. 49.

TABLE 10 Chiral Resolutions of Some Drugs on 3-Chloro-4-methyl (CSPs 8 and 15, Bonded to Silica via the Narrow Opening of CD) and 3-Chloro-2-methyl (CSPs 13 and 16, Bonded to Silica via the Wider Opening of CD) Phenylcarbamate Derivatives of β-CD

	CSPs[b]							
	8		15		13		16	
Solutes[a]	*k*	α	*k*	α	*k*	α	*k*	α
I	1.18	1.11	1.45	1.05	1.23	1.00	1.90	1.00
II	0.67	1.00	0.80	1.00	0.55	1.00	0.90	1.00
III	1.50	1.00	1.07	1.00	1.17	1.00	1.45	1.00
IV	0.95	1.00	1.70	1.00	1.17	1.00	2.10	1.00
V	1.87	1.00	1.70	1.00	1.17	1.00	2.10	1.00
VI	1.40	1.00	1.57	1.00	1.30	1.00	1.77	1.00
VII	2.07	1.84	2.20	2.04	0.90	1.11	1.80	1.11
VIII	2.50	1.05	2.96	1.06	2.05	1.04	3.70	1.02
IX	12.33	1.35	1.26.02	1.50	1.58	1.09	2.13	1.13

[a]Solute structures are given in Figure 15.
[b]CSP structures are given in Figure 16.
Source: Ref. 49.

In another study, the effects on chiral recognition of CD type, the nature and position of the substituents on the phenyl ring, and the binding mode were investigated in detail [49]. No marked change of the chiral recognition abilities was established when the CD binding side was reversed, that is, it made no difference whether the opening of the cone-shaped CD to silica gel was the narrower (primary) one or the wider (secondary) side. Furthermore, the authors concluded that chiral recognition of some compounds by these CSPs critically depends on the type of CD used [49]. Lévêque et al. [115], who synthesized mono-2-*O*-pentenyl- and mono-6-*O*-pentenyl β-CD-based CSPs, concluded that these CD derivatives, linked to silica gel, were not equivalent in chiral discrimination and that the orientation of CDs grafted onto silica gel played an important role in the chiral resolution.

3.4.8 Other Parameters

In addition to the parameters already discussed, other factors contribute the chiral resolution on these cyclodextrin phases. These include injection amount, silica gel particle size, spacers between the CD and the silica gel, and column dimensions. In 1985 Feitsma et al. [26] studied variations in chiral resolution by varying

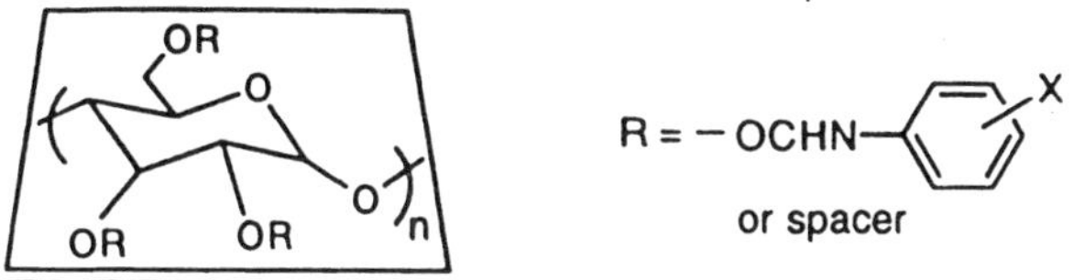

OR, O, OR, OR, n

R = –OCHN–(C6H4)–X or spacer

CSP	n	X	Immobilization position	Spacer[a]
1	6 (α-CD)	4-Cl-3-CH_3	2.3	A
2	7 (β-CD)	4-Cl-3-CH_3	2.3	A
3	8 (γ-CD)	4-Cl-3-CH_3	2.3	A
4	7 (β-CD)	3,4-$(CH_3)_2$	2.3	A
5	7 (β-CD)	3,4-Cl_2	2.3	A
6	7 (β-CD)	2,5-$(CH_3)_2$	2.3	A
7	7 (β-CD)	2,5-Cl_2	2.3	A
8	7 (β-CD)	3-Cl-4-CH_3	2.3	A
9	7 (β-CD)	5-Cl-2-CH_3	2.3	A
10	7 (β-CD)	2-Cl-5-CH_3	2.3	A
11	7 (β-CD)	4-Cl-2-CH_3	2.3	A
12	7 (β-CD)	2-Cl-4-CH_3	2.3	A
13	7 (β-CD)	3-Cl-2-CH_3	2.3	A
14	7 (β-CD)	2-Cl-6-CH_3	2.3	A
15	7 (β-CD)	3-Cl-4-CH_3	6	A
16	7 (β-CD)	3-Cl-2-CH_3	6	A
17	7 (β-CD)	4-Cl-3-CH_3	2.3	B

[a] A: Hexamethylene diisocyanate; B: 4,4′-diphenylmethane diisocyanate.

FIGURE 15 The chemical structures of some CSP-based α- and β-cyclodextrin derivatives. (From Ref. 49.)

the amount of cyclohexylphenylglycolic acid injected. The authors reported a decrease in the retention times of the two enantiomers when the injection amount was increased. Similarly, Lemer et al. [112] studied the effects of the amount of β-CD-based CSP loaded on the chiral resolution of ephedrine, methamphetamine, and selegiline. The resolution factor values decreased drastically when the amount injected was increased. The findings of these two research groups are shown in Figure 17.

Stalcup et al. [44] reported an interesting effect of spacers on the chiral resolution of some drugs, namely, there was no chiral resolution without a spacer between the CD and the silica gel. Chankvetadze et al. [49] studied the effects of such spacers on the chiral resolution of certain drugs. Two types of spacer, 4,4′-diphenylmethane diisocyanate and hexamethylene diisocyanate, were used to

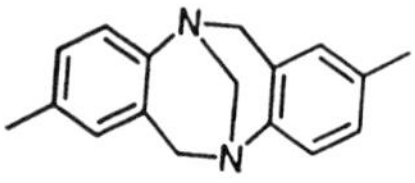

I

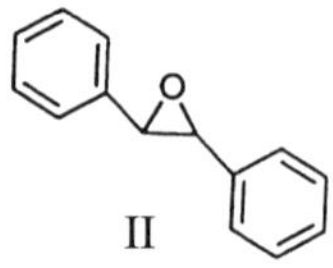

II

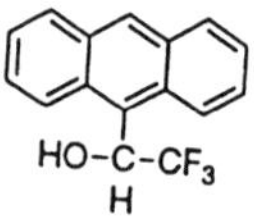

III

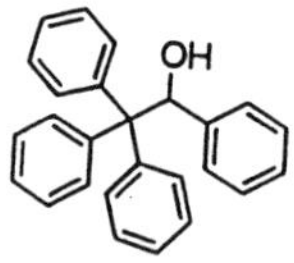

IV

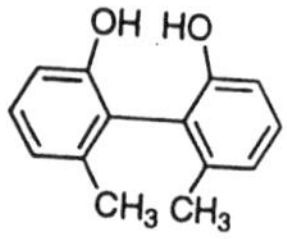

V

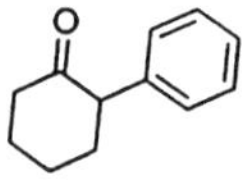

VI

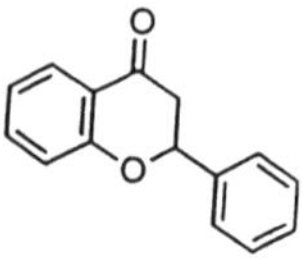

VII

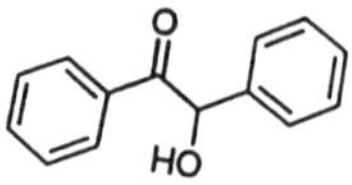

VIII

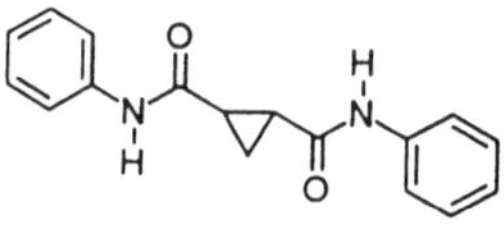

IX

FIGURE 16 The chemical structures of some drugs resolved on different CSPs based on α- and β-cyclodextrin derivatives (Fig. 15): Tröger's base (I), *trans*-2,3-diphenyloxirane (II), 1-(9-anthryl)-2,2,2-trifluoroethanol (III), 1,2,2,2-tetraphenylethanol (IV), 2,2′-dihydroxy-6,6-dimethylbiphenyl (V), 2-phenylcyclohexanone (VI), flavanone (VII), benzoin (VIII), and *trans*-cyclopropanedicarboxylic acid anilide (IX). (From Ref. 49.)

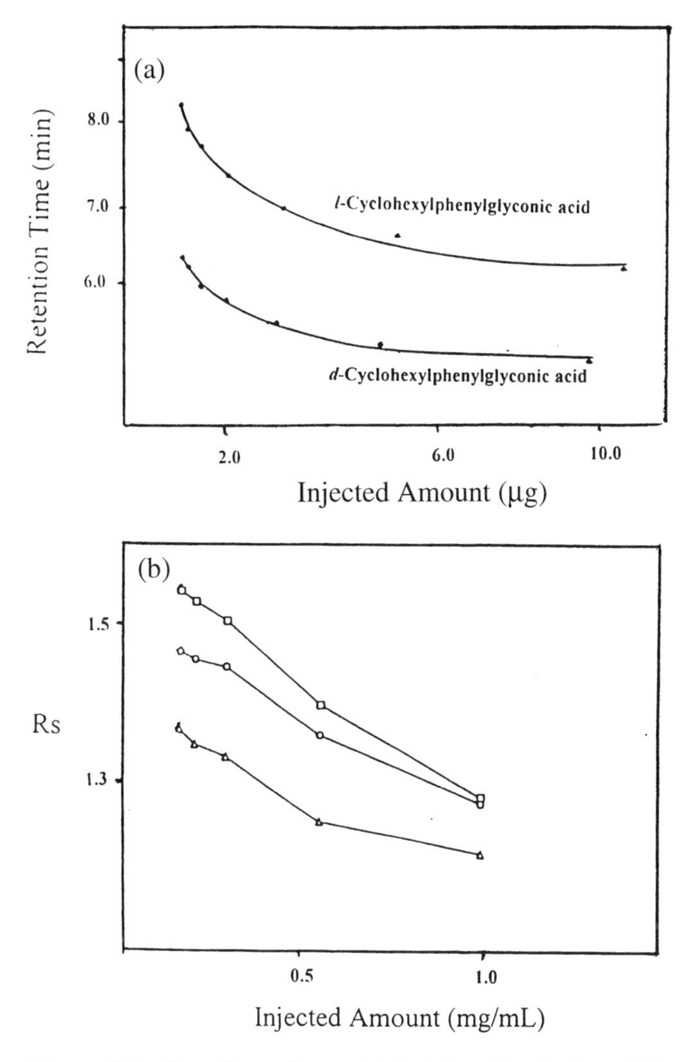

FIGURE 17 The effects of amount injected on the chiral resolution of (a) cyclohexylphenylglyconic acid and (b) ephedrine (△), methamphetamine (□), and selegiline (○) on β-CD-based CSPs. (From Refs. 26,112.)

bind the same chiral selector onto a silica surface. CSPs having the first diisocyanate showed better chiral recognition than CSPs containing the second. The most interesting spacer effect was the reversal of the elution order for flavanone. Similar results were observed with benzoin when the spacer was changed [116]. Hargitai and Okamoto [110] studied the influence of pore diameter of silica gel on the chiral resolution of some drugs. Pore diameters of 60, 100, and 300 Å were used for this purpose. The enantioselectivities of these silica-based CSPs varied from one drug to another. In addition, enantiorecognition on these CSPs may be optimized by using columns of different dimensions.

3.5 CHIRAL RECOGNITION MECHANISM

The primary C-6 hydroxyl groups, free to rotate, can partially block the CD cavity from one end. The mouth of the opposite end of the CD cavity is enriched owing to the presence of C-2 and C-3 secondary hydroxyl groups. This arrangement favors complex formation with a variety of compounds. The presence of stereospecific glucopyronose units, as well as the restricted conformational freedom and orientation of the secondary hydroxyl groups, are assumed to be responsible for the chiral recognition capacities of CDs [117]. Therefore, different diastereoisomeric inclusion complexes are formed by the two enantiomers (Fig. 18). This is known from data obtained from crystallography, NMR, and mass spectrometric studies [10,42,114,118–122]. Basically, in the chiral recognition process, enantiomers enter the cavities of CD molecules and diastereoisomeric complexes are formed; hence these units are called inclusion complexes. Therefore, even a slight variation in the chromatographic conditions, such as a

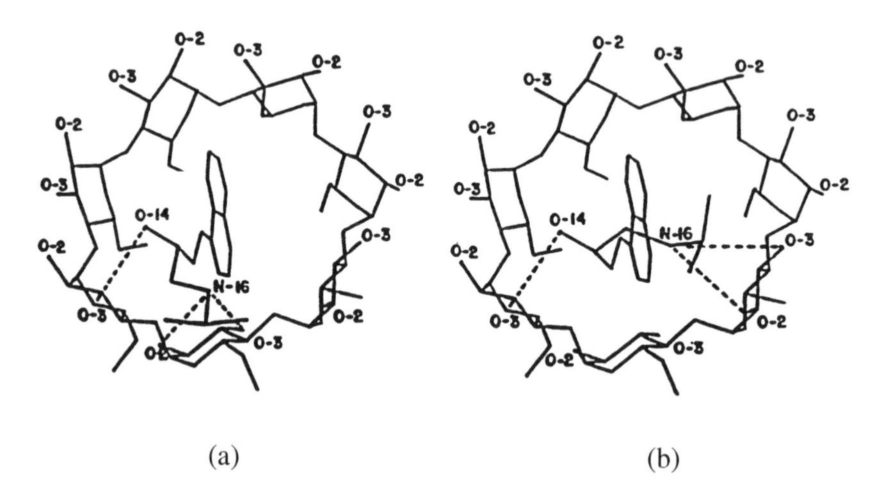

FIGURE 18 Computer-projected structures of the diastereoisomeric inclusion complexes of (a) *d*-propranolol and (b) *l*-propranolol in β-CD; dashed lines, hydrogen bondings. (From Ref. 9.)

change in the mobile phase or the derivatization of the CDs, affects the chiral resolution greatly. Because of this, many research groups have synthesized different derivatives for use in the chiral resolution of different molecules. Therefore, the selection of the CD phase is an important factor in the formation of a stable complex with a specific racemic compound. This aspect has been discussed (Sects. 3.4.6 and 3.4.7).

Lipkowitz and Stoehr [114] carried out an NMR study of methyl mandelate on β-CD. The authors conducted molecular dynamics simulations and predicted the guest–host (i.e., the interactions between the methyl mandelate molecule and the CD). It was observed that short-range dispersion forces rather than long-range Coulombic forces were responsible for both complexation and enantiodiscrimination. To explain the chiral resolution on CDs, several models of and approaches to guest–host formation were proposed and discussed [59,120,123,124]. Inclusion complex stability is controlled by a number of interactions between the enantiomers and the CSPs. The most important interactions are hydrogen bonding, interactions between dipoles and induced dipoles, and those of π–π forces [4,11,114,125]. Steric effect also plays a crucial role in the chiral resolution, since complexation is governed by the type of groups on the enantiomers [48,57]. However, other forces, such as van der Waals, ionic interactions, and solvation effect, also have roles in chiral resolution. These different types of interaction, which occur between the CD hydroxyl groups and various groups of enantiomers, may be enhanced by introducing electronegative or phenyl groups into CDs (derivatization). This is why the CSPs of CD derivatives are better than those obtained from native CDs.

In summary, the enantiomers fit in stereogenically different ways into the chiral cavity of CD, and these arrangements are stabilized at different magnitudes by different interactions as discussed earlier. Manipulation of these differences, in turn, results in the chiral separation of enantiomers. It has also been reported that the magnitudes of these interactions vary in the normal, polar organic, and reversed-phase modes, and hence the chiral recognition mechanism differs slightly in all the three mobile phase modes. However, the exact mechanism of chiral resolution on these phases is still obscure. A schematic representation of chiral resolution on these phases is shown in Figure 19.

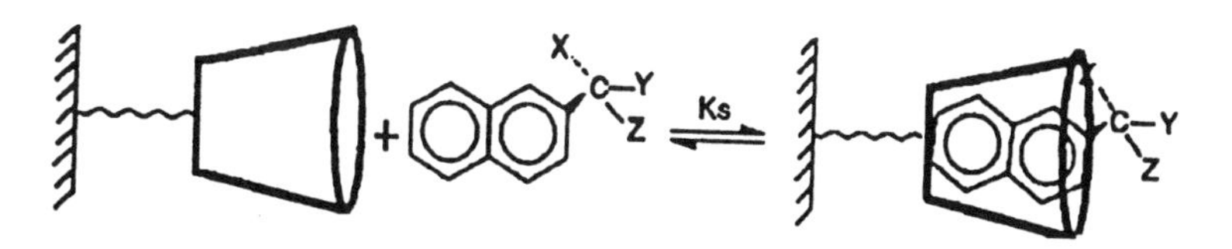

FIGURE 19 Schematic representation of solute inclusion in a cyclodextrin cavity.

3.6 APPLICATIONS IN SUB- AND SUPERCRITICAL FLUID CHROMATOGRAPHY

Because of the wide range of applications of CD-based CSPs, these phases have been evaluated for the chiral resolution of enantiomers in sub- and supercritical FC modes. In 1987 Macaudiere et al. [126] reported the first chiral resolution of phosphine oxides and amides on native cyclodextrins-based CSPs under sub-critical conditions. Later, several reports were published on the chiral resolution of various compounds using CD-based CSPs in sub- and supercritical FC models [127–131]. Work on the chiral separation by sub-FC and SFC was reviewed recently [130,132]. Shen and coworkers [128] carried out a detailed comparison of the SFC performance of packed and open tubular columns containing the same cyclodextrins-modified polymers. They observed that the retention factors could differ drastically within a certain range of pressure, with no significant change in chiral selectivity. Enantiomeric separation of a variety of drugs and related compounds was achieved on an (*S*)-naphthylethylcarbamoylated-β-cyclodextrin (S-NEC-CD) chiral stationary phase (CSP) using sub-FC and SFC. The mobile phase was a simple carbon dioxide–methanol mixture. Resolution of cromakalim, which is not possible on an S-NEC-CD column in LC, was readily accomplished in SFC. The importance of modifier, temperature, and pressure was also assessed in relation to retention, selectivity, and resolution. The nature of the modifier and the modifier concentration were found to be crucial parameters [133].

Schurig et al. [134] bonded permethylated β-CD via an octamethylene bridge to polydimethylsiloxane, resulting in the chiral polymer Chiralsil-Dex. This polymer was immobilized onto the inner surface of a capillary or onto a silica particle by cross-linking or bonding, and the resulting CSPs were investigated in SFC. Duval et al. [135] linked covalently mono-2- and mono-6-*O*-pentenyl-β-cyclodextrins (mono-2-pent-β-CD and mono-6-pent-β-CD), to mercaptopropyl silica gel (thiol-Si) through thioether or sulfone linkage, and the CSPs developed revealed different enantioselectivities in the separation of piperidine-2,6-dione-related drugs, namely, aminoglutethimide and thalidomide, under supercritical fluid conditions. The authors discussed the impact on chiral discrimination of the position of the pentenyl moiety on one of the glucopyranosidics of the CD cage [135]. Additionally, the role of the heteroatom present in the spacer arm between the CD and the silica gel, in this case a thioether or sulfone functionality, was also addressed in terms of enantioselectivities [135].

Verillon and Boutant [136] presented a method of separating the enantiomers of some β-adrenergic blockers at the preparative scale. SFC modified with carbon dioxide and standard HPLC columns having bores of 4 to 20 mm could produce from 10 to 500 mg of purified enantiomers per hour. The authors described the practical rules of their method and its performance. In another study, Lévêque and Duval [137] described the fast and easy transportation of

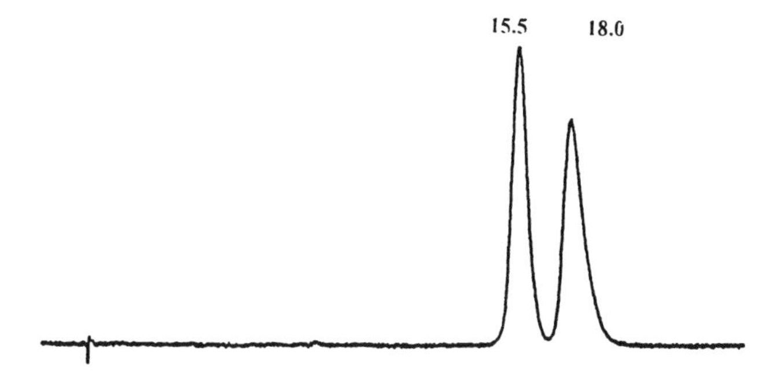

FIGURE 20 Chromatograms of the chiral resolution of tropicamide on Cyclobond I SN CSP using SFC. (From Ref. 130.)

chiral separation conditions of racemates from HPLC to SFC. The HPLC and SFC modes of liquid chromatography were used to compare chiral resolution on naphthylethylcarbamoylated β-CDs phases [130]. Williams et al. [138] reported that the separations obtained in HPLC (using all three mobile phase modes) could be replicated with a simple carbon dioxide–methanol eluent in SFC. Armstrong et al. [139] resolved the enantiomers of 1-aminoindanone on a β-CD-based phase. Figure 20, which shows chromatograms of the chiral resolution of tropicamide on Cyclobond I SN CSP, illustrates the kind of chiral resolution on CD-based CSPs that can be obtained by using SFC. Some of the applications of CD-based CSPs using sub- and supercritical fluid chromatography are given in Table 11.

TABLE 11 The Enantiomeric Resolution of Racemic Compounds on Cyclodextrin-Based CSPs by Means of Subcritical and Supercritical Fluid Chromatography

Racemic compounds	CSPs	Refs.
Alkanes	β-CD	130,134
Amide	β-CD	126,130
Ketone	β-CD	130,134
Lactones	β-CD	130,134
Cromakalim	β-CD derivatives	133
Aminoglutethimide and thalidomide	β-CD derivatives	135
β-Adrenergic blockers	β-CD derivatives	136
1-Aminoindane	β-CD derivatives	139
Phosphine oxides	β-CD	126

3.7 APPLICATIONS IN CAPILLARY ELECTROCHROMATOGRAPHY

Native CDs and their derivatives have been packed into capillaries columns of different dimensions and the CSPs developed used for the chiral resolution of different compounds by capillary electrochromatography (CEC) [140,141]. The CD-based chiral selectors were used in open tubular [142–151], packed [152,153], and monolithic columns [153–157]. Schurig et al. [134] linked permethylated β-CD via an octamethylene bridge to polydimethylsiloxane, resulting in the chiral polymer Chiralsil-Dex. The polymer was immobilized onto the inner surface of a capillary by cross-linking or bonding, and the CSP was used in CEC chiral resolution. Pesek et al. [149] etched the capillary wall and bonded it with β-CD as an achiral selector. The authors observed that the bonded organic moieties on the etched inner wall of the capillary provided the solutes sufficient interactions to influence retention through a capillary by electroosmosis or a combination of electroosmotic and electrophoretic forces. Koide and Ueno [156] achieved the enantiomeric resolution of cationic compounds by CEC on β-CD-bonded, negatively charged polyacrylamide gels. The columns used were capillaries filled with a negatively charged polyacrylamide gel, a so-called monolithic stationary phase, covalently bonded to an allyl-carbamoylated β-CD derivative. High efficiencies ($\leq$150,000 plates/m) were obtained. The applications of CEC using CD-based CSPs are summarized in Table 12 [140,142–147,150,154–158]. The chiral recognition mechanism in sub- and supercritical fluid chromatography and in capillary electrochromatography is similar to that in HPLC separations.

TABLE 12 The Enantiomeric Resolution of Racemic Compounds on Cyclodextrin-Based CSPs by Means of Capillary Electrochromatography

Racemic compounds	CSPs	Refs.
Profens and barbiturates	Native and derivatized CDs in open tubular columns	140,142–147
Ephedrine	Native and derivatized CDs in open tubular columns	150
Benzoin, mandelates, praline derivatives, barbiturates, and glutethimide	Native and derivatized CDs in packed capillary columns	152
Dansyl amino acids, hydroxy acids, and warfarin	Organic monoliths of CD	154–157
Cationic and neutral compounds	β-CD	158

3.8 CONCLUSION

The applications of CD-based CSPs for chiral resolution in liquid chromatography has increased exponentially in the last decade owing to the high potential of these phases as chiral selectors in all three mobile phase modes. Moreover, the synthesis of CD derivatives has increased the demand of these phases because they have unique selectivities for a larger number of racemic compounds. Besides, the chiral resolution strategies on these phases may be predicted theoretically because of their well-known structures and sizes. This facility makes experiments inexpensive and less laborious. Thus, liquid chromatographic approaches using CD-based chiral stationary phases are the method of choice for the chiral resolution of a wide range of racemic compounds. Although these phases present no serious limitation, the evaluation of the exact chiral recognition mechanism is demanding. Therefore, the chiral recognition mechanism should be established in detail, to permit the use of these CSPs in a more scientific way in all modes of chromatography.

REFERENCES

1. Villiers A, C R 112: 536 (1891).
2. Schardinger F, Zentr Bacteriol Parasitenk Abt II 29: 188 (1911).
3. Gröger M, Katharina E, Woyke A, Cyclodextrins, in "Science Forum an der Universität Siegen," Siegen, Germany (2001).
4. Han SM, Armstrong DW, HPLC separation of enantiomers and other isomers with cyclodextrin bonded phases: rule for chiral recognition, in Chiral Separations by HPLC (Krstulovic AM, Ed.), Ellis Horwood, Chichester, p. 208 (1989).
5. Michon J, Rassat A, J Am Chem Soc 101: 4337 (1979).
6. Han SM, Atkinson WM, Purdie N, Anal Chem 56: 2827 (1984).
7. Lightner DA, Gawronski JK, Gawronska J, J Am Chem Soc 107: 2456 (1985).
8. Ihara Y, Nakashini E, Mamuro K, Koga J, Bull Chem Soc Jpn 59: 1901 (1986).
9. Armstrong DW, Ward TJ, Armstrong RD, Beesley TE, Science 222: 1132 (1986).
10. Hamilton JA, Chen L, J Am Chem Soc 110: 5833 (1988).
11. Stalcup AM, Cyclodextrin bonded chiral stationary phases in enantiomer separations, in A Practical Approach to Chiral Separations by Liquid Chromatography (Subramanian G, Ed.), VCH Verlag, Weinheim, Germany, p. 95 (1994).
12. Armstrong DW, Nome F, Spino LA, Golden TD, J Am Chem Soc 108: 1418 (1986).
13. Bender ML, Komiyama M, "Cyclodextrin," Springer-Verlag, Berlin (1978).
14. Szejtli J, Cyclodextrins and Their Inclusion Complexes, Akademia Kiado, Budapest (1982).
15. Hinze WL, Applications of cyclodextrins in chromatographic separations and purification methods, in Separations and Purification Methods (Van Oss C, Ed.), Vol. 10, p. 159, Marcel Dekker, New York (1981).
16. Jones D, Am Lab 72 (1987).

17. Harada A, Furue M, Nozakura S, J Polym Sci 16: 187 (1978).
18. Zsadon B, Szilasi M, Otta KH, Tudos F, Fenyvesi E, Szejtli J, Acta Chim Acad Sci Hung 100: 265 (1979).
19. Zsadon B, Szilasi M, Tudos F, Fenyvesi E, Szejtli J, Stark 31: 11 (1979).
20. Zsadon B, Szilasi F, Tudos F, Szejtli J, J Chromatogr 208: 109 (1981).
21. Zsadon B, Decsei L, Szilasi M, Tudos F, J Chromatogr 270: 127 (1983).
22. Fujimura K, Ueda T, Ando T, Anal Chem 55: 446 (1983).
23. Kawaguchi Y, Tanaka M, Nakae M, Funazo K, Shono T, Anal Chem 55: 1852 (1983).
24. Tanaka M, Kawaguchi Y, Shono T, Uebori M, Kuge Y, J Chromatogr 301: 345 (1984).
25. Feitsma KG, Drenth BFH, De Zeeuw RA, J High Resolut Chromatogr 7: 147 (1984).
26. Feitsma KG, Bosmann J, Drenth BFH, De Zeeuw RA, J Chromatogr 333: 59 (1985).
27. Fujimura K, Kitagawa M, Takayanagi H, Ando T, J Liq Chromatogr 9: 607 (1986).
28. Fisher CM, Chromatogr Int 5: 10 (1984).
29. Armstrong DW, US Patent 4,539,399 (1985).
30. Armstrong DW, Alak A, DeMond W, Hinze WL, Riehl TE, J Liq Chromatogr 8: 261 (1985).
31. Fisher CM, Chromatogr Int 8: 38 (1985).
32. Beesley TE, Am Lab 78 (1985).
33. Chang CA, Abdel H, Melchoir N, Wu K, Pennell KH, Armstrong DW, J Chromatogr 347: 51 (1985).
34. Ward TJ, Armstrong DW, J Liq Chromatogr 9: 407 (1986).
35. Isasaq HJ, J Liq Chromatogr 9: 229 (1986).
36. Weaver DE, van Lier RBL, Anal Biochem 154: 590 (1986).
37. Issaq HJ, Weiss D, Ridlon C, Fox SD, Muschik GM, J Liq Chromatogr 9: 1791 (1986).
38. Issaq HJ, McConnell JH, Weiss DE, Williams DG, Saavedra JE, J Liq Chromatogr 9: 1783 (1986).
39. Issaq HJ, Glennon M, Weiss DE, Chmurny GN, Saavedra JE, J Liq Chromatogr 9: 2763 (1986).
40. Abidi SL, J Chromatogr 362: 33 (1986).
41. Armstrong DW, Li W, Chromatography 2: 43 (1987).
42. Street Jr KW, J Liq Chromatogr 10: 655 (1987).
43. Connelly JA, Siehl DL, Methods Enzymol 142: 422 (1987).
44. Stalcup AM, Chang S, Armstrong DW, J Chromatogr 513: 181 (1990).
45. Armstrong DW, Stalcup AM, Hilton ML, Duncan JD, Faulkner JH, Chang SC, Anal Chem 62: 1610 (1990).
46. Stalcup AM, Chang SC, Armstrong DW, J Chromatogr 540: 113 (1991).
47. Stalcup AM, Gahm KH, Anal Chem 68: 1369 (1996).
48. Armstrong DW, Chang SC, Wang Y, Ibrahim H, Reid III GR, Beesley TE, J Liq Chromatogr Relat Technol 20: 3279 (1997).
49. Chankvetadze B, Yashima E, Okamoto Y, Chirality 8: 402 (1996).

50. Blondel F, Peulon V, Prigent Y, Duval R, Combret Y, Eighth International Symposium on Chiral Discrimination, Edinburgh, Scotland, June 30–July 3, 1996, p. 1.5 (1996).
51. Elfakir CC, Dreux M, Bourgeaux E, Lévêque H, Duval R, 214th ACS National Meeting, Las Vegas, Sept. 7–11, 1997 (1997).
52. Caron I, Elfakir C, Dreux M, Ninth International Symposium on Cyclodextrins, Santiago de Compostela, May 31–June 3, 1998 (1998).
53. Beesley TE, Scott RPW, Chiral Chromatography, John Wiley & Sons, New York, p. 265 (1998).
54. Armstrong DW, Yang X, Han SM, Menges RA, Anal Chem 59: 2594 (1987).
55. Han SM, Armstrong DW, J Chromatogr 389: 256 (1987).
56. Han SM, Han YI, Armstrong DW, J Chromatogr 441: 376 (1988).
57. Seeman JI, Secor HV, Armstrong DW, Timmons KD, Ward TJ, Anal Chem 60: 2120 (1988).
58. Stalcup AM, Jin HL, Armstrong DW, J Liq Chromatogr 13: 473 (1990).
59. Armstrong DW, Chang CD, Lee SH, J Chromatogr 539: 83 (1991).
60. Armstrong DW, Chen S, Chang C, Chang S, J Liq Chromatogr 15: 545 (1992).
61. Chen S, Li Y, Armstrong DW, Borrell JI, Martinez-Teipel B, Matallana JL, J Liq Chromatogr 18: 1495 (1995).
62. Ekborg-Ott KH, Armstrong DW, Chirality 8: 49 (1996).
63. Chang SC, Wang CR, Armstrong DW, J Liq Chromatogr 15: 1411 (1992).
64. Abidi SL, J Chromatogr 404: 133 (1987).
65. Fukushima Y, Murayama K, Santa T, Hana H, Imai K, Biomed Chromatogr 12: 1 (1998).
66. Stammel W, Woesle B, Thomas H, Chirality 7: 10 (1995).
67. Haginaka J, Wakai J, Anal Chem 62: 997 (1990).
68. Han SM, Biomed Chromatogr 11: 259 (1997).
69. Lee SH, Berthod A, Armstrong DW, J Chromatogr 603: 83 (1992).
70. Fujimura K, Suzuki S, Hayashi K, Masuda S, Anal Chem 62: 2198 (1990).
71. Stalcup AM, Faulkner JR Jr, Tang Y, Armstrong DW, Levy LW, Regalado E, Biomed Chromatogr 5: 3 (1991).
72. Haynes JL III, Shamsi SA, O'Keefe F, Darcey R, Warner IM, J Chromatogr A 803: 261 (1998).
73. Ryu JW, Kim DW, Lee KP, Pyo D, Park JH, J Chromatogr A 814: 247 (1998).
74. Armstrong DW, Hilton DW, Coffin L, LC-GC 9: 647 (1992).
75. Tazerouti F, Badjah-Hadj-Ahmed AY, Meklati BY, Franco P, Minguillon C, Chirality 14: 59 (2002).
76. Chang SC, Reid III GL, Chen S, Chang CD, Armstrong DW, Trends Anal Chem 12: 144 (1993).
77. Matchett MW, Branch SK, Jefferies TM, Chirality 8: 126 (1996).
78. Zukowski J, Chirality 10: 362 (1998).
79. Péter A, Kámán J, Fülöp F, van der Eycken J, Armstrong DW, J Chromatogr A 919: 79 (2001).
80. Stalcup AM, Williams KL, J Liq Chromatogr 15: 29 (1992).
81. Armstrong DW, Duncan JD, Lee SH, Amino Acids 1: 97 (1991).

82. Kanazawa H, Matsushima Y, Okubo S, Mashige F, J Chromatogr A 871: 181 (2000).
83. West RR, Cardellina JH, J Chromatogr 539 15 (1991).
84. Shaw CJ, Sanfilippo PJ, McNally JJ, Park SA, Press JB J Chromatogr A 631: 173 (1993).
85. A Guide to Using Cyclodextrin Bonded Phases for Liquid Chromatography, Advanced Separation Technologies, Inc., Whippany, NJ (1996).
86. Beesley TE, Scott RPW, "Chiral Chromatography," John Wiley & Sons, New York, p. 375 (1998).
87. Armstrong DW, Han SM, Han YI, Anal Biochem 167: 261 (1987).
88. Armstrong DW, Han YI, Han SM, Anal Chim Acta 208: 275 (1988).
89. Maguire JH, J Chromatogr 387: 453 (1987).
90. Armstrong DW, Alak A, Bui K, DeMond W, Ward TJ, Reihl TE, Hinze WL, J Incl Phenom 2: 533 (1984).
91. McClanahan JS, Maguire JH, J Chromatogr 381: 438 (1986).
92. Florance J, Galdes A, Konteatis Z, Kosaryeh Z, Langer K, Martucci C, J Chromatogr 414: 313 (1987).
93. Armstrong DW, Ward TJ, Czech A, Czech BP, Bartsch RA, J Org Chem 50: 5556 (1985).
94. Armstrong DW, Demond DW, Czech BP, Anal Chem 57: 481 (1985).
95. Krstulovic AM, Gianviti JM, Burke JT, Mompon B, J Chromatogr 426: 417 (1988).
96. Pham-Huy, C, Radenen B, Sahui-Gnassi A, Claude J, J Chromatogr B 665: 125 (1995).
97. Camilleri P, Reid CA, Manallack DT, Chromatographia 11/12: 771 (1994).
98. Barderas AV, Duprate F, J Liq Chromatogr 17: 1709 (1994).
99. Furuta R, Nakazawa H, J Chromatogr 625: 231 (1992).
100. Kuijpers PH, Gerding TK, de Jong GJ, J Chromatogr 625: 223 (1992).
101. Plesek J, Bruner B, J Chromatogr 626: 197 (1992).
102. Lim TY, Kim HJ, J Chromatogr A 933: 99 (2001).
103. Li S, Purdy WC, J Chromatogr 543: 105 (1991).
104. Aboul-Enein HY, Islam MR, Bakr SA, J Liq Chromatogr 11: 1485 (1988).
105. Armstrong DW, Bertrand GL, Ward KD, Ward TJ, Anal Chem 62: 332 (1990).
106. Thuaud N, Sebille B, Deratani A, G. Lelievre G, J Chromatogr 555: 53 (1991).
107. Weems H, Zamani K, Chirality 4: 268 (1992).
108. Piperaki S, Parissi-Poulou M, Chirality 5: 258 (1993).
109. Piperaki S, Kakoulidou-Tsantili A, Parissi-Poulou M, Chirality 7: 257 (1995).
110. Hargitai T, Okamoto Y, J Liq Chromatogr 16: 843 (1993).
111. Hilton ML, Chang SC, Gasper MP, Pawlowska M, Armstrong DW, Stalcup AM, J Liq Chromatogr 16: 127 (1993).
112. Lemer K, Jirovsky D, Seveik J, J Liq Chromatogr Relat Technol 19: 3173 (1996).
113. Chen L, Zhang LF, Ching CB, Ng SC, J Chromatogr A 950: 65 (2002).
114. Lipkowitz KB, Stoehr CM, Chirality 8: 341 (1996).
115. Lévêque H, Adam N, Duval R, Seventh International Symposium on Preparative and Industrial Chromatography and Allied Techniques, Sept. 23–25, 1998, Strasbourg, France (1998).

116. Hargitai T, Kaida Y, Okamoto Y, J Chromatogr A 628: 11 (1993).
117. Lipkowitz KB, Green K, Yang JA, Chirality 4: 205 (1992).
118. Boeham RE, Martire DE, Anal Chem 60: 522 (1988).
119. Linnemayr K, Rizzi A, Günther A, J Chromatogr A 791: 299 (1997).
120. Morin N, Guillaumine YC, Rouland JC, Chromatographia 48: 388 (1998).
121. Redondo J, Blazquez MA, Torrens A, Chirality 11: 694 (1999).
122. Lipkowitz KB, J Chromatogr A 906: 417 (2001).
123. Armstrong DW, Yang X, Han SM, Menges RA, Anal Chem 59: 316 (1987).
124. Berthod A, Chang SC, Armstrong DW, Anal Chem 64: 395 (1992).
125. Torr MA, Nelson G, Patonay G, Warner IM, Anal Lett 21: 843 (1988).
126. Macaudiere P, Caude M, Rosset R, Tambute A, J Chromatogr 405: 135 (1987).
127. Jung M, Mayer S, Schurig V, LC-GC 12: 458 (1994).
128. Shen Y, Chen Z, Owen NL, Li W, Bradshaw JS, Lee ML, J Microcol Sep 8: 249 (1996).
129. Williams KL, Sander LC, Wise SA, J Chromatogr A 746: 91 (1996).
130. Phinney KW, Sub- and supercritical fluid chromatography for enantiomer separations, in Chiral Separation Techniques: A Practical Approach (Subramanian G, Ed.), VCH Verlag, Weinheim, Germany, p. 299 (2001).
131. Röder W, Ruffing FJ, Schomburg G, Pirkle WH, J High Resolut Chromatogr 10: 665 (1987).
132. Terfloth G, J Chromatogr A 906: 301 (2001).
133. Salvador A, Herbreteau B, Lafosse M, Dreux M, Analusis 25: 263 (1997).
134. Schurig V, Mayer S, Jung M, Fluck M, Jakubetz H, Glausch A, Negura S, Chem Anal 142: 401 (1997).
135. Duval R, Lévêque H, Prigent Y, Aboul-Enein HY, Biomed Chromatogr 15: 202 (2001).
136. Verillon F, Boutant R, Proceedings of the Fifth Meeting on Supercritical Fluids, Materials, and Natural Products Processing, March 23–25, 1998, Nice, France (1998).
137. Lévêque H, Duval R, Proceedings of the Fifth Meeting on Supercritical Fluids, Materials, and Natural Products Processing, March 23–25, 1998, Nice, France (1998).
138. Williams KL, Sander LC, Wise SA, Chirality 8: 325 (1996).
139. Armstrong DW, Tang Y, Ward T, Nichols M, Anal Chem 65: 1114 (1993).
140. Schurig V, Wistuba D, Electrophoresis 20: 2313 (1999).
141. Fanali S, Catarcini P, Blaschke G, Chankvetadze B, Electrophoresis 22: 3131 (2001).
142. Mayer S, Schurig V, J High Resolut Chromatogr 15: 129 (1992).
143. Mayer S, Schurig V, J Liq Chromatogr 16: 915 (1993).
144. Schurig V, Jung M, Mayer S, Fluck M, Negura S, Jakubetz H, J Chromatogr A 694: 119 (1994).
145. Schurig V, Jung M, Mayer S, Negura S, Fluck M, Jakubetz H, Angew Chem 106: 2265 (1994).
146. Jakubetz H, Czesla H, Schurig V, J Microcol Sep 9: 421 (1997).
147. Mayer S, Schleimer M, Schurig V, J Microcol Sep 6: 43 (1994).

148. Armstrong DW, Tang Y, Ward T, Nichols M, Anal Chem 65: 2265 (1993).
149. Pesek JJ, Matyska MT, Menezes S, J Chromatogr A, 853: 151 (1999).
150. Szeman J, Ganzler K, J Chromatogr A 668: 509 (1994).
151. Schurig V, Wistuba D, Electrophoresis 20: 50 (1999).
152. Wistuba D, Czesla H, Roeder M, Schurig V, J Chromatogr A 815: 183 (1998).
153. Wistuba D, Schurig V, Electrophoresis 21: 3152 (2000).
154. Koide T, Ueno K, Anal Sci 14: 1021 (1998).
155. Koide T, Ueno K, Anal Sci 16: 1065 (2000).
156. Koide T, Ueno K, J Chromatogr A 893: 177 (2000).
157. Koide T, Ueno K, J High Resolut Chromatogr 23: 59 (2000).
158. Koide T, Ueno K, Anal Sci 15: 791 (1999).

4

Macrocyclic Glycopeptide Antibiotics–Based Chiral Stationary Phases

Macrocyclic antibiotics are one of the newest and perhaps the most varied class of chiral selectors [1]. The concept of utilizing macrocyclic glycopeptide as the chiral stationary phase (CSP) for high-performance liquid chromotography (HPLC) was introduced by Armstrong in 1994 [2]. Since then, their use for chiral resolution by liquid chromatography is increasing exponentially [3,4]. The antibiotics have been found to have a very good potential for chiral resolution. It may be due to their specific structures and the possibility of their use in a wide range of mobile phases. In addition, their relatively small size and the fact that their structures are known allows basic studies on chiral recognition to be done easily and exactly. They are often complimentary in the type of compounds they can separate. For example, rifamycin B, an ansamycin, is enantioselective for many positively charged analytes, whereas vancomycin can resolve a variety of chiral compounds containing free carboxylic acid functional groups. The antibiotics used for chiral resolution are vancomycin, vancomycin aglycon, teicoplanin, teicoplanin aglycon, ristocetin A, thiostrepton, rifamycin, fradiomycin, streptomycin, kanamycin, and avoparcin. However, the most commonly used antibiotics are vancomycin, teicoplanin, teicoplanin aglycon, and ristocetin A. This chapter describes the structures, properties, applications, effect of various chromatographic factors on chiral resolution, and the chiral recognition mechanism. The use of these antibiotic-based chiral selectors in subcritical and supercritical fluid chromatography, capillary electrochromatography, and thin-layer chromatography is also discussed.

4.1 STRUCTURES AND PROPERTIES

Only 11 antibiotics have been used as chiral selectors for the enatiomeric resolution of a variety of racemic compounds. It is very interesting to note that all of them contain ionizable groups at different pHs in their structures. The

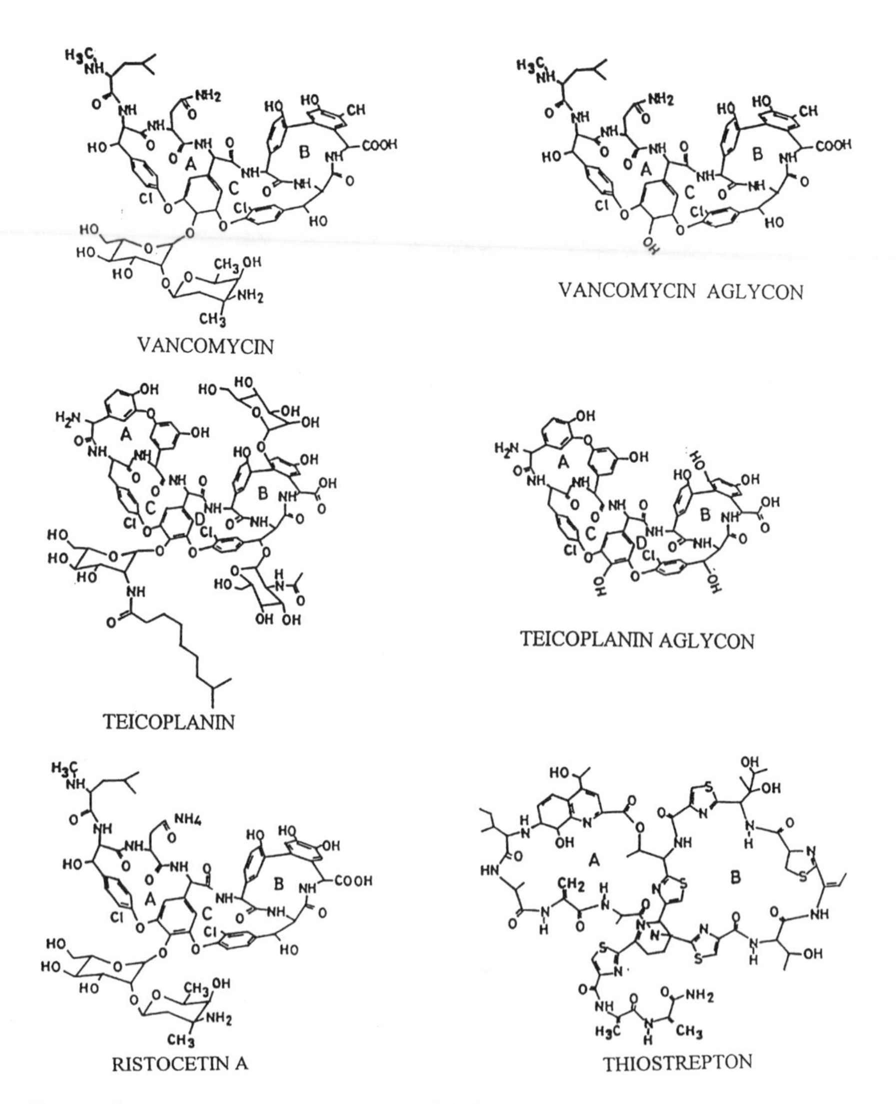

FIGURE 1 The chemical structures of antibiotics.

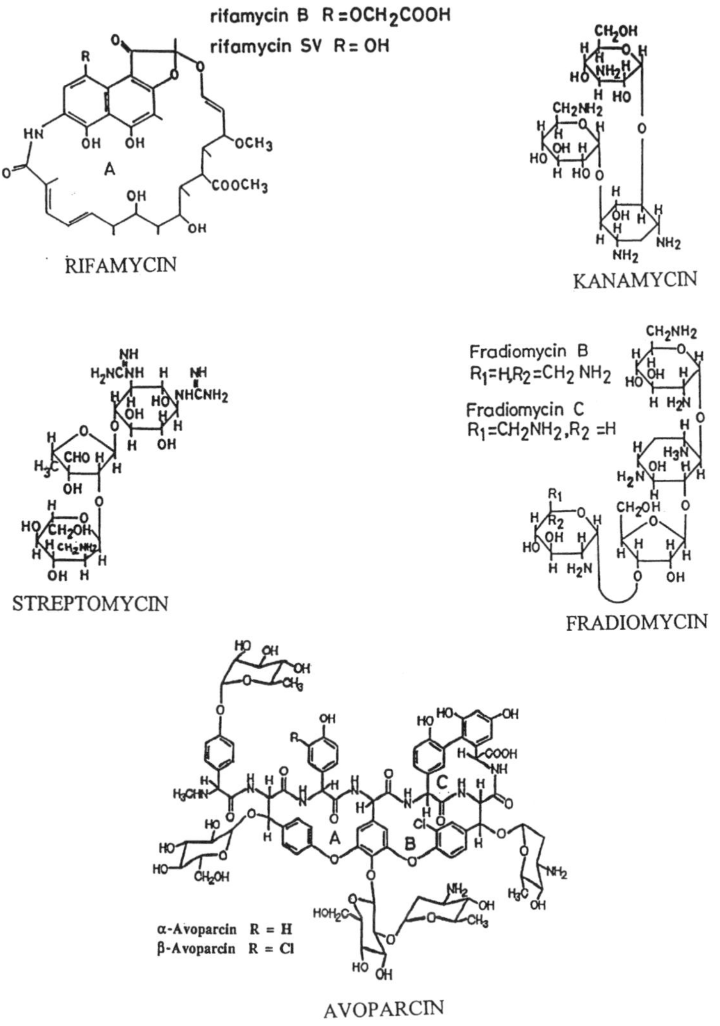

rifamycin B $R = OCH_2COOH$
rifamycin SV $R = OH$

RIFAMYCIN

KANAMYCIN

STREPTOMYCIN

Fradiomycin B
$R_1 = H, R_2 = CH_2NH_2$
Fradiomycin C
$R_1 = CH_2NH_2, R_2 = H$

FRADIOMYCIN

α-Avoparcin R = H
β-Avoparcin R = Cl

AVOPARCIN

FIGURE 1 Continued.

chemical structures of these antibiotics are shown in Figure 1. The physical and chemical properties of these antibiotics are discussed in the following subsections.

4.1.1 Vancomycin

Vancomycin has been used widely for enantiomeric resolution as it is very effective for the enantiorecognition of anionic compounds, particularly those containing carboxylic groups in their structure. The selectivity toward these compounds is due to the presence of amine groups. It is synthesized by *Streptomyces orientalis* bacterium and is very soluble in water and polar aprotic solvents and less soluble or insoluble in higher alcohols and less polar organic solvents [5]. In an aqueous solution, vancomycin dimerizes depending on the solvent type and its concentration. The solution of vancomycin is stable if stored at low temperature and in a buffer of pH value 3–6 [5,6]. The pK values are 2.9, 7.2, 8.6, 9.6, 10.4, and 11.7. The isoelectric point is 7.2. There are 18 chiral centers in this molecule with 3 cavities (viz. A, B, and C in Fig. 1). Five aromatic ring structures and hydrogen donor and acceptor sites are readily available close to the ring structures. Several functional groups such as carboxylic, hydroxyl, amino, and amido are mainly responsible for the ionization of these antibiotics in buffer with various pHs and compositions and, therefore, it is enantioselective in nature. The relative stability of the aqueous solution of vancomycin is about 1–2 weeks. Without the sugar component of vancomycin, the antibiotic is called a vancomycin aglycon (Fig. 1). The physical and chemical properties of vancomycin aglycon are similar to vancomycin.

4.1.2 Teicoplanin and Teicoplanin Aglycon

Teicoplanin is a macrocyclic glycopeptide antibiotic that is structurally related to vancomycin and ristocetin A. Its solubility in water is poor in comparison to vancomycin. Teicoplanin shows a different behavior, exhibiting a very slight anionic character even at an acidic pH (p$I \sim 3.8$). It is obtained from fermentation of *Actinoplanes teichomyceticus* bacterium. There are 23 chiral centers in this molecule with four cavities (viz. A, B, C, and D in Fig. 1). Three sugar moieties are also present. The hydrogen donor and acceptor sites are readily available close to the ring structures. It also contains a hydrophobic acyl side chain attached to a 2-amino-2-deoxy-β D-glucopyranosyl moiety, which activates its surface and enables [7] the formation of micellar aggregates [critical micellar concentration (CMC) about 0.18 mM in unbuffered aqueous solutions]. The relative stability of the aqueous solution of teicoplanin is about 2–3 weeks. The above-mentioned characteristics render teicoplanin highly stereo-specific in nature. Teicoplanin without a sugar part is called a teicoplanin aglycon antibiotic (Fig. 1). This antibiotic contains almost all of the groups and cavities as in teicoplanin and, therefore, has similar chemical and physical properties.

4.1.3 Ristocetin A

Ristocetin A has a structure very similar to teicoplanin and vancomycin. It is produced as the fermentation product of *Nocardia lurida* bacterium and it is characterized by an aglycon portion with 4 fused macrocyclic rings, 1 tetrasaccharide moiety and 6 monosaccharides, together with 38 stereogenic centers. It has also 21 hydroxyl groups, 2 amine groups, 6 amido groups, and 1 methyl ester. This antibiotic also contains three cavities (viz. A, B, and C in Fig. 1). The pI of the antibiotic is 7.5. The relative stability of the aqueous solution of ristocetin A is about 3–4 weeks. It has been used to resolve anionic compounds and those enantiomers not separated by vancomycin [7].

4.1.4 Thiostrepton

Thiostrepton is a macrocyclic polypeptide antibiotic that is structurally different from vancomycin and teicoplanin. It is soluble in water and acetic acid. It is obtained from *Streptomyces azureus* bacterium. There are 17 chiral centers in this antibiotic, with 2 large cavities (A and B in Fig. 1). Five thiazole rings, 1 quinoline ring, 5 hydroxyl groups, 10 amide linkages, and 1 secondary amine make the molecule stereo-specific in nature.

4.1.5 Rifamycin

Different members of the rifamycin family are used for enantiomeric resolution. They are very soluble in water and in low-molecular-weight alcohols and acetone. The members of this class can be used either in the positive, negative, or neutral mode. Among these members, rifamycin B was used commonly for enantio-resolution of a variety of compounds. It is obtained from *Nocardia mediterranei* bacterium. The pK values are 2.8 and 6.7, respectively. There are nine chiral centers in this molecule with one large cavity (A in Fig. 1). A napthohydro-quinone chromophore with a long aliphatic bridge, four hydroxyl groups, one carboxylic moiety, and one amide bond make the molecule stereo-specific in nature.

4.1.6 Kanamycin

Kanamycin is an aminoglycoside antibiotic. It is obtained from *Streptomyces kanamyceticus* bacterium and is soluble in water. There are 16 chiral centers in this molecule and no cavity. It contains two pyranose rings, six hydroxyl groups, and three amine groups. This characteristic makes the molecule capable of enantiorecognition of different antipodes.

4.1.7 Streptomycin

Streptomycin is also an aminoglycoside antibiotic that is soluble in water. It is obtained from *Streptomyces griseus* bacterium. There are 15 chiral centers in this molecule and no cavity. It has one pyranose and one furanose ring, six hydroxyl groups, two primary amine groups, and four secondary amine groups. This arrangement makes the molecule stereo-specific in nature.

4.1.8 Fradiomycin

The fradiomycin is structurally similar to kanamycin. It contains one pyranose, two furanose, and one cyclohexane ring. There are five primary amine and seven hydroxyl groups. Various alkyl groups are also present. The presence of these different types of group makes the antibiotics enantioselective at different pHs. Fradiomycin does not have a cavity.

4.1.9 Avoparcin

Avoparcin [8] is an antibiotic produced by *Streptomyces candidus* bacterium and is commonly used in animal feedstuffs. Commercially, it is available as a product called Avotan that contains 10% avoparcin, which is used to promote animal growth. Unfortunately, the large-scale use of avoparcin appears to give rise to a vancomycin-resistant enterococci in the feces of chickens and pigs. There are two forms of avoparcin: the unsubstituted α-avoparcin and the chlorinated β-avoparcin. The structures of these two forms are shown in Fig. 1. The molecular weights of these forms are 1909 and 1944, respectively. The ratio of α- to β-avoparcin is 1 : 4. The mixture is a white, amorphous solid and is hygroscopic, with no clearly defined melting point. These two forms are soluble in methanol, water, dimethylformamide, and dimethyl sulfoxide. The aglycon portion of avoparcin contains three connected semirigid macrocyclic rings (one 12 membered and two 16 membered). These two forms of avoparcin contain 32 stereogenic centers with 4 carbohydrate chains, 16 hydroxy groups, 1 carboxylic group, 2 primary amines, 1 secondary amine, 6 amide linkages and chlorine atoms (two for β-avoparcin and one for α-avoparcin). There are three cavities (A, B and C in Fig. 1) in this molecule.

4.2 PREPARATION AND COMMERCIALIZATION

Almost all of the antibiotics used for chiral resolution are produced by different bacteria and are commercially available. The preparation of antibiotic CSPs only involves the bonding of these antibiotics on a suitable support. Although the antibiotics can be bonded to a number of different substrates, the material used most commonly in liquid chromatography is silica gel. The reagents may be organosilanes that are terminated by carboxylic groups, amine groups, epoxy groups, and isocyanate groups. Examples of such reagents are 2-(carboxymethoxy)-ethyltrichlorosilane, 3-aminopropyldimethylethoxysilane, (3-glycidoxypropyl)-trimethoxysilane and 3-isocyanatopropyltriethoxysilane. Thus, the linkage attaching these peptide antibiotics to silica gel could be ether, thioether, amine, amide, carbamate, or urea [9].

To make the procedure clear, an example of a reaction sequence that could be used to attach a peptide to silica employing 3-isocyanatopropyltriethoxysilane as the bonding agent is presented in Figure 2. The silica gel (spherical silica 5 μm in particle diameter and mean pore size of 100 Å) is added to the toluene containing a silyl reagent. The mixture is refluxed, and the ethanol produced in the reaction with silanol groups is distilled off periodically. The silica gel is then

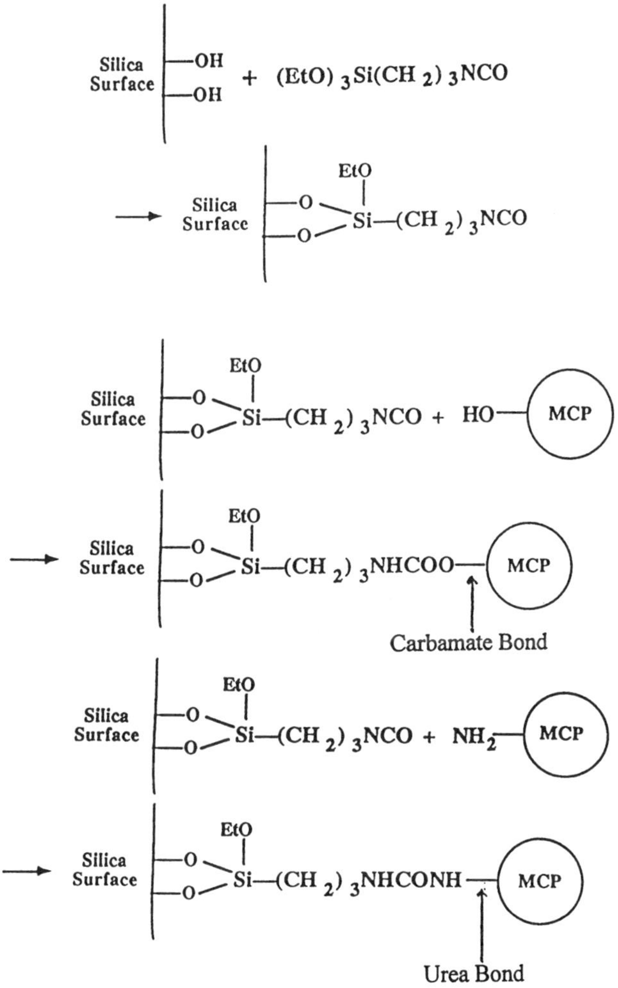

Silica Surface –OH –OH + $(EtO)_3Si(CH_2)_3NCO$ → Silica Surface –O–O– Si(EtO)–$(CH_2)_3NCO$

Silica Surface –O–O– Si(EtO)–$(CH_2)_3NCO$ + HO–MCP → Silica Surface –O–O– Si(EtO)–$(CH_2)_3NHCOO$–MCP

Carbamate Bond

Silica Surface –O–O– Si(EtO)–$(CH_2)_3NCO$ + NH_2–MCP → Silica Surface –O–O– Si(EtO)–$(CH_2)_3NHCONH$–MCP

Urea Bond

FIGURE 2 The chemical pathway for covalent bonding of antibiotic with silica gel. (From Ref. 8.)

filtered, washed with hot toluene, and dried. In the next step, the peptide antibiotic is allowed to react with an isocyanate group on the silica gel. The particular bond used to attach the peptide antibiotic to the carbamate group will depend largely on the nature of the peptide antibiotic used. Prior to the reaction, the peptide antibiotic may be derivatized and such groups as 3,5-dimethylphenyl may be added to the macromolecule to improve or change the interactive nature of the material with regard to chromatographic retention and selectivity. The reaction is carried out by adding the peptide (dried in vacuum at 100°C over phosphorus pentoxide) dropwise to a dispersion of silica particles, in anhydrous N,N-dimethylformamide. The reaction is carried out at 90–95°C under nitrogen atmosphere. If the isocyanate reacts with a hydroxyl group on the peptide antibiotic, then a carbamate bond is formed. Alternatively, if the isocyanate reacts with an amino group present on the peptide antibiotic structure, then the bond formed is a urea bond [9]. Very recently, Gasparrini et al. [10–13] developed a novel and effective "one pot" synthetic strategy to covalently link teicoplanin and teicoplanin aglycon antibiotics to aminopropyl functionalized silica gel via a bifunctional aliphatic isocyanates. It has been observed that all the above-mentioned antibiotics are not suitable and economic enough to be used as chiral selectors and, therefore, only some of them were commercialized. The commercial antibiotic CSPs are available from Advanced Separation Technologies Inc. (Whippaney, NJ, USA). The most commonly used antibiotics are vancomycin, teicoplanin, teicoplanin aglycon, and ristocetin A. These antibiotics-based CSPs are sold under the names Chirobiotic V, Chirobiotic R, Chirobiotic T, Chirobiotic TAG, which contain vancomycin, ristocetin A, teicoplanin, and teicoplanin aglycon antibiotics, respectively. These CSPs are available in the form of columns of different sizes and dimensions.

4.3 APPLICATIONS

4.3.1 Analytical Separations

The application of antibiotics as chiral selectors has resulted in the successful resolution of almost all types of neutral, acidic, and basic racemic molecule. These antibiotics have been used for the enantiomeric resolution of amino acids, their derivatives, peptides, alcohols, and other pharmaceuticals. The selectivities of the most commonly used antibiotic-based (vancomycin, teicoplanin, and ristocetin A) CSPs varied from one racemate to another and are given in Table 1. Vancomycin was used for the chiral resolution of amino acids, amines, amides, imides, cyclic amines, amino alcohols, hydantoins, barbiturates, oxazolidinones, acids, profens, and other pharmaceuticals. Teicoplanin was found to be excellent chiral selector for the enantiomeric resolution of amino acids, amino alcohols, imides, peptides, hydantoins, α-hydroxy and halo acids, and oxazolidinones, whereas ristocetin A is capable of chiral resolution of amino acids, imides, amino

TABLE 1 The Selectivities of Antibiotic CSPs in the Three Mobile Phase Modes

Mobile phase	Vancomycin	Teicoplanin	Ristocetin A
Normal phase	Hydantoins, barbiturates imides, oxazolidinones	Hydantoins, imides	Imides, hydantoins, amino acids
Reversed phase	Amines, amides, imides, acids, profens	α-Hydroxy acids, oxazolidinones, amino acids, peptides	α-Hydroxy acids, acids, profens, amino acids, amino esters, hydantoins, peptides
New polar organic phase	Cyclic amines, amino alcohols	Amino alcohols, amino acids	α-Hydroxy and halo acids, organic acids, profen, amino acids

esters, peptides, α-hydroxy acids, hydantoins, acids, and profens. The different racemates resolved on the antibiotic CSPs are summarized in Table 2. These antibiotics behave as excellent chiral selectors for a variety of molecules in normal, reversed, and new modified organic polar mobile phases. The properties of these antibiotics make them superb chiral selectors in liquid chromatography. Because of their particular characteristics as chiral selectors, these macrocyclic antibiotics can be used for the resolution of the racemates, which have not been reported as separated on any other chiral stationary phase [2].

Chen et al. [20] demonstrated the complementary use of the two macrocyclic CSPs (i.e., vancomycin and teicoplanin) for the chiral resolution of substituted pyridones. The chromatograms are shown in Figure 3. It was observed that although one type of CSP failed to achieve the separation, the other can often succeed. It can be seen from Figure 3 that all three of the enantiomeric pairs could be resolved either on the vancomycin or on teicoplanin-based CSPs. However, the resolution times of the different solutes varied widely from about 5 to 16 min. Armstrong et al. [28] resolved the enantiomers of alkoxysubstituted esters of phenylcarbamic acids using vancomycin and teicoplanin CSPs. Lehotay and coworkers [29] compared the chiral resolution of amino acids on vancomycin and teicoplanin antibiotics. The macrocyclic antibiotics, vancomycin and teicoplanin, were used for the enantioselective separation of semisynthetic ergot alkaloids in reversed-phase modes [30]. Aboul-Enein and Serignese [17] resolved several cyclic imides on vancomycin under normal phase and reversed-phase modes. Recently, the reversal order of elution of oxazolidiones and dansyl amino acids on Chirobiotic V, Chirobiotic T, and Chirobiotic R has been reported by Armstrong et al. [31]. A mechanistic study on the chiral resolution of 26 sulfoxides was carried out using vancomycin, vancomycin aglycon, teicoplanin, teicoplanin

TABLE 2 Enantiomeric Resolution of Different Racemates on Antibiotics as CSPs

Compounds resolved	Antibiotics (CSP)	Ref.
Bendroflumethiazide, benoxaprofen, benzoin methyl ether, 4-benzyl-2-oxazolidinone, biotin, bromacil, bupivacaine, chloroquine, citalopram, clenbuterol, coumachlor, coumafuryl, devrinol, fenoprofen, fenoterol, fluoxetine, flurbiprofen, folinic acid, hydrochloroquine, ibuprofen, isopromethazine, labetalol, luciferin, mephobarbital, α-methylbenzyl amine 3,5-DNB derivative, methylphenidate (ritalin), 5-methyl-5-phenylhydantoin, metoprolol, naproxen, phensuximide, pindolol, proglumide, promethazine, promethazine sulfoxides, propanolol, pyridoglutethimide, sotalol, terbutaline, 3*a*,4,5,6-tetrahydrosuccinimido (3,4-*b*) acenaphthen-10-one, and warfarin	Vancomycin	5
Amino acids, amino alcohols, bendroflumethiazide, bezoin methyl ester, *N*-benzoylleucine, bromacil, 3-methyl-5-cyano-6-methoxy-3,4-dihydro-2-pyridone,3-benzylphthalide, 1-1-binapthyl-2,2′-diylhydrogen phosphate, coumafuryl, devrinol, bupivacaine, coumachlor, 5-(-4-hydroxyphenyl)-5-phenylhydantoin, indapamide, indoprofen, methsuximide, 5-methyl-5-phenylhydantoin, norverapamil, proglumide, pyridoglutethimide, temazepam, verapamil, and warfarin	Vancomycin	1
Amino acids, AQC-3aminopiperidine dihydrochloride, bendrolumethiazide, coumachlor, indoprofen, warfarin, AQC (6-aminoquinoline-*N*-hydroxysuccinimidyl carbamate)	Vancomycin	7
Tetralone derivatives	Vancomycin	14
4- or 5-Substituted 2-methoxy-6-oxa-1,4,5,6-tetrahydropyridine-3 carbonitriles	Vancomycin	15
Aryloxypropanols	Vancomycin	16
Cyclic imides	Vancomycin	17
Sulfoxides	Vancomycin and vancomycin aglycon	18
4-Aryldihydropyrimidine-5-carboxylates, aza analogs of nifedipine-type dihydropyridine calcium channel modulators	Vancomycin	19
4- or 5-Substituted pyridones	Vancomycin	20

(*continued*)

TABLE 2 Continued

Compounds resolved	Antibiotics (CSP)	Ref.
Amino acids, albuterol, arotinolol, atenolol, bendroflumethiazide, biotin, bromacil, clenbuterol, cyclophosphamide, folinic acid, ifosfamide, isoproterenolol, α-methyl-, α-phenyl-succinimde, metoprolol, oxazepam, oxprenolol, pindolol, propanolol, sotalol, and terbutaline	Teicoplanin	5
Amino acids, dipeptides, organic acids, althiazide, bendoroflumethiazide, 4-benzyl-2-methoxy-6-oxo-1,4,5,6-tetrahydropyridine-3-carbonitrile, 4-benzyl-2-oxalidinone, 1-benzoyl-2-tert-buytl-3-methyl-4-imidazolidinone, bromacil, bupivacaine, carprofen, coumachlor, coumafuryl, 4-cyclohexyl-2-methoxy-6-oxo-1,4,5,6-tetrahydropyridine-3-carbonitrile, devrinol, flurbiprofen, buprofen, indoprofen, ketoprofen, methsuximide, 4-methyl-2-methoxy-6-oxo-1,4,5,6-tetrahydropyridine-3-carbonitrile, 5-methyl-5-phenylhydantoin, 5-(4-methylphenyl)-5-phenylhydantoin, α-methyl-α-phenyl-succinamide, phesuxamide, 3-phenylphthalide, 4-phenyl-2-methoxy-6-oxo-1,4,5,6-tetrahydropyridine-3-carbonitrile, proglumide, suprofen, tropicamide, and warfarin	Teicoplanin	6
Native amino acids, peptides, α-hydroxyl carboxylic acids, cyclic amines, and amides	Teicoplanin	21
Amino acids, carboxylic acids, organometalic complexes, β-blockers, nonsteroidal anti-inflamatory drugs, and carnitines	Teicoplanin and teicoplanin aglycon	10
Sulfoxides	Teicoplanin and teicoplanin aglycon	18
Amino acids containing 1,2,3,4-tetrahydroisoquinoline, tetraline, 1,2,3,4-tetrahydro-2-carboline, cyclopentane, cyclohexane, heptane and bicyclo [2.2.1] heptane skeleton	Teicoplanin	22
Amino acids and imino acids	Teicoplanin and teicoplanin aglycon	23
Phenyl glycine	Teicoplanin	24
Amino acids and carboxylic acids	Teicoplanin	25
Aryloxypropanols	Teicoplanin	16
Native amino acids and peptides	Teicoplanin	26
Usual amino acids and their derivatives	Teicoplanin	27

(continued)

TABLE 2 Continued

Compounds resolved	Antibiotics (CSP)	Ref.
4- or 5-Substituted 2-methoxy-6-oxa-1,4,5,6-tetrahydropyridine-3-carbonitriles	Teicoplanin	15
4-Aryldihydropyrimidine-5-carboxylates, aza analogs of nifedipine-type dihydropyridine calcium channel modulators	Teicoplanin	19
4- or 5-Substituted pyridones	Teicoplanin	20
Tetralone derivatives	Ristocetin A	21
Thioridazine and 2,2,2-trifluoro-1(9-anthyryl)ethanol	Thiostrepton	1
Idazoxam	Rifamycin B	1

aglycon, and ristocetin A antibiotic CSPs using a wide range of mobile phases [18]. It has also been observed that the sugar moieties of vancomycin and teicoplanin antibiotics did not participate in the chiral resolution of these sulfoxides. Recently, Armstrong et al. [32] compared the chiral resolution of plant growth regulators on vancomycin, teicoplanin, and ristocetin A CSPs using the reversed-phase mode.

Wainer et al. [24] resolved the enantiomers of albuterol in plasma on teicoplanin CSP. Jandera et al. [33] studied the retention behavior of underivatized phenylglycine on the Chirobiotic T column packed with amphoteric glycopeptide teicoplanin covalently bonded to the silica surface. A teicoplanin-based CSP was tested for enantioseparation of underivatized amino acids and their *N-tert*-butyloxycarbonyl (*t*-Boc) derivatives [34]. Peter et al. [35] studied the resolution of amino acids on teicoplanin CSP. Aboul-Enein and Serignese [36,27] have resolved clenbuterol enantiomers qualitatively and quantitatively in blood on teicoplanin CSP. Berthod et al. [38] studied 26 racemates (amino acids, carnitine, and bromacil) on teicoplanin and teicoplanin aglycon CSPs using seven reversed phases and two polar organic mobile phases. They observed a difference in the enantioselective free energy ranging from 0.3 to 1 kcal/mol for amino acid enantiomers. The chromatograms of 3,4-dihydroxyphenyl alanine (DOPA) and 4-hydroxymandelic acid on teicoplanin and teicoplanin aglycon CSPs are shown in Figure 4. Enantiomeric resolution of cyclic β-substituted quaternary α-amino acids was carried out on teicoplanin CSP [39]. Tesova et al. [40] used teicoplanin for the chiral resolution of *N-tert*-butyloxycarbonyl amino acids. Peyrin and co-workers [41] described a wide range of mobile phases for the chiral resolution of dansyl amino acids on teicoplanin CSP. An approach based on the development of various equilibria was carried out in order to describe the retention behavior of the solute in the chromatographic system. The equilibrium constants were used to explain the retention and separation factors observed. Gasparrini et al. [10] developed a novel linkage between teicoplanin and teicoplanin aglycon

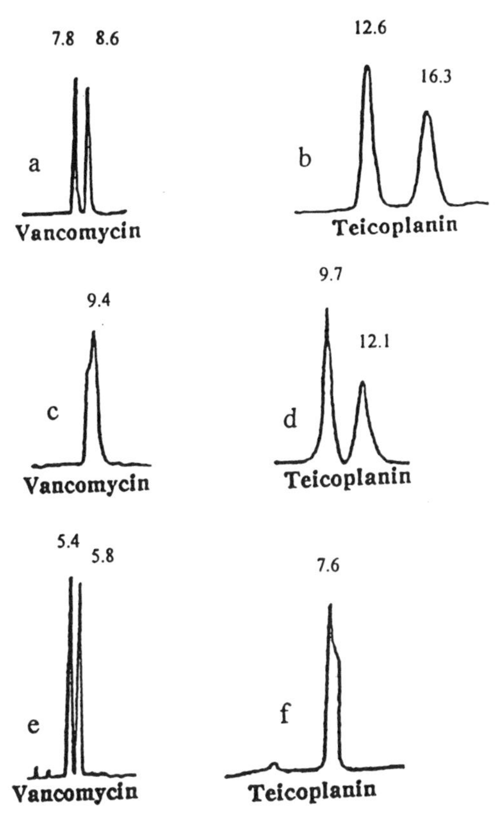

FIGURE 3 Chromatograms of substituted pyridones on vancomycin (a, c, e) and teicoplanin (b, d, f) CSPs using methanol–1% triethylammonium acetate buffer, pH 4.1 (10 : 90, v/v). (From Ref. 20.)

antibiotics with amino silica gel, and the developed CSPs were tested for the resolution of carboxylic acids, organometalic complexes, amino acids, β-blockers, anti-inflammatory drugs, and carnitines.

In addition to the common use of vancomycin and teicoplanin, the use of other antibiotics as chiral selectors in HPLC were limited. Only a few reports are available in the literature dealing with these antibiotics as HPLC CSPs. Armstrong et al. [42] resolved about 230 racemates on the ristocetin A antibiotic covalently bonded to silica gel. The resolution was carried out using the three modes of mobile phases. The results were complimentary to those obtained on vancomycin and teicoplanin CSPs. In another study, the effect of selector coverage and mobile phase composition on enantiomeric resolution with ristocetin A CSP was carried out by Armstrong et al. [43]. Thiostrepton-based CSP was also used to resolve the enantiomers of thioridazine, 2,2,2-trifluoro-1-

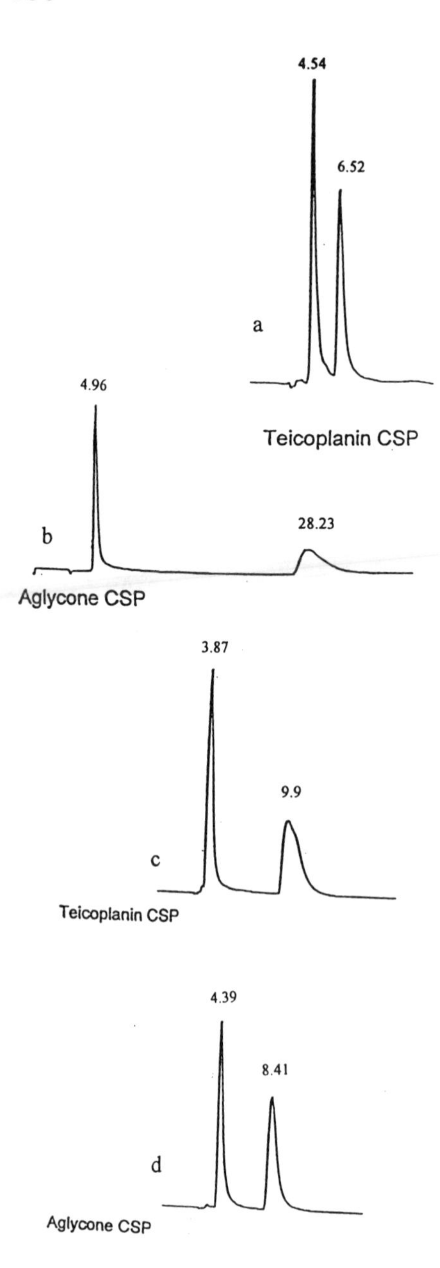

FIGURE 4 Chromatograms of DL-DOPA and 4-hydroxymandelic acid on teicoplanin (a, c) and teicoplanin aglycon (b, d) CSPs using methanol–water (60 : 40, v/v, pH 4.1) for DL-DOPA and methanol-water (85 : 15, v/v, pH 4.1) for 4-hydroxymandelic acid. (From Ref. 38.)

(9-anthryl)ethanolalthiazide, and 4-benzyl-2-oxazolidinone [1]. In addition, the enantiomers of idazoxan, coumafuryl, laudanosine, and 1-[5-chloro-2-(methyl-amino)phenyl]-1,2,3,4-tetrahydrosoquinoline have been resolved on rifamycin B antibiotic [1]. Ristocetin A [44] has been used for the chiral resolution of 28 amino acids. Excellent resolution was achieved by using the reversed phase and the polar organic mobile phase [44]. Recently, Aboul-Enein and Ali [14] used ristocetin A CSP for the chiral resolution of tetralone derivatives. Avoparcin, a costly antibiotic, has been tested for the chiral resolution of a variety of racemates such as verapamil, thyroxine, mephenytoin, and 2-imidazolidone-4-carboxylic acid [8]. Figure 5 presents the resolution of tetralone derivatives 3-amino-3-phenyl propionic acid, midodrine, *N*-g-fluorenylmethyl oxycarbonyl (*N*-FMOC)-glycyl-arginine, thyroxine, trihexyphenidyl and verapamil on ristocetin A- and avoparcin-based CSPs.

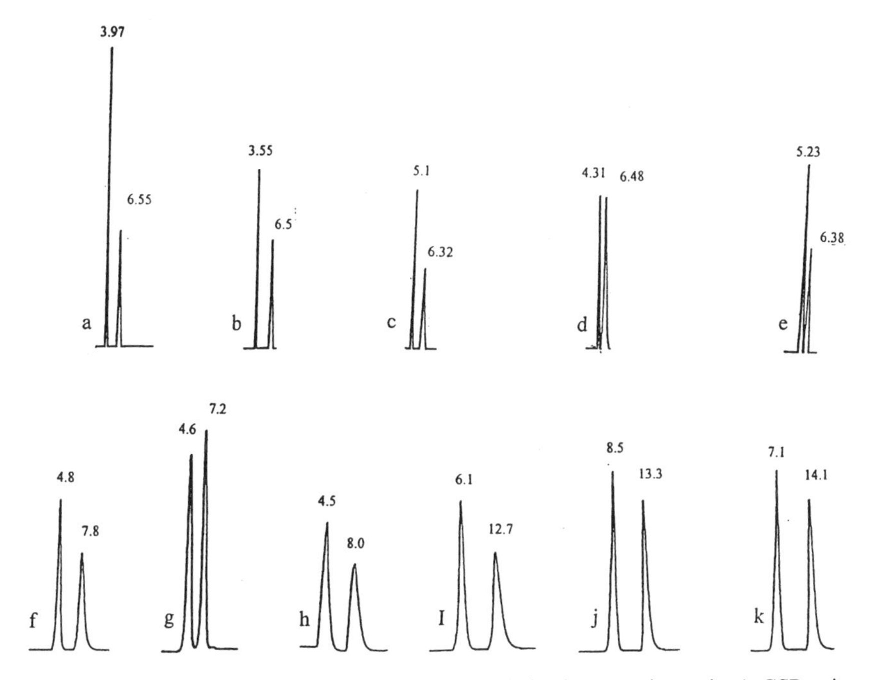

FIGURE 4.5 Chromatograms of (a–e) tetralone derivatives on ristocetin A CSP using hexane-ethanol-TEA (12 : 8 : 02, v/v/v), (f) 3-amino-3-phenyl propionic acid [methanol–water (60 : 40, v/v)], (g) midodrine [methanol–1% triethyl ammonium acetate buffer, pH 7 (20 : 80, v/v)], (h) *N*-FMOC-glycyl-arginine [methanol–water (60 : 40, v/v)], (i) thyroxine [methanol–1% triethyl ammonium acetate buffer, pH 7 (20 : 80, v/v)], (j) trihexyphenidyl [methanol–1% triethyl ammonium acetate buffer, pH 7 (20 : 80, v/v)] and (k) verapamil [methanol–1% triethyl ammonium acetate buffer, pH 7 (20 : 80, v/v)] on avoparcin CSP (From Refs. 8 and 14.)

4.3.2 Preparative Separations

Armstrong et al. [1,21] and Beesley and co-workers [8,45] reported the chiral preparative separation at the milligram scale on antibiotic CSPs. Columns up to 2 in. in diameter are available and can effectively operate up to 500 mL/min. It has been reported that the high flow rates are suitable for the increasing sample throughput. The preparative chiral separations on antibiotic CSPs have been reported using reversed-phase and normal phase modes [2]. Although resolution is better in the reversed-phase mode for the analyte (e.g., 5-methyl-5-phenyl-hydantoin on Chirobiotic V), the solubility is much greater in the normal phase solvent, allowing for greater column loads. The most efficient processing of this sample has been described by the AmeriChrom recycle and shave systems [2]. It was observed that the retention times with the vancomycin column are significantly shorter than those reported with many other CSPs. Therefore, throughput reported as grams per hour is more illustrative than grams per injection. The Chirobiotic T column was used for the preparative chiral separation of chloro-kynurenine using the reversed-phase mode [2]. Recently, Francotte [46] has reviewed the chiral preparative separation on antibiotics CSPs. The author described the limitations of these CSPs in the preparative scale.

4.4 OPTIMIZATION OF HPLC CONDITIONS

Recently, Aboul-Enein and Ali reviewed the chiral resolution on antibiotic CSPs by HPLC [3,47]. It was observed that chiral resolution on antibiotic CSPs is governed by various HPLC parameters. The antibiotic CSPs may be used in normal, reversed, and new modified polar organic phase modes. The most important parameters which control the chiral resolution on antibiotic CSPs by HPLC are mobile phase composition, pH of the mobile phase, flow rate, temperature, structures of solutes, structures of antibiotics, and other parameters. These parameters are discussed herein.

4.4.1 Mobile Phase Compositions

Because of the complex structure of the antibiotics, most of the antibiotics function equally well in reversed, normal, and modified polar organic phases. All of the three solvent modes generally show different selectivity with different analytes. Sometimes equivalent separations were obtained in both normal and reversed phases. This ability to operate in two different solvent modes is an advantage in determining the best preparative methodology where sample solubility is a key issue. In normal phase chromatography, the most common solvents used are hexane, ethanol, methanol, and so on. The optimization of chiral resolution is achieved by adding some other organic solvents such as acetic acid, tetrahydrofuran (THF), diethylamine (DEA) and triethylamine (TEA) [42]. The protocols for the development and optimization of mobile phases on vancomycin, teicoplanin, and ristocetin A CSPs are shown in Schemes 1 to 3, respectively. The effect of the composition of mobile phases on the enantiomeric resolution of different racemates on a normal phase column is shown in Figure 6. It can be concluded from this figure that lower concentrations of alcohols favor the chiral resolution.

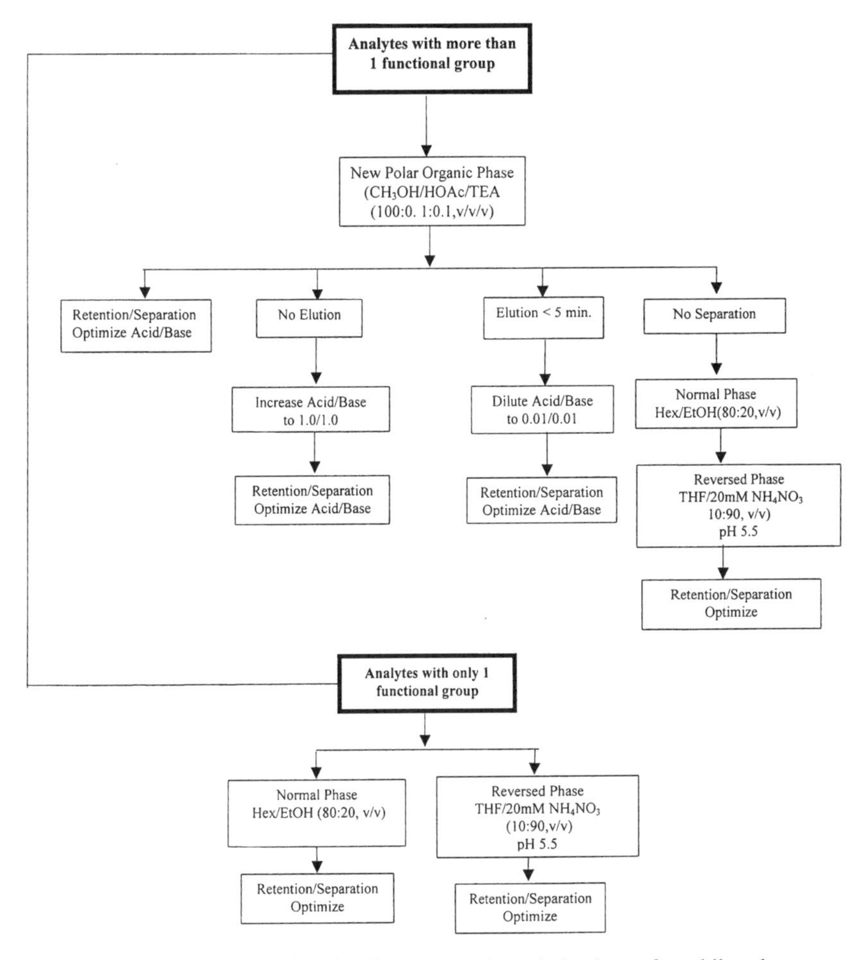

SCHEME 1 Protocol for the development and optimization of mobile phases on vancomycin-based CSP.

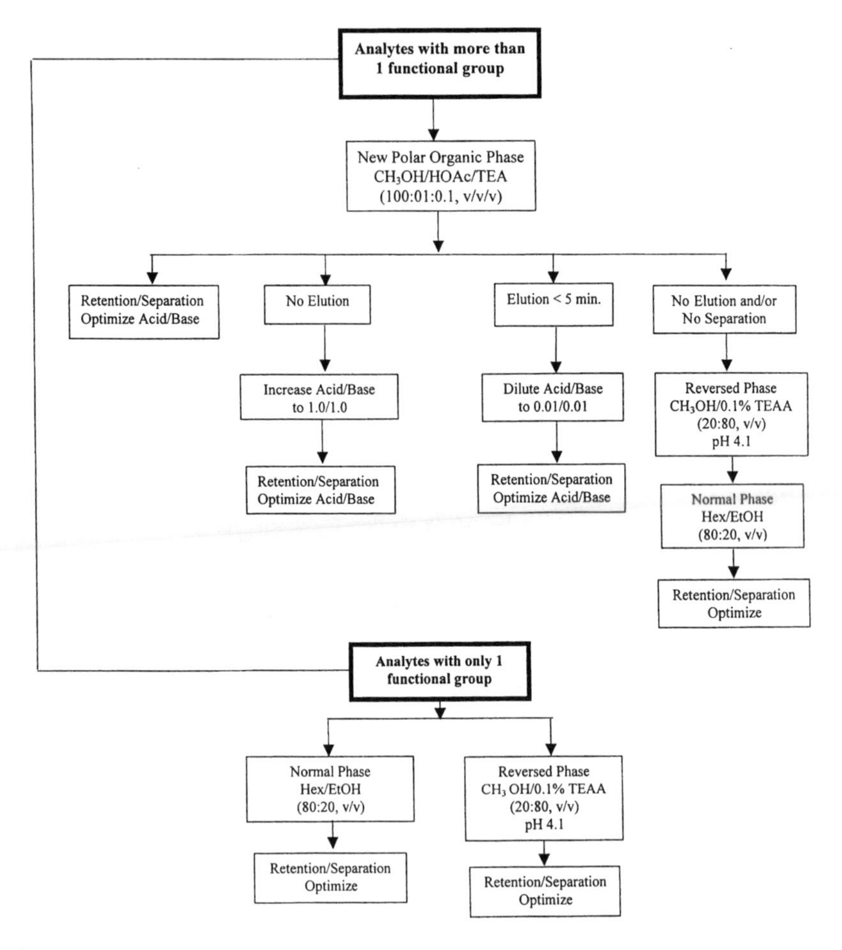

SCHEME 2 Protocol for the development and optimization of mobile phases on teicoplanin-based CSP.

In the reversed phase system, buffers are used most often as the mobile phases with small amount of organic modifiers. The use of buffers as the mobile phases has increased the efficiency of the resolution. Ammonium nitrate, triethylammonium acetate (TEAA), and sodium citrate buffers have been used very successfully. A variety of organic modifiers have been used to alter selectivity [2,5,22]. Acetonitrile, methanol, ethanol, 2-propanol, and THF have shown good selectivities for various analytes. In the reversed-phase mode, the amount of organic modifiers is typically low, usually of the order of 10–20%. The typical starting composition of the mobile phase is an organic modifier–buffer

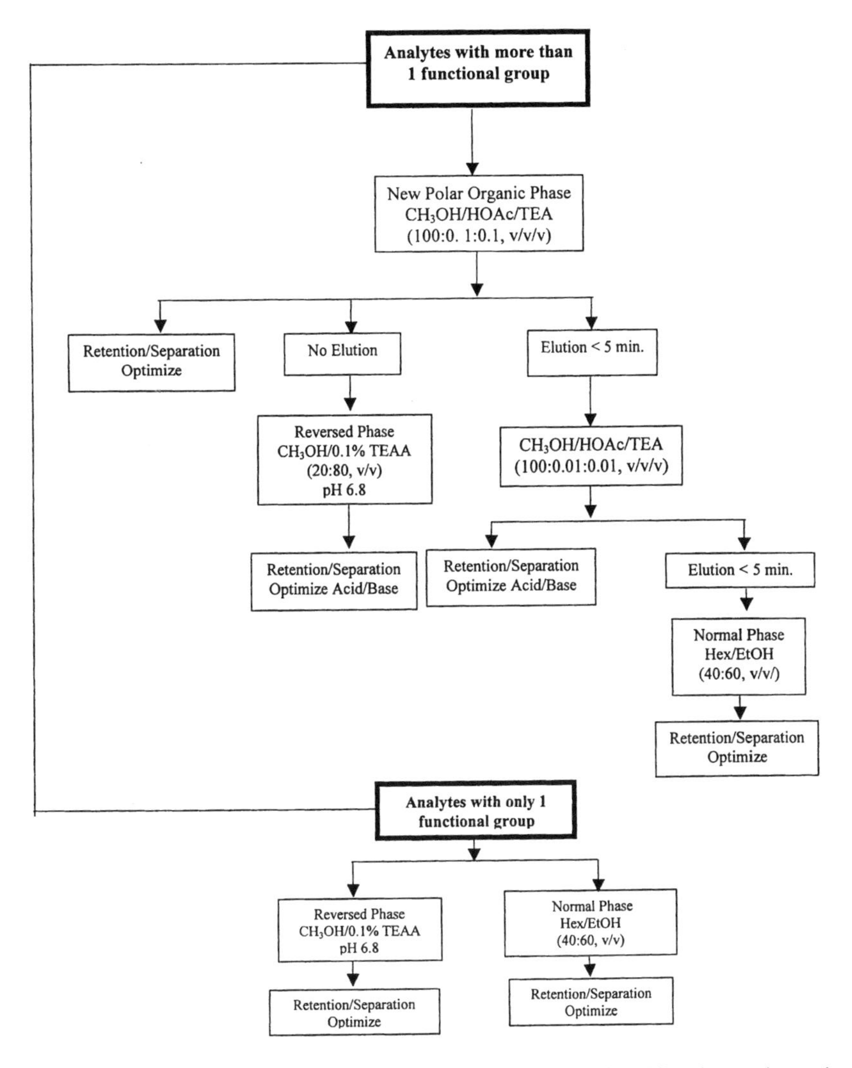

SCHEME 3 Protocol for the development and optimization of mobile phases ristocetin-based CSP.

(pH 4.0–7.0) (10 : 90). Using alcohols as organic modifiers generally requires higher starting concentrations (i.e., 20% for comparable retention when using acetonitrile or tetrahydrofuran in starting concentration of 10%). The effect of organic solvents on the enantioselectivities also depends on the type of antibiotic. In fact, better recognition is obtained at acidic buffer pH values below or close to

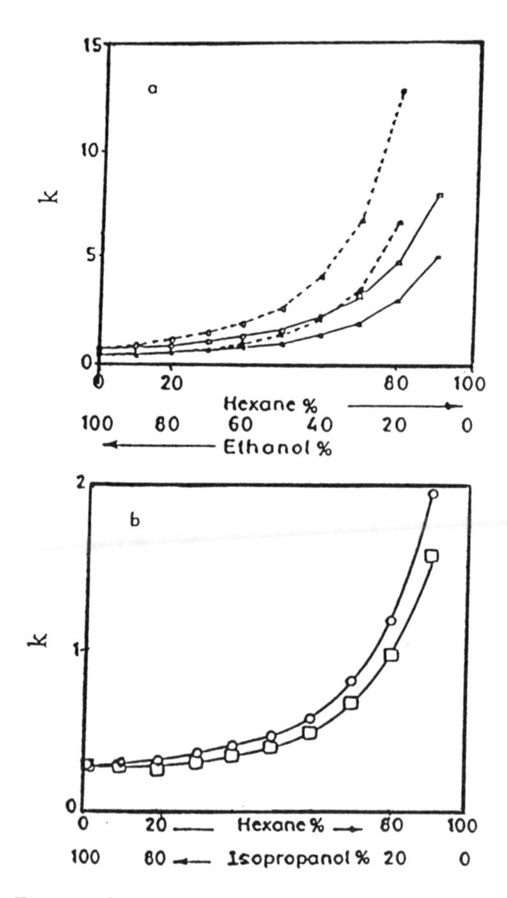

FIGURE 6 Effect of mobile phase composition on the resolution of enantiomers of different racemates in normal phase HPLC on antibiotic CSPs. (a) First (■) and second (□) eluted enantiomers of γ-phenyl-γ-butyrolactone and first (●) and second (○) eluted enantiomers of 4-phenyl-2-methoxy-6-oxo-2,4,5,6-tetrahydropyridine-3-carbonitrile on Chirobiotic T column and (b) first (□) and second (○) eluted enantiomers of mephenytoin on Chirobiotic V column. (From Refs. 1 and 21.)

the isoelectric point of the antibiotics, especially for vancomycin. For vancomycin, a low concentration of organic solvents did not significantly influence the separation; however, enantioresolution is improved for some compounds with ristocetin A and teicoplanin [45], even at low concentrations of organic modifiers. Figure 7 shows the effect of the organic modifiers on the enantioselectivities of different enantiomers in reversed-phase liquid chromatography. The effect of organic modifiers on chiral resolution varies from racemate to racemate [34,42].

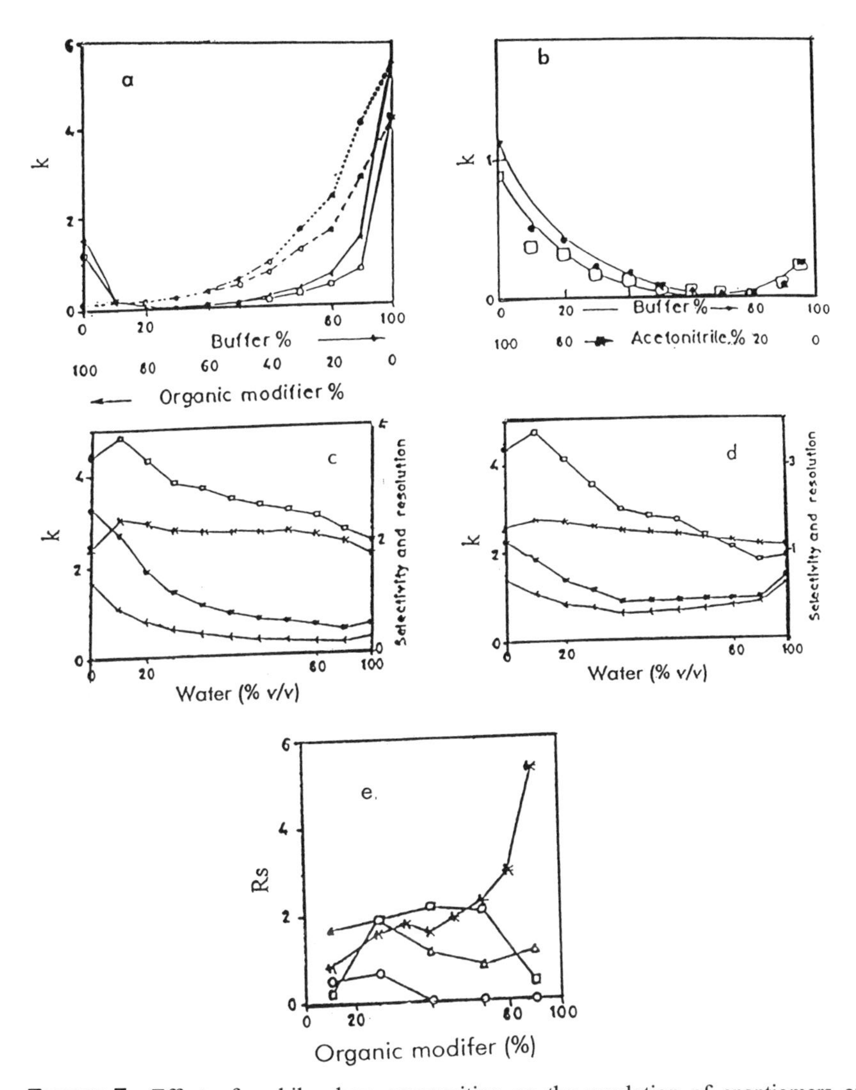

FIGURE 7 Effect of mobile phase composition on the resolution of enantiomers of different racemates in reversed-phase HPLC on antibiotic CSPs. (a) First (□, ○) and second (◆, ●) enantiomers of 5-methyl-5-phenylhydantoin on a Chirobiotic T column using an acetonitrile–triethylammonium acetate buffer (—) and a methanol–triethylammonium acetate buffer (- - - -) as mobile phases; (b) first (□) and second (●) enantiomers of 5-methyl-5-phenylhydantoin on a Chirobiotic V column, (c, d): first (+) and second (●) eluted enantiomers, and the resolution (□) and selectivity (×) for methionine (c) and phenylalanine (d) on a Chirobiotic T column; (e) dependence of the resolution of the enantiomers of fluoxetine (Prozac) on the amount of methanol (□), 2-propanol (×), acetonitrile (○), and THF (△) as organic modifiers. (From Refs. 1 and 2.)

The presence of triethylamine in the aqueous portion of the mobile phase was shown to be important for the separation of enantiomers of *t*-Boc-amino acids, whereasa native amino acids were almost not affected by the addition of TEAA [34]. Lee and Beesley [48] observed that a change in the ionic interactions occurred when changing the acid/base ratio in reversed and polar organic mobile phases or changing the buffer pH in the reversed-phase mode. Consequently, the retention, selectivity, and peak shape is affected.

A simplified approach has proven very effective for the resolution of a broad spectrum of racemate analytes. The first consideration in this direction is the structure of the analytes. If the compound has more than one functional group capable of interacting with the stationary phase and at least one of those groups is on or near the stereogenic center, then the first mobile phase choice would be the new polar organic phase. Because of the strong polar groups present in the macrocyclic peptides, it was possible to convert the original mobile phase concept to 100% methanol with the acid/base added to effect selectivity. The key factor in obtaining complete resolution is still the ratio of acid to base. The actual concentration of the acid and base only affects the retention. Therefore, starting with a 1 : 1 ratio, some selectivity is typically observed, then different ratios of 1 : 2 and 2 : 1 are applied to note the change in resolution indicating the trend. If the analyte is eluting too fast, the concentration of acid/base is reduced. Conversely, if the analyte is retained too well, the acid/base concentration is increased. The parameters for concentration are between 1% and 0.001%. Above 1%, the analyte is too polar and indicates a typical reversed-phase system, whereas below 0.001% indicates a normal phase system. Both trifluoroacetic acid (TFA) and acetic acid have been used as the acid component, with ammonium hydroxide as the base. For an analyte that has only one functional group or for reasons of solubility, typical normal phase solvents (hexane/ethanol) or reversed-phase solvents (THF/buffer) are employed. The typical chromatograms, using the new polar organic modified phases, of the chiral resolution of trimipramine, chlorthalidone, and naproxen on vancomycin, teicoplanin, and ristocetin A CSPs, respectively, are shown in Figure 8.

4.4.2 pH of the Mobile Phase

The pH is an important controlling factor of enantiomeric resolution in normal, reversed, and new organic phases. Generally, buffers are used in the mobile phases to control the pH in HPLC. The pH of the buffers ranges from 4.0 to 7.0 in the reversed-phase system. In general, the analytes interact more favorably at a pH at which they are not ionized. Therefore, retention and selectivity of the molecules that possess ionizable acidic or basic functional groups can be effected by altering the pH. A strategy is to take advantage of a difference in pK_a values, (i.e., keeping the analyte of interest neutral and strongly interacting while keeping other components ionized and poorly retained). Because of the complexities of

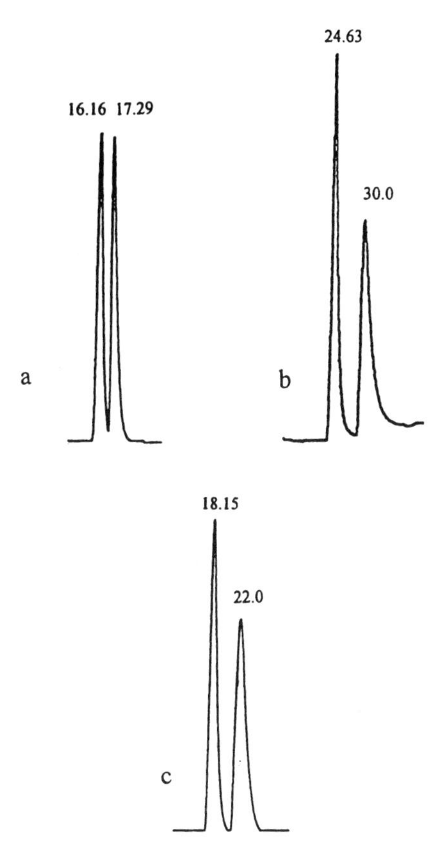

FIGURE 8 Chromatograms of (a) trimipramine on vancomycin CSP using methanol–acetic acid–TEA (100 : 0.3 : 0.2, v/v/v), (b) chlorthalidone on teicoplanin CSP using methanol–TEAA, pH 4.1 (10 : 90, v/v) and (c) naproxen on ristocetin A CSP using methanol–EAA (30 : 70, v/v) mobile phases. (From Ref. 2.)

these interactions, it is necessary to observe the retention and resolution as the function of pH, usually testing at pH 4.0 and pH 7.0 or 0.50 pH units above and below the pK_a value. It has been observed that with increasing the pH values, the values of R_s (resolution factor), k (retention factor), and α (separation factor) decrease in most of the cases. Therefore, the safest and suitable values of pH in reversed-phase systems vary from 4.0 to 7.0 [1,2]. The enantiomeric resolution by normal phase and new modified polar organic phases has been achieved below pH 7.0. To show the effect of pH on the chiral resolution of some racemates on teicoplanin and vancomycin CSPs, the values of k, α, and R_s are given in Tables 3 and 4, respectively. It can be observed from these tables that pH 4.0 and 7.0 are

TABLE 3 Effect of pH on the Chiral Resolution of Several Racemates on Chirobiotic T CSP Using Methanol–1% Triethylammonium Acetate Buffer (20 : 80, v/v) as the Mobile Phase

pH	k_1	α	R_s
5-(4-Hydroxyphenyl)-5-phenylhydantoin			
7.0	4.29	1.30	1.12
6.0	3.78	1.35	1.25
5.0	3.38	1.38	1.36
4.1	3.10	1.40	1.70
3.6	1.82	1.31	1.10
5-Methyl-5-phenylhydantoin			
7.0	0.97	2.34	2.58
6.0	0.95	2.30	2.64
5.0	0.92	2.28	2.72
4.1	0.87	2.11	2.87
3.6	0.70	1.87	2.18
3-Phenylphthalide			
7.0	2.66	2.06	2.93
6.0	2.88	2.01	3.04
5.0	2.94	1.96	3.26
4.1	2.98	1.78	3.69
3.6	2.20	1.47	2.24
1-Benzoyl-2-*tert*-butyl-3-methyl-4-imidazolidone			
7.0	1.41	2.28	2.80
6.0	1.83	2.20	3.05
5.0	1.97	2.11	3.22
4.1	2.09	1.85	3.40
3.6	1.69	1.64	1.97
N-Formylphenylalanine			
7.0	0.89	2.52	2.28
6.0	1.45	2.36	2.32
5.0	1.56	2.27	2.43
4.1	1.61	2.16	2.53
3.6	1.10	2.05	2.33
Mandelic acid			
7.0	0.14	6.57	2.75
6.0	0.28	4.96	2.89
5.0	0.36	4.34	2.98
4.1	0.46	3.28	3.17
3.6	0.40	2.35	2.43

Source: Ref. 45.

TABLE 4 Effect of pH on Chiral Resolution of Several Racemates on Chirobiotic V CSP Using Acetonitrile-1% Triethylammonium Acetate Buffer (10 : 90, v/v) as the Mobile Phase

pH	k_1	α	R_s
Coumachlor			
7.6	1.00	1.69	1.60
6.2	1.21	1.64	2.00
5.5	2.00	1.60	2.70
4.5	3.00	1.64	4.10
3.6	1.65	1.42	3.90
Devrinol			
7.6	1.15	1.80	3.20
6.2	1.16	1.77	3.30
5.5	1.36	1.76	3.50
4.5	1.46	1.70	3.60
3.6	1.19	1.41	3.70

Source: Ref. 1.

suitable for chiral resolution. Recently, Schlauch et al. [39] studied the effect of pH on the chiral resolution of cyclic substituted amino acids. They observed two different enantiomeric and diastereomeric discrimination mechanisms based on different interactions with the stationary phase. The effect of the pH on the chiral resolution of mandelic acid on teicoplanin is studied by Jandera et al. [25]. These

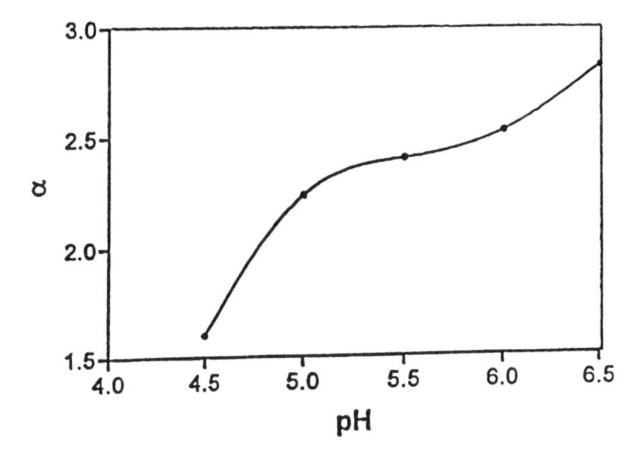

FIGURE 9 Effect of mobile phase pH on the chiral resolution of mandelic acid on Chirobiotic T CSP using ethanol–water–TEA (17.5 : 70.2 : 12.3, v/v/v) as the mobile phase. (From Ref. 25.)

findings are shown in Figure 9, which indicates that pH 7 is the optimum pH. Briefly, the interactions of enantiomers with CSP are effected by the pH of the mobile phase and, hence, the enantioselectivity is pH dependent.

4.4.3 Flow Rate

Armstrong et al. [21] studied the effect of flow rate on the resolution of 4-hydroxy-5-cyano-6-methoxy-3,4-dihydro-2-pyridone and α-methyl-α-phenyl succinamide on the teicoplanin column. It has been observed that the flow rate does not affect the enantioselectivity (α) but does affect the separation efficiency. This is reflected by the inverse relationship between R_s and the flow rate. Decreasing the flow rate from 2.0 to 1.0 mL/min enhanced the resolution by 20–30%. Further, the decrease in the flow rate did not produce any increase in the resolution. No increase in resolution was obtained below the flow rate of 0.5 mL/min in any separation mode. The loss of efficiency at higher flow rates has also been observed with most of the chiral antibiotics. The enantioselectivities of some of the derivatives of amino acids on the teicoplanin phase have been enhanced by simply lowering the flow rates [1]. The effect of the flow rate on the normal phase enantiomeric resolution of 3*a*,4,5,6-tetrahydrosuccinamido[3,4-*b*]acenapthen-10-one on vancomycin has been studied by Armstrong and co-workers [1]. They have varied the flow rates from 0.5 to 2.0 mL/min and observed that the values of α decreased from 1.31 to 1.29 and the values of R_s decreased from 1.28 to 1.11. The effect of flow rate on the enantioselectivity of various racemates is given in Table 5. Recently, Peter et al. [44] optimized the chiral resolution of synthetic amino acids on ristocetin A CSP by controlling the flow rate. The flow rate up to 500 mL/min has been used for preparative chiral resolution [2].

4.4.4 Temperature

Temperature is also an important parameter for controlling the resolution of enantiomers in HPLC. The enthalpy and entropy control of chiral resolution on antibiotic CSPs is similar to the case of polysaccharide-based CSPs (Chapter 2). Armstrong et al. [1] have studied the effect of temperature on the resolution behavior of proglumide, 5-methyl-5-phenylhydantoin and *N*-carbamyl-D-phenylalanine on the vancomycin column. The experiments were carried out from 0°C to 45°C. These results are given in Table 6 for three chiral compounds. It has been observed that the values of k, α, and R_s for the three studied molecules have decreased with the increase in temperature, indicating the enhancement of chiral resolution at low temperature. In another work, the same workers [22] have also studied the effect of temperature on the resolution of certain amino acid derivatives on the teicoplanin chiral stationary phase. They further observed poor resolution at ambient temperature, whereas the resolution increased at low

TABLE 5 Effect of Flow Rate on Enantiomeric Resolution on Chirobiotic V and Chirobiotic T CSPs

Flow Rate (mL/min)	α	R_s
3*a*,4,5,6-Tetrahydrosuccinimido [3,4-*b*] acenaphthen-10-one[a]		
0.50	1.31	1.28
0.75	1.31	1.19
1.00	1.27	1.14
1.50	1.30	1.13
2.00	1.29	1.11
4-Hexyl-5-cyano-6-methoxy-3,4-dihydro-2-pyridone[b]		
0.50	1.4	2.4
1.0	1.4	2.3
1.5	1.4	2.0
2.0	1.4	1.6
α-Methyl-α-phenylsuccinamide[b]		
0.5	1.3	1.9
1.0	1.3	1.8
1.5	1.3	1.6
2.0	1.3	1.4

[a] Chirobiotic V using 2-propanol-hexane (50 : 50, v/v) as the mobile phase. (*Source:* Ref. 1.)
[b] Chirobiotic T using ethanol–hexane (30 : 70, v/v) as the mobile phase. (*Source:* Ref. 21.).

and high temperatures. The increase in the resolution at lower temperature may be the result of the increase in efficiency of the column. It has also been observed that the change in temperature has a greater effect on the retention of solutes in the normal phase in comparison to the reversed phase. It might be due to the fact that the binding constant of a solute to the macrolide involves several interactive mechanisms that dramatically change with temperature. Inclusion complex formation is effectively prevented for most solutes in the temperature range of 60–80°C. The lower temperature enhances the weaker bonding forces and the net result is that the chromatographers have an additional powerful means to control selectivity and retention. The effect of temperature in HPLC on enantioselectivity of a variety of racemates is given in Figure 10 [22]. This figure indicates that the effect of temperature on enantioselectivity varies from antipode to antipode. Peter et al. [44] optimized the chiral resolution of synthetic amino acids on ristocetin A CSP by controlling the experimental temperature. Peyrin and co-workers [41] calculated the thermodynamic parameters for the chiral resolution of dansyl amino acids on the teicoplanin stationary phase. The authors observed that the driving forces of the solutes on CSP varied with temperature, indicating the control of chiral recognition by the interactions of solutes with teicoplanin CSP. Peter et al. [49] studied the effect of temperature on the chiral separation of amino

TABLE 6 Effect of the Temperature on Chiral Resolution of Some Racemates on Vancomycin CSP Using Acetonitrile–1% Triethylammonium Acetate, pH 4.1 (10 : 90, v/v) as the Mobile Phase

Temp. (°C)	k	α	R_s
Proglumide			
0	1.33	2.27	3.60
5	1.33	2.11	3.30
15	1.31	1.87	2.40
22	1.18	1.75	2.10
35	0.93	1.57	1.80
45	0.76	1.44	1.60
5-Methyl-5-phenylhydantoin			
0	0.35	1.38	1.50
5	0.27	1.36	1.00
22	0.24	1.34	1.00
35	0.24	1.30	0.90
45	0.19	1.32	0.70
N-Carbamyl-D,L-phenylalanine			
0	0.51	1.39	1.50
5	0.39	1.34	1.30
15	0.38	1.23	1.00
22	0.31	1.20	0.80
35	0.27	1.11	0.70
45	0.22	1.00	0.00

acids on teicoplanin CSP using the reversed-phase mode. The retention and selectivity factors of all amino acids decreased with the increase in temperature. The natural logarithms of the retention factors ($\ln k$) of the investigated compound depended linearly on the inverse of the temperature ($1/T$). van't Hoff plots afforded thermodynamic parameters such as enthalpy (H°), entropy (S°), and Gibbs free energy (G°) changes for the transfer from the mobile to the stationary phase. The thermodynamic constants (H°, S°, and G°) were calculated in order to promote an understanding of the thermodynamic driving forces, for the retention in the chromatographic system.

4.4.5 Structures of Solutes

Figure 1 shows a variety of groups on antibiotic CSP that can take part in the bonding with the different groups of the racemates. Therefore, different structures of the racemates provide different types of bondings and, therefore, the different patterns of chiral resolution observed. Amino acids is the best class of com-

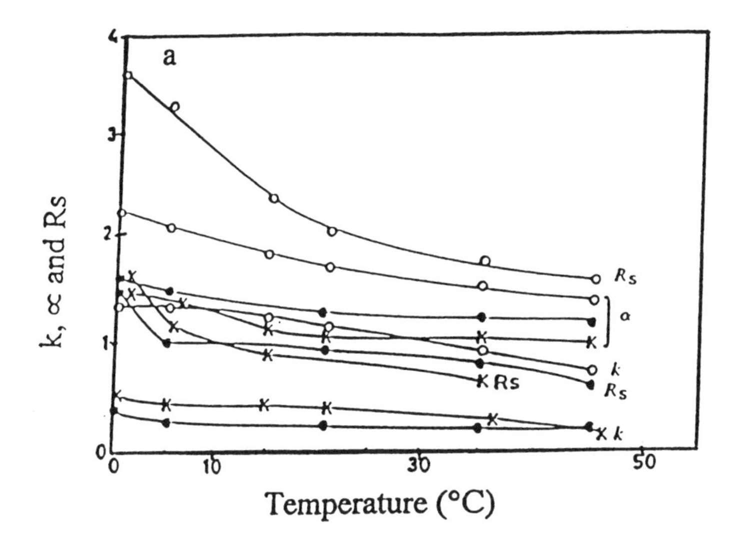

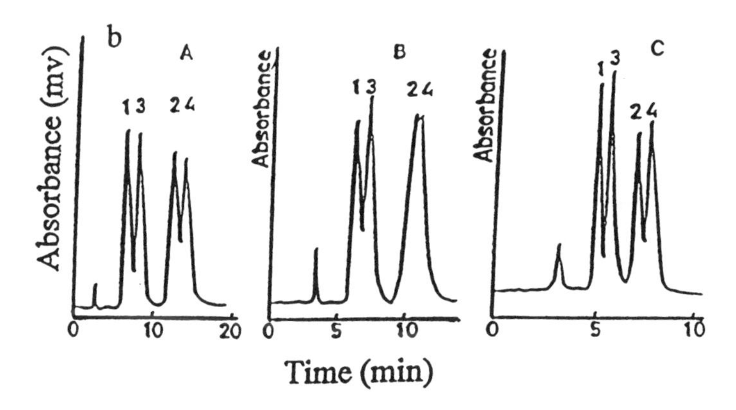

FIGURE 10 Effect of temperature on enantiomeric resolution on antibiotic CSPs. (a): k, α, and R_s for proglumide (○), 5-methyl-5-phenylhydantoin (●) and *N*-corbyl-DL-phenylalanine (×) on Chirobiotic V column using acetonitrile–1% triethylammonium acetate buffer (10 : 90, v/v) as the mobile phase and (b) separation of enantiomers of β-methyl phenylalanine on the Chirobiotic T column using water–methanol (10 : 90, v/v) as the mobile phase at (A) 1°C, (B) 20°C, (C) 50°C. 1 = *erythro*-L; 2 = *erythro*-D; 3 = *threo*-L; 4 = *threo*-D. (From Refs. 1 and 22.)

pounds by which effect of the substituents on chiral resolution can be studied. A number of reports are available for the chiral resolution of amino acids on antibiotics CSPs [1–7,22,23,25–27]. The survey of these reports indicated the different enantioselectivities of all the amino acids on a particular CSP reflecting the effect of substituents on chiral resolution. Aboul-Enein and Ali [14] have observed different chiral recognition patterns of certain tetralone derivatives on vancomycin, teicoplanin, and ristocetin CSPs. Generally, the resolution of these tetralone derivatives was in the order of I > II > IV > V > III (for structures of tetralone derivatives, see Fig. 18 of Chap. 2). This could be the result of the different types of bonding of different magnitudes between these derivatives and the chiral stationary phases. The resolution of II and IV derivatives was better than III and V analytes, which could be due to the different steric effects among these (II–IV) analytes. The different steric effects among these compounds might be due to the different positions of the substituent groups on the phenyl ring.

Lee et al. [48] achieved the chiral resolution of a variety of ionizable racemates on vancomycin, teicoplanin, and ristocetin A CSPs. The authors discussed the results of the chiral resolution due to the structures of the solutes and the CSPs. Ananieva and co-workers [50] resolved several substituted amino acids on teicoplanin and teicoplanin aglycon CSPs. The authors observed different patterns of retention and selectivity of these amino acids on the reported CSPs. Retention factors for the amino acids on teicoplanin were in the order of bezoyl- < *o*-phthaldialdehyde- < phthalyl- < carbobenzyloxy < 2,4-dinitrophenyl- < dansyl < 9-fluorenylmethyloxycarbonyl- and those on teicoplanin aglycon were *o*-phthaldialdehyde- < benzoyl- < phtalyl- < carbobenzyloxy- < 2,4-dinitrophenyl- < dansyl- < 9-fluorenylmethyloxycarbonyl-. This type of chiral recognition behavior might be the result of the different structures of the solutes. Armstrong et al. [16] studied the chiral resolution of alkylamino derivatives of aryloxypropanols on vancomycin and teicoplanin CSPs. The authors observed that steric effect interactions were contributing to enantioselectivity. The substituents in the 2-position of the aromatic ring influence the asymmetric carbon atom environment and have a negative effect on the resolution of the enantiomers. The substituents in the 3-position of the aromatic ring have no significant effect on the enantioseparation. It seems that the type of nitrogen substituent in the hydrophilic part of molecule has the dominant influence on the resolution of the enantiomers. Further, Armstrong et al. [32] studied the chiral resolution of substituted tryptophan compounds on vancomycin, teicoplanin, and ristocetin A CSPs. The authors noted the effect of electron-donating (methyl) and electron-withdrawing (fluoro) groups (on the indole moiety of the tryptophan molecule) on enantioselectivity. The fluoro substituents (relative to the methyl group in the same position) decrease retention and increase enantioselectivity. However, the resolution of the fluorotryptophan enantiomers was somewhat less than those of the corresponding methyl tryptophan.

4.4.6 Structures of Antibiotics

It is clear from Figure 1 that all of the reported antibiotics have different structures with various groups and, hence, all of the antibiotics cannot be used for the chiral resolution of any racemic compound. Therefore, the optimization of chiral resolution on these CSPs may be achieved by selecting the appropriate antibiotic for a particular racemate. There are a number of reports indicating that all of the antibiotics cannot be used for the chiral resolution of all the antipodes [2,9–13,14,17,18,20,28–31]. For example, Aboul-Enein and Ali [14] studied the chiral resolution of tetralone derivatives on vancomycin, teicoplanin, and ristocetin A CSPs. It was observed that the best resolution of these tetralone derivatives was achieved on vancomycin CSP. The retention of reported tetralone derivatives was in the order vancomycin > ristocetin A > teicoplanin columns. It may be due to the different steric effects among the studied tetralone derivatives and CSPs. From these findings, it may be concluded that the strength of the steric effect was in the order teicoplanin > ristocetin A > vancomycin. Therefore, vancomycin and ristocetin A provide a stronger bonding with enantiomers than the bonding provided by teicoplanin. The role of the carbohydrate moieties of teicoplanin on chiral resolution was studied extensively by Armstrong et al. [38]. They have observed that carbohydrate units on teicoplanin are not needed for the chiral resolution of common amino acids. Furthermore, Armstrong et al. [18] observed that the aglycon portion of vancomycin and teicoplanin has a negative effect on the chiral resolution of a number of racemic compounds. On the other hand, sometimes the carbohydrate units were found suitable for the chiral resolution of certain racemates. Briefly, the suitability of the antibiotics CSPs varies from analyte to analyte.

4.4.7 Other HPLC Parameters

Apart from the above-discussed parameters for HPLC optimization of chiral resolution on antibiotic CSPs, some other HPLC conditions may be controlled to improve chiral resolution on these CSPs. The effect of the concentrations of antibiotics (on stationary phase) on enantioresolution varied depending on the type of racemates. The effect of the concentrations of teicoplanin has been studied on the retention (k), enantioselectivity (α), resolution (R_s), and theoretical plate number (N) for five racemates [21]. An increase in the concentration of teicoplanin resulted in an increase of α and R_s values. The most surprising fact is that the theoretical plate number (N) increases with the increase in the concentration of teicoplanin. It may be the result of the resistance of mass transfer resulting from analyte interaction with free silanol and/or the linkage chains (antibiotics linked with silica gel). This would tend to trap an analyte between the silica surface and the bulky chiral selector adhered to it. This is somewhat analogous to the effect of stationary phase adsorption on the efficiency of achiral separation in micellar chromatography [27]. A denser surface coverage of the selector could prevent this deep penetration by steric means. This would tend to limit the analyte interactions to surface interaction with the chiral selector alone, thereby enhancing efficiency and possibly affecting selectivity in some cases. It is apparent that the surface coverage and the orientation of the stationary phase are important for these separations. However, there are few reports in the literature indicating the effect of surface coverage for any chiral stationary phase [45]. The effect of the concentrations of antibiotics on the retention (k), enan-tioselectivity (α), resolution (R_s), and theoretical plate number (N) is given in Table 7. A perusal of Table 7 indicates that 1.5 g was found suitable for good chiral resolution of the reported racemates. In addition, Berthod et al. [51] used the entirely new approach for the chiral resolution of underivatized amino acids. They used copper metal ions as the mobile phase additives. Furthermore, the authors observed a complex formation between copper ions and teicoplanin that resulted in an improved resolution. Wang et al. [52] introduced a new approach to optimize the chiral resolution on these CSPs. They immobilized vancomycin and teicoplanin antibiotics on the same silica gel (mixed CSP) and it was tested for the chiral resolution of profens and β-blockers. The results were compared with those obtained by individual antibiotic CSPs or antibiotic CSPs coupled in series. The authors reported that the mixed CSP was not suitable. However, the coupling of these CSPs in series resulted in an improved resolution in comparison to the results obtained with individual CSPs.

TABLE 7 The Effect of the Teicoplanin Concentration (X), on the Column, on Enantioselectivity in HPLC Using Acetonitrile–Triethylammonium Acetate Buffer of pH 4.1 (10 : 90, v/v) as the Mobile Phase

Racemates	X(g)	k_1	α	R_s	N (m^{-1})
Bromacil	1.5	1.72	1.3	1.7	32,300
	1.0	0.94	1.2	1.5	26,900
Dansyl methionine	1.5	3.6	1.4	2.4	40,680
	1.0	2.10	1.3	1.7	29,300
Mandelic acid	1.5	0.40	2.3	2.3	31,100
	1.0	0.37	1.6	2.0	21,500
5-Methyl-5-phenylhydantoin	1.5	0.78	1.7	2.3	34,500
	1.0	0.46	1.4	1.9	28,600
3-Phenylphthalide	1.5	2.87	1.3	2.3	25,100
	1.0	1.57	1.2	1.8	20,900

Source: Ref. 45.

4.5 CHIRAL RECOGNITION MECHANISM

The macrocylic antibiotics are stable in a 0–100% organic modifier and exhibit high sample capacity. The selectivities of antibiotics in normal, reversed, and new polar organic mobile phase modes make them ideally suitable chiral selectors. The macrocyclic antibiotics stationary phases have similarities and differences with both the cyclodextrin and protein phases from the mechanism point of view. All of the antibiotics contain ionizable groups and their charge and, possibly, their conformation can vary with the pH of the mobile phase. The three-dimensional complex structures and different spatial stereochemical arrangements of the functional groups of the antibiotics containing different chiral centers, inclusion cavities, phenyl rings, pyranose, furanose, quinoline and thiazole rings, several hydrogen donor and acceptor sites, sugar moieties, and other groups are responsible for their surprising chiral selectivities in different modes. This allows for an excellent potential to resolve a greater variety of racemates. The possible interactions involved with the use of antibiotics as chiral selectors for chiral recognition are as follows:

1. $\pi-\pi$ Complexation
2. Hydrogen-bonding
3. Inclusion complexation
4. Dipole interactions
5. Steric interactions
6. Anionic and cationic bindings

The functions of these interactions are well known [2,53,54] and a detailed discussion of them is beyond the scope of this book. However, these interactions take place individually or in combinations that can result in the very high chiral recognition capacities for these antibiotics. The strength of these interactions depends on the type of mobile phase used. The reversed-phase condition favors ionic interactions, hydrophobic inclusion, hydrogen-bonding, and steric interaction. The normal phase favors $\pi-\pi$ complexation, hydrogen bonding, dipole stacking, and steric interaction. On the other hand, the new polar organic phase mode enhances hydrogen-bonding, dipole stacking, $\pi-\pi$ interaction, and steric interaction. Vancomycin, teicoplanin, and ristocetin A are supposed to be the best chiral selectors because of the presence of an aglycon (fused macrocyclic rings) portion which can exhibit different morphological characteristics such as the openness of the aglycon cavity and the degree of helical twist. The twist degree does not seem to depend on the molecular size; in fact, vancomycin, which possesses the smallest macrocyclic ring, has the highest twist degree.

The importance of the amino group of amino acids in the interaction mechanism was evaluated. Native amino acids have better possibility for interaction with teicoplanin; they are more retained on the CSP and better enantioresolved than the blocked amino acids [34]. Recently, Lee and Beesley [48] presented the different interactions, such as $\pi-\pi$, hydrogen-bonding, dipole–dipole, and steric repulsion, between the racemates and the glycopeptide antibiotics CSPs (vancomycin, teicoplanin, and ristocetin A) under aqueous versus organic mobile phases. Finally, the electronegative atoms such as oxygen, nitrogen and halogens of racemates form hydrogen-bondings and dipole–dipole-induced interactions with certain groups in the antibiotics. In addition, $\pi-\pi$ interactions also occur between the phenyl ring of aromatic racemates and the CSP. During the chiral resolution, the enantiomers fit stereogenically in different ways into the chiral cavities of the CSP, which is stabilized by various types of bonding (as discussed earlier) of different magnitudes and, hence, the resolution of enantiomers occur. In addition to these bondings, the steric effect also governs the chiral resolution on antibiotic CSPs. In addition, some other achiral weak bondings like van der Waal forces and ionic bondings may also contribute in the chiral resolution. A graphical representation of chiral recognition mechanism of phenylalanine amino acid on teicoplanin CSP is presented in Figure 11. It may be observed from this figure that the D-enantiomer of phenylalanine is more strongly bonded to CSP (because of lower steric hindrance) than the L-enantiomer and, hence, the L-enantiomer is eluted first, followed by the D-enantiomer.

4.6 MISCELLANEOUS APPLICATIONS

The use of antibiotic-based CSPs in supercritical fluid chromatography (SFC) is under development stage and are not extensively used in sub-SFC and SFC modes of chromatography. Only a few reports are available on chiral resolution on these CSPs by SFC and only vancomycin and teicoplanin were used for this purpose. Medvedovici et al. [55] resolved some racemates on vancomycin and teicoplanin CSPs, covalently bonded to silca gel. A high concentration of organic modifiers was essential in SFC. However, a complex mixture of methanol, water and glycerol was used for chiral separation on teicoplanin CSP. The separation of coumachlor enantiomers on vancomycin CSP was achieved by Phinney [56]. The chromatograms are shown in Figure 12. Donneck et al. [57] used vancomycin, immobilized on aldehyde functionalized silica support and packed capillary, for the chiral resolution of basic racemates (β-blockers) [57]. Various modifiers were tested, but methanol as the modifier resulted in successful results. The effect of temperature on the chiral resolution was also studied and resulted in nonlinear van't Hoff plots. Large differences in enantioselectivity in a homologous series of local anesthetic compounds indicated that the fit of a host–guest interaction between analyte and vancomycin could be very precise. Recently, vancomycin was used to resolve the enantiomers of anticoagulants, aryloxypropionic acid and arylpropionic acid, using SFC [58]. Duret et al. [59] used vancomycin as the

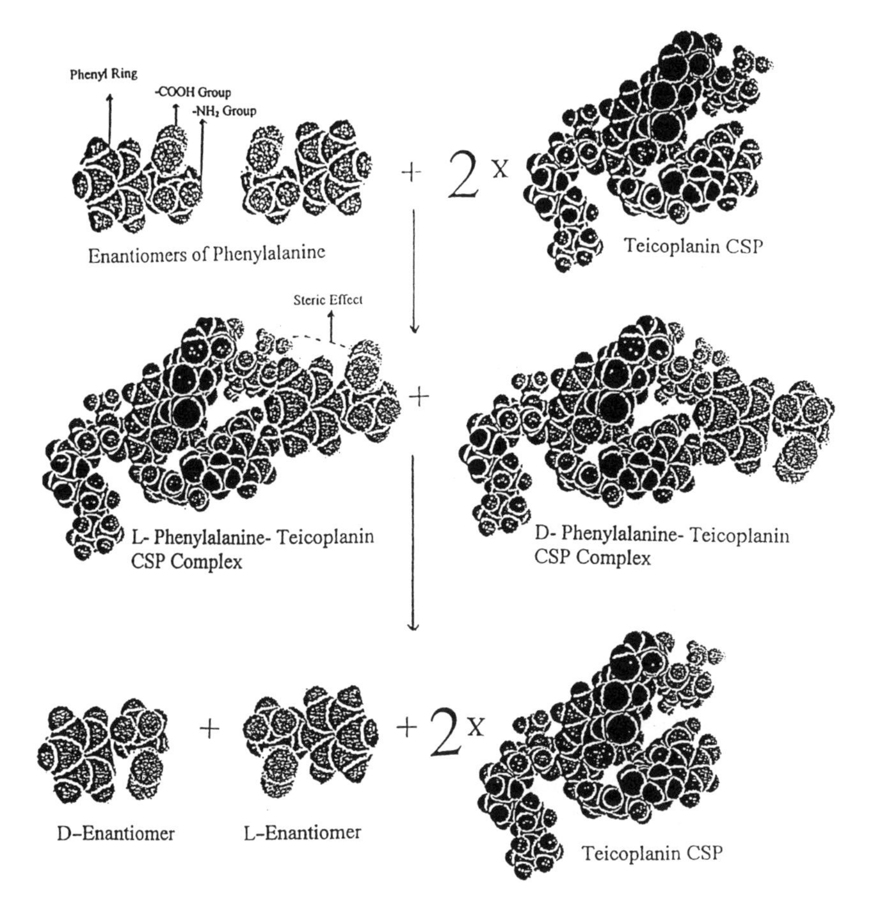

FIGURE 11 Graphical representation of chiral recognition mechanism of phenylalanine on teicoplanin CSP.

chiral selector for the enantiomeric resolution of dansyl norleucine by centrifugal partition chromatography. The authors reported exceptionally high values of selectivity factors ($\alpha > 30$).

The use of antibiotic-based CSPs has been reported in capillary electrochromatography (CEC) for chiral resolution [60]. Teicoplanin CSP covalently bonded to silica gel was used to resolve the enantiomers of tryptophan and dinitrobenzoyl leucine by CEC [61]. Good levels of enantioselectivity were obtained with optimized separations. Vancomycin covalently bonded to silica gel was also evaluated in CEC for the chiral resolution of thalidomide and β-adrenergic blocking agents under all the three mobile phase modes. The

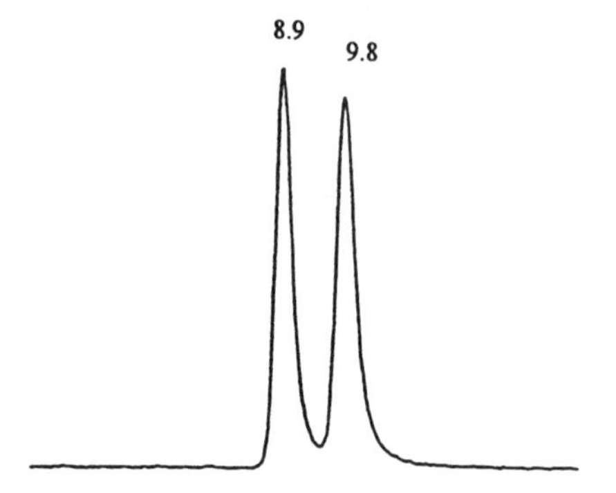

FIGURE 12 Chromatograms of the chiral resolution of coumachlor on vancomycin CSP using 15% methanol in carbon dioxide under the SFC mode. (From Ref. 56.)

selectivities of the CSP were found to be very good. Different ratios of methanol and acetonitrile played a significant role in controlling both the resolution and efficiency [62]. Enantiomeric resolution of alprenolol, metoprolol, and coumachlor was reported on teicoplanin CSP under reversed and new polar organic modified phases [63]. A statical experimental design was used to investigate the effects of nonaqueous polar organic mobile phase parameters on the CEC electroosmotic flow, resolution, and peak efficiency. Results indicated that higher efficiency and resolution values could be attained at higher methanol contents, which is similar to findings obtained on this CSP in HPLC. Recently, the enantiomers of venlafaxine and its main active metabolite *O*-desmethylvenlafaxine were resolved by CEC on vancomycin CSP [64]. The mobile phase used was 100 mM ammonium acetate buffer (pH 6)–water–acetonitrile (5 : 5 : 90, v/v/v). The acetonitrile concentration was found to modulate the elution time, efficiency, and selectivity.

Bhushan and Parsad [65] resolved dansyl amino acids on erythromycin impregnated thin-layer chromatographic (TLC) silica plates. The mobile phase used was different ratios of 0.5 M aqueous NaCl–acetonitrile–methanol. Further, Bhushan and Thiong'o [66] achieved the chiral resolution of dansyl amino acids on silica TLC plates impregnated with vancomycin chiral selector. The mobile phase used for this study was acetonitrile–0.5 M aqueous NaCl (10 : 4 and 14 : 3, v/v). The chiral recognition mechanisms of antibiotic CSPs in sub-SFC, SFC, CEC, and TLC modes of chromatography were found to be similar to HPLC.

4.7 CONCLUSION

The development of the macrocyclic antibiotics as chiral selectors has resulted in an inexpensive, easy, reproducible and fast enantiomeric resolution method for a wide variety of racemates both for analytical and preparative scales. There is a good potential to use these antibiotics for a large-scale preparative resolution of drug racemates that can be resolved to provide the desired active single enantiomer in a pure form. In addition, these CSPs are multimodal in nature and can be switched from one mobile phase to another without any deleterious effect, which makes column coupling an efficient and economical methodology. By using the method development procedures and optimizing chromatographic parameters, these CSPs can be used to simulate further investigations and understanding of the chiral recognition for a large number of racemates. Research work carried out on these CSPs and discussed herein indicated no serious limitation of these antibiotics. However, the high ultraviolet background of these macrocyclic antibiotics could prevent operation at low wavelengths and generally reduces sensitivity compared to other chiral selectors. We believe that in the near future, antibiotic-based CSPs will become the method of choice for the chiral resolution of a wide variety of racemic compounds.

REFERENCES

1. Armstrong DW, Tang Y, Chen S, Zhou Y, Bagwill C, Chen R, Anal Chem 66: 1473 (1994).
2. Chirobiotic Handbook, Guide to Using Macrocyclic Glycopeptide Bonded Phases for Chiral LC Separations, 2nd ed., Advance Separation Tech. Inc., Whippany, NJ, (1999).
3. Aboul-Enein HY, Ali I, Chromatographia 52: 679 (2000).
4. Maier NM, Franco P, Linder W, J Chromatogr A 906: 3 (2001).
5. Armstrong DW, Rundlett KL, Chen JR, Chirality 6: 496 (1994).
6. Ward TJ, LC-GC Int 14: 428 (1996).
7. Gasper MP, Berthod A, Nair UB, Armstrong DW, Anal Chem 68: 2501 (1996).
8. Ekborg-Ott KH, Kullman JP, Wang X, Gahm K, He L, Armstrong DW, Chirality 10: 627 (1998).
9. Beesley TE, Scott RPW (Eds.), Chiral Chromatography, John Wiley & Sons, New York (1998).
10. Gasparrini F, D'Acquarica I, Villani, C, Misiti D, Pierini M, 13th International Symposium on Chirality 2001, Orlando, FL, Abstract P-234: 65 (2001).
11. D'Acquarica I, Gaspirrini F, Misiti D, Villani C, Carotti A, Cellamare S, Muck S, J Chromatogr A 857: 145 (1999).
12. Berthod A, Yu T, Kullman JP, Armstrong DW, Gasparrini F, D'Acquarica I, Misiti D, Carotti A, J Chromatogr A 897: 113 (2000).
13. D'Acquarica I, Gasparrini F, Misiti D, Zappia G, Cimarelli C, Palmieri G, Carotti A, Cellamare S, Villani C, Tetrahedron: Asymm 11: 2375 (2000).
14. Aboul-Enein HY, Ali I, Arch Pharm 334: 258 (2001).
15. Vespalec R, Billiet HAH, Frank J, Bocek P, Electrophoresis 17: 1214 (1996).
16. Hrobonova K, Lehotay J, Cizmarikova R, Armstrong DW, J Liq Chromatogr Relat Technol 24: 2225 (2001).
17. Aboul-Enein HY, Serinese V, Chirality 10: 358 (1998).

18. Xiao TL, Jenks W, Armstrong DW, 13th International Symposium on Chirality 2001, Orlando, FL, Abstract P-243, p. 68 (2001).
19. Kleidernigg OP, Kappe CO, Tetrahedron: Asymmetry 8: 2057 (1997).
20. Chen S, Liu Y, Armstrong DW, Borrell JI, Martinez-Teipel B, Matallana JL, J Liq Chromatogr 18: 1495 (1995).
21. Armstrong DW, Liu Y, Ekborg-Ott K, Chirality 7: 474 (1995).
22. Peter A, Torok G, Armstrong DW, J Chromatogr A 793: 283 (1998).
23. Berthod A, Liu Y, Bagwill C, Armstrong DW, J Chromatogr A 731: 123 (1996).
24. Fried KM, Koch P, Wainer IW, Chirality 10:484 (1998).
25. Jndera P, Skavrada M, Klemmova K, Backovska V, Guiochon G, J Chromatogr A 917: 123 (2001).
26. Desiderio D, Fanali S, J Chromatogr A 807: 37 (1998).
27. Armstrong DW, Ward TJ, Berthord A, Anal Chem 58: 579 (1986).
28. Hroboova K, Lehotay J, Iungelova J, Eimarik J, Armstrong DW, 13th International Symposium on Chirality 2001, Orlando, FL, Abstract P-128, p. 38 (2001).
29. Lehotay J, Hrobonova K, Krupcik J, Cizmarik J, Pharmazie 53: 863 (1998).
30. Tesaova E, Zaruba K, Flieger M, J Chromatogr A 844: 137 (1999).
31. Zhang B, Xiao L, Lee JT, Hui F, Armstrong DW, 13th International Symposium on Chirality 2001, Orlando, FL, Abstract P-147, p. 44 (2001).
32. Hui F, Ekborg-Ott KH, Armstrong DW, J Chromatogr A 906: 91 (2001).
33. Jndera P, Backovska V, Felinger A, J Chromatogr A 919: 67 (2001).
34. Tesaova E, Bosakova Z, Pacakova V, J Chromatogr A 839: 121 (1999).
35. Peter A, Olajos E, Casimir R, Tourwe D, Broxterman QB, Armstrong DW, J Chromatogr A 871: 105 (2000).
36. Aboul-Enein HY, Serinese V, J Liq Chromatogr Relat Technol 22: 2177 (1999).
37. Aboul-Enein HY, Serinese V, Biomed Chromatogr 13: 520 (1999).
38. Berthod A, Chen X, Kullman JP, Armstrong DW, Gasparrini F, DiAcquarica, Villani C, Anal Chem 72: 1767 (2000).
39. Schlauch M, Frahm AW, J Chromatogr A 868: 197 (2000).
40. Tesova E, Bosakova Z, Zuskova I, J Chromatogr A 879: 147 (2000).
41. Peyrin E, Ravelet C, Nicolle E, Villet A, Grosset C, Ravel A, Alary J, J Chromatogr A 923: 37 (2001).
42. Ekborg-Ott, KH, Liu Y, Armstrong DW, Chirality 10: 434 (1998).
43. Ekborg-Ott KH, Wang, X, Armstrong DW, Microchemical J 62: 26 (1999).
44. Peter A, Torok G, Armstrong DW, Toth G, Tourwe D, J Chromatogr A 904: 1 (2000).
45. Wu, DR, Lohse K, Beesley TE, 23rd International Symposium on HPLC (HPLC 99), Canada (1999).
46. Francotte ER, J Chromatogr A 906: 379 (2001).
47. Aboul-Enein HY, Ali I, IL Farmaco 57: 513 (2002).
48. Lee JT, Beesley TE, 13th International Symposium on Chirality 2001, Orlando, FL, Abstract P-225, p. 62 (2001).
49. Peter A, Torok G, Armstrong DW, Toth G, Tourwe D, J Chromatogr A 828: 177 (1998).
50. Ananieva IA, Davankov VA, Armstrong DW, 13th International Symposium on Chirality 2001, Orlando, FL, Abstract P-257, p. 72 (2001).

51. Berthod A, Tizon, V, Leonce E, Caussignac C, Valleix A, Armstrong DW, 13th International Symposium on Chirality 2001, Orlando, FL, Abstract P-203, p. 53 (2001).
52. Wang AX, Lee JT, Beesley TE, LC-GC 18: 629 (2000).
53. Allenmark S., Chromatographic Enantioseparation: Methods and Applications, 2nd Ed., Ellis Horwood, New York (1991).
54. Yashima, E, Okamoto Y, Chiral recognition mechanisms of polysaccharides chiral stationary phases, in The Impact of Stereochemistry on Drug development and Use, Aboul-Enein HY, Wainer IW (Eds.), John Wiley & Sons, New York (1997).
55. Medvedovici A, Sandra P, Toribio L, David F, J Chromatogr A 785: 159 (1997).
56. Phinney KW, Sub- and supercritical fluid chromatography for enantiomer separations, in Chiral Separation Techniques: A Practical Approach, Subramanian G (Ed.), Wiley–VCH, Weinheim, Germany, p. 299 (2001).
57. Donneck J, Svensson LA, Gyllenhaal O, Karlsson KE, Karlsson A, Vessman J, J Microcolumn Sep 11: 521 (1999).
58. Selditz U, Copinga S, Grol CJ, Frank JP, De Zeeuw RA, Wikstroem H, Gyllenhall O, Pharmazie 54: 183 (1999).
59. Duret P, Foucault A, Margraff R, J Liq Chromatogr & Relat Technol 23: 295 (2000).
60. Dermaux A, Lyen F, Sandra P, J High Resolut Chromatogr 21: 575 (1998).
61. Caster-Finch AS, Smith NW, J Chromatogr A 848: 375 (1999).
62. Wikstrom H, Svensson LA, Torstensson A, Owens PK, J Chromatogr A 869: 395 (2000).
63. Karlsson C, Wikstrom H, Armstrong DW, Owens PK, J Chromatogr A 897: 349 (2000).
64. Fanali S, Rudaz S, Veuthey JL, Desiderio C, J Chromatogr A 919: 195 (2001).
65. Bhushan R, Parshad V, J Chromatogr A 736: 235 (1996).
66. Bhushan R, Thiong'o GT, J Planar Chromatogr Mod TLC 13: 33 (2000).

5

Pirkle-Type and Related Chiral Stationary Phases

Generally, all of the chiral stationary phases (CSPs) have some specific structures such as polysaccharides, proteins are polymeric, cyclodextrins and crown ethers have cyclic features, antibiotic-based CSPs contain chiral cavities, and ligand-exchange phases comprise metal ions coordinated with suitable ligand exchangers. However, in 1976, Mikeš et al. [1] introduced a new concept by attaching a small chiral molecule to silica gel. In this CSP, the organic groups of the chiral molecule remain directed away from the silica gel, appearing in the form of a brush; hence, it is called a brush-type phase. Later, Pirkle and co-workers developed these types of CSP extensively and now these CSPs are popularly called Pirkle-type CSPs [2–12]. Normally, the chiral molecule attached to the silica gel contains a π-electron donor or a π-electron receptor or both types of group. Therefore, these CSPs are classified into three groups [i.e., π-acidic (with π-electron-acceptor groups), π-basic (with π-electron-donor groups), and π-acidic–basic (with π-electron-acceptor and -donor groups)]. The reciprocality concept put forth by Pirkle allowed several generations of these types of CSP [4,7]. The main advantage of these types of phase is having the chiral molecule attached to the silica gel. A specific and required chiral molecule (to attach onto silica gel) can be selected by the reciprocality concept and bonded to the silica gel; hence, the chiral resolution of a wide variety of racemic compounds can be obtained easily and successfully. Recently, some chiral molecules having specific groups, other than π donor or π acceptors, such as polar and polarizable groups have been grafted onto the silica gel surface. These types of CSP were found to have great potential for the chiral resolution of different racemic compounds. The high stability of these phases at high pressure in normal phase or reversed-phase modes make them ideally suitable for the chiral resolution of large number of racemic compounds. It is also very interesting to note that these phases have been used successfully for the chiral resolution at analytical and preparative scales. In view of this, the present chapter describes the art of the chiral resolution by liquid chromatography using Pirkle-type CSPs.

5.1 STRUCTURES AND PROPERTIES

Basically, these CSPs contain two parts (i.e., silica gel and the chiral molecule attached to silica gel). A general structure outlining these CSPs is shown in Figure 1. Normally, the chiral molecules are bonded to silica gel through ionic or covalent bonds. The phenyl groups of the chiral molecules are supposed to have the tendency to donate the π-electrons, whereas the phenyl groups containing electronegative groups/atoms are π-electron deficient and have the tendency to accept π-electrons.

The first ionically bonded phase was presented by Pirkle; it contained (*R*)-3,5-dinitrobenzoyl phenylglycine [13]. The most commonly used π-acid moiety is the 3,5-dinitrophenyl group introduced by the reaction of 3,5-dinitrobenzoyl chloride (DNB-Cl) on chiral selectors such as amino acids, amino alcohols, and amines. In addition, pentafluorobenzoyl derivatives have also been reported [14,15]. The π-basic phases are complimentary to the π-acidic phases. These CSPs include the presence of phenyl- or alkyl-substituted phenyl groups. Macaudiere et al. [9] designed a CSP containing both π-acidic and π-basic

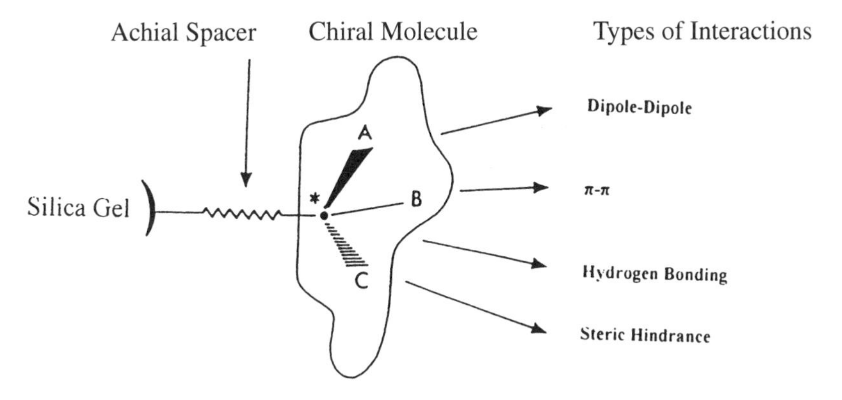

FIGURE 1 Graphical representation of a Pirkle-type CSP.

groups on the same chiral molecule. The main precursors of these CSPs are amino acids, amino alcohols, and amines. However, other chiral molecules have also been used as the precursors for these CSPs.

Normally, these CSPs bear π-electron-donating or π-electron-accepting groups closer to the chiral center and, therefore, $\pi-\pi$ interactions with the complimentary groups of the racemic molecules resulted into the charge-transfer complexes, which is an essential feature of the chiral resolution on these CSPs. These CSPs are very effective for the chiral resolution of racemic compounds containing both π-electron-donating and π-electron-accepting groups. The structures of some of these phases are shown in Figure 2.

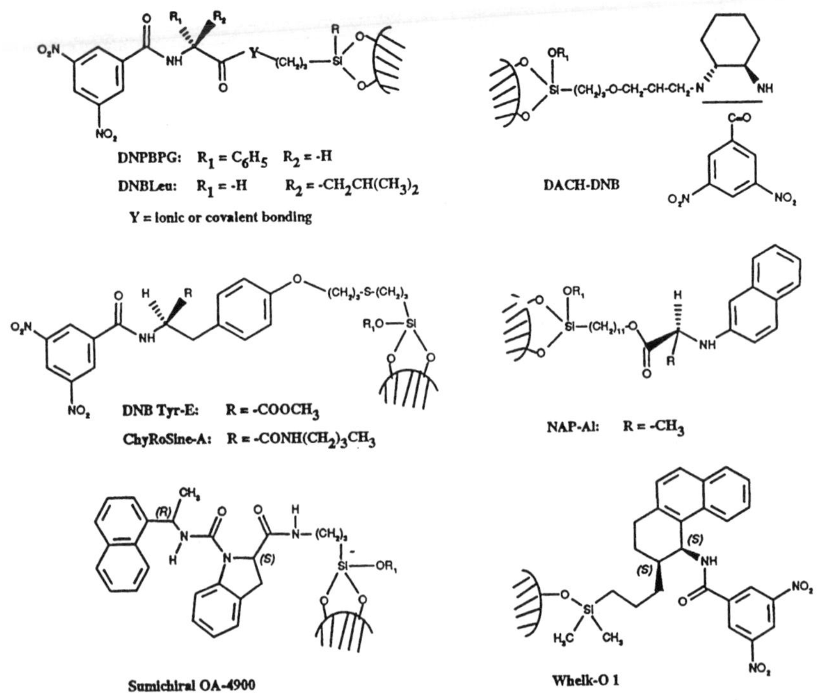

DNPBPG: $R_1 = C_6H_5$ $R_2 = -H$

DNBLeu: $R_1 = -H$ $R_2 = -CH_2CH(CH_3)_2$

Y = Ionic or covalent bonding

DACH-DNB

DNB Tyr-E: $R = -COOCH_3$

ChyRoSine-A: $R = -CONH(CH_2)_3CH_3$

NAP-Al: $R = -CH_3$

Sumichiral OA-4900

Whelk-O 1

DNB: Dinitrobenzyl, DNBPG: Dinitrobenzylphenylglycine and DACH: Diaminocyclohexane

FIGURE 2 The chemical structures of some Pirkle-type CSPs.

5.2 PREPARATION AND COMMERCIALIZATION

Pirkle phases contain small chiral molecules and, hence, are easy to prepare in comparison to other phases. These CSPs are prepared either by ionic or covalent bondings of the chiral molecule to the silica gel (normally of 5 μm particle size). The preparation of these CSPs containing ionic bonding involves a three-step procedure. The silica gel is allowed to react with some suitable silyl reagent such as 3-aminopropyldimethoxyethoxysilane and the product obtained is 3-aminopropyl silica gel. In the next step (step II), the chiral molecule [e.g., (*R*)-phenylglycine] is treated with 3,5-dinitrobenzoyl chloride to produce the benzoyl derivative, which was then allowed to react with the 3-aminopropyl group that was linked to silica gel via one or more siloxyl bonds. A solution of (*R*)-3,5-dinitrobenzyol phenylglycine in tetrahydrofuran (THF) is allowed to react with a suspension of the 3-aminopropyl silica gel and the mixture is stirred well (step III), simply producing the ionically bonded stationary phase. These types of phases are quite effective in the normal phase mode, but they can be somewhat labile in the aqueous phase; therefore, attempts were made to bind the chiral molecule to silica gel by covalent bonds.

In covalent bonding, silica gel is derivatized with a suitable silyl reagent. There are numerous silyl reagents available that contain aminopropyl, hydroxypropyl, and isocyanatopropyl groups which can be bonded to silica gel. The chiral molecule is also derivatized with some suitable derivatizing reagent. Finally, the derivatized chiral molecule is allowed to bind onto the surface of derivatized silica gel. For example, the synthetic steps of the (*R*)-3,5-dinitrobenzoyl phenylglycine CSP are shown in Figure 3. It is a three-step process including the derivatization of silica gel with 3-aminopropyltriethoxysilane to produce 3-aminopropyl silica gel (step I). In step II, (*R*)-phenylglycine is derivatized with 3,5-dinitrobenzoyl chloride. In the third step, the derivatized phenylglycine is covalently bonded to the surface of 3-aminopropyl silica gel. Contrary to the preparation of ionic bonding phases, the mixture of (*R*)-3,5-dinitrobenzoyl phenylglycine and 3-aminopropyl silica gel is refluxed for 2–3 hr in toluene. These types of CSP are quite stable in the reversed-phase mode even at extreme values of pH and, hence, have been used successfully for the chiral resolution of a wide variety of racemic compounds using different types of mobile phase. With the development of the research in this area, different types of chiral molecule have been grafted onto silica gel, and the new CSPs were developed and commercialized by different trade names such as Whelk-O1, Ulmo, Kromasil Chiral TBB, Kromasil Chiral DMB, Chiris, and so forth. The commercialized CSPs are summarized in Table 1.

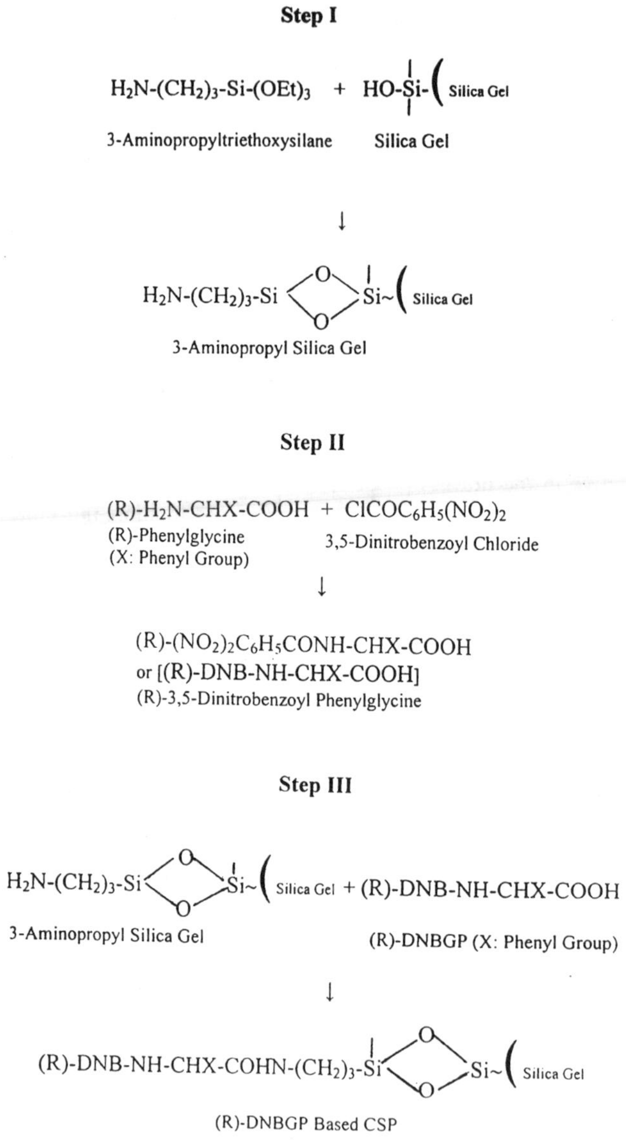

Step I

$H_2N\text{-}(CH_2)_3\text{-}Si\text{-}(OEt)_3$ + HO-Si-(Silica Gel

3-Aminopropyltriethoxysilane Silica Gel

↓

$H_2N\text{-}(CH_2)_3\text{-}Si$ O O Si~(Silica Gel

3-Aminopropyl Silica Gel

Step II

(R)-H_2N-CHX-COOH + $ClCOC_6H_5(NO_2)_2$

(R)-Phenylglycine 3,5-Dinitrobenzoyl Chloride
(X: Phenyl Group)

↓

(R)-$(NO_2)_2C_6H_5$CONH-CHX-COOH
or [(R)-DNB-NH-CHX-COOH]
(R)-3,5-Dinitrobenzoyl Phenylglycine

Step III

$H_2N\text{-}(CH_2)_3\text{-}Si$ O O Si~(Silica Gel + (R)-DNB-NH-CHX-COOH

3-Aminopropyl Silica Gel (R)-DNBGP (X: Phenyl Group)

↓

(R)-DNB-NH-CHX-COHN-$(CH_2)_3$-Si O O Si~(Silica Gel

(R)-DNBGP Based CSP

FIGURE 3 The chemical pathway for the preparation of 3,5-dinitrobenzoyl phenylglycine CSP (Pirkle type).

TABLE 1 Various Commercial Pirkle-Type CSPs

Commercialized CSPs	Companies
Opticrown Chiralhyun-Leu-1 and Opticrown Chiralhyun-PG-1	Usmac Corp., Glenview, IL, USA
Whelk-O1, Whelk-O2, Leucine, Phenylglycine, β-Gem 1, α-Burke 1, α-Burke 2, Pirkle 1-J, Naphthylleucine, Ulmo, and Dach	Regis Technologies, Austin, TX, USA
Nucleosil Chiral-2	Macherey-Nagel, Duren, Germany
Sumichiral OA	Sumika Chemical Analysis Service, Konohana-ku Osaka, Japan
Kromasil Chiral TBB and Kromasil Chiral DMB	Eka Chemicals Separation Products, Bohus, Sweden
Chirex Type I	Phenomenex, Torrance, CA, USA
Chiris series	IRIS Technologies, Lawrence, KS, USA

5.3 APPLICATIONS

The Pirkle-type chiral stationary phases are quite stable and exhibit good chiral selectivities to a wide range of solute types. These CSPs are also popular for the separation of many drug enantiomers and for amino acid analysis. Primarily, direct chiral resolution of racemic compounds were achieved on these CSPs. However, in some cases, prederivatization of racemic compounds with achiral reagents is required. The applications of these phases are discussed considering π-acidic, π-basic, and π-acidic–basic types of CSP. These CSPs have also been found effective for the chiral resolution on a preparative scale. Generally, the normal phase mode was used for the chiral resolution on these phases. However, with the development of new and more stable phases, the reversed phase mode became popular.

5.3.1 Analytical Separations

The applications of π-acidic chiral stationary phases include the resolution of α-blockers and β-blockers, amines, arylacetamine, alkylcarbinols, hydantoins, barbiturates, naphthols, benzodiazapines, carboxylic acids, lactams, lactones, phthaldehydes selenoids, and phosphorus compounds. Hyun et al. [16] achieved a chiral resolution of a homologous series of N-acyl-α-(1-naphthyl)ethylamine on N-(3,5-dinitrobenzoyl-(R)-phenylglycine and N-(3,5-dinitrobenzoyl)-(S)-leucine CSPs. The authors used hexane-2-propanol (80 : 20, v/v) as the mobile phase. Similarly, the scope of π-basic CSPs comprises the chiral resolution of β-blockers, amino acids, amines, diamines, amino phosphonates, naphthols, benzadiazapines, carboxylic acids, hydroxy acids, dipeptides, tripeptides, diols, glycerols, lactones, lactams, sulfoxides, and thiols. Uray et al. [17] studied the chiral recognition behavior of 3,5-dinitrobenzoylated (DNB) diphenylalkane-amine-derived CSPs and their four racemic precursors using the normal phase mode. In 13 cases, successful enantioseparation of these π-electron-acceptor analytes was achieved on the π-acceptor CSPs. The authors reported that the stereochemistries of the DNB-modified stereogenic centers of the chiral selector and the higher retained enantiomer of the corresponding precursor were identical.

Currently, π-acidic–basic-type CSPs are very effective and common in use, as they have a wide applicability. The most common applications of these CSPs include the resolution of α-blockers, β-blockers, amino acids, amines, amino phosphonates, arylacetamine, alkylcarbinols, hydantoins, barbiturates, naphthols, benzodiazapines, carboxylic acids, hydroxy acids, dipeptides, tripeptides, lactams, lactones, phthaldehydes, phosphorus compounds, diols, glycerols, sulfoxides, thiols, and selenoids. Imai et al. [18] derivatized phenylalanine and leucine amino acids with 4-fluoro-7-nitro-2,1,3,-benzoxadiazole (afluorogenic reagent) and separated them on a Sumichiral OA 2500 (S) column using 20 mM ammonium acetate in methanol as the mobile phase. In other studies, the same authors [19–21] derivatized amino acids with different fluorogenic reagents and, again, achieved the chiral resolution of these derivatives on a series of Sumichiral OA CSPs using different mobile phases. Siddiqui et al. [22] used the phenylglycine-type CSP to access the enantiomeric purity of different intermediates in the synthetic route of (1′*R*, 2′*S*, 5′*R*)-menthyl-(5*R*)-acetoxy-1,3-oxathiolan-(2*R*)-carboxylates [22]. Caccamese et al. [23] studied the chiral resolution of 17 α-hydroxybenzoylphosphonate diethylesters on Whelk-O1 CSP using hexane-2-propanol (70 : 30, v/v) as the mobile phase. Quaglia et al. [24] obtained the enantioseparation of 3-benzoylchroman-4-ones, the potent antiviral agents, on Whelk-O1 CSP with different ratios of hexane and 2-propanol as the mobile phases. In another study, the same group [25] used (*R*,*R*)-Whelk-O1 CSP for the chiral resolution of 2-cyclopentylthio-6-[1-(2,6-difluorophenyl)ethyl]-3,4-dihydro-5-methylpyrimidin-4(3*H*)-one, a potent inhibitor of human immunodeficiency virus (HIV)-1. The mobile phase used was hexane-2-propanol (90 : 10, v/v). Baeyens et al. [26] compared the chiral resolution of 2-arylpropionic acids on the Whelk-O1 column using a variety of normal mobile phases. Uray and Kosjek [27] obtained the chiral resolution of 4-(2′, 4′-difluorobiphenyl-4-yl)-4-oxo-2-methylbutanoic acid (flobufen) on Whelk-O1 and Ulmo CSPs using different ratios of hexane, 2-propanol, and trifluoroacetic acid. The authors reported the best resolution on Whelk-O1 CSP. Again, the same authors [28] prepared and evaluated the several brush-type CSPs based on undecanoyl- or butanoyl-bound (*R*,*R*)-1,2-diphenylethane-1,2-diamine as chiral selectors. A benzoylated analog of the commercially available Ulmo CSP was shown to be very effective in separating the enantiomers of *N*-acyl amino acids. In this way, the CSPs containing both π-acidic and π-basic moieties proved very effective and successful for the chiral resolution of different groups of racemic compounds. The chromatograms of the chiral resolution on some of the commonly used CSPs are shown in Figure 4. The chiral resolution on these CSPs is summarized in Table 2.

5.3.2 Preparative Separations

The Pirkle-type CSPs are also effective for the chiral resolution of racemic compounds at the preparative scale. Some reports have been published on the chiral resolution using these phases at the preparative scale [53,70,94,131,132]. The Whelk-O1 CSP has been found quite good for preparative purposes. Even using a conventional analytical column (4.6-mm inner diameter×25-cm length), small samples of a racemate can often be resolved. In one study, 12.6 mg of naproxen was resolved within 20 min on the Whelk-O1 analytical column [131]. In 1994, Möller et al. [133] developed new chiral stationary phases for preparative purposes. The chiral selectors were *N,N′*-diallyltartardiamide derivatives and a multifunctional hydrosilanes. The authors reported the separation of various racemic compounds at milligram levels. Jageland et al. [134] presented the methodology to improve the chiral resolution at the preparative scale using Kromasil chiral columns. Pirkle-1J has been reported as a useful CSP for the chiral resolution at the semipreparative scale [127]. The production of large quantities of the enantiomers has been claimed for Chirex columns [125]. Eka Chemical (Marietta, GA, USA) has also commercialized Kromasil columns for preparative chiral separations [126]. α-Burke 1 CSP was reported to be useful for the chiral resolution of various racemic compounds at the preparative level [127]. Sumichiral columns were claimed as useful CSPs for preparative purposes [135]. Recently, Pirkle and Koscho [136] reported the preparative chiral separation of various *N*-acetyl proline anilides on Pirkle-1J and Burke 2 CSPs.

5.4 OPTIMIZATION OF HIGH-PERFORMANCE LIQUID CHROMATOGRAPHIC CONDITIONS

The chiral resolutions on π-acidic and π-basic CSPs were carried out under the normal phase mode. However, some reports are also available dealing with the use of reversed-phase eluents, but the prolonged use of the reversed-phase mobile phase is not recommended. With the development of the more stable and new CSPs, the use of the reversed-phase mode came into existence on these CSPs. Currently, both modes of mobile phases (i.e., normal and reversed) are in use. Therefore, the optimization of the chiral resolution on these phases can be achieved by varying the concentration of the mobile phases, including the use of organic modifiers. In addition, the temperature, structures of solutes, and CSPs are also important parameters that control the chiral resolution on these CSPs.

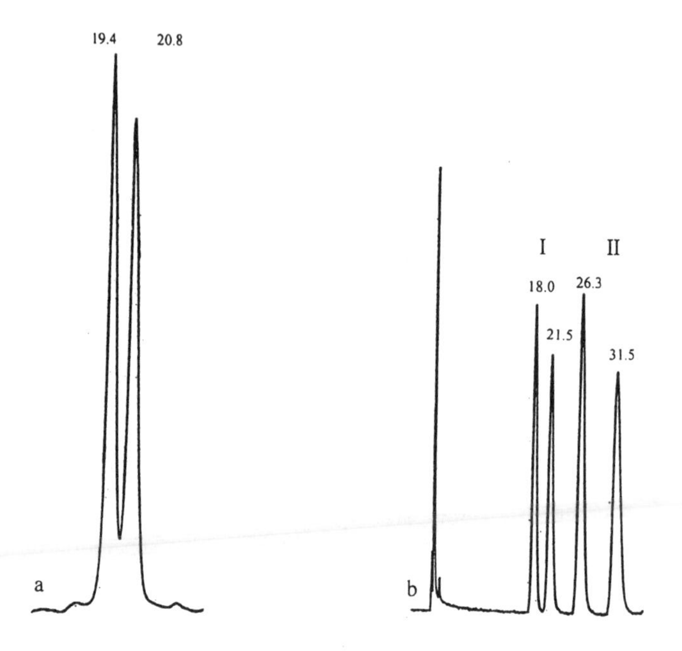

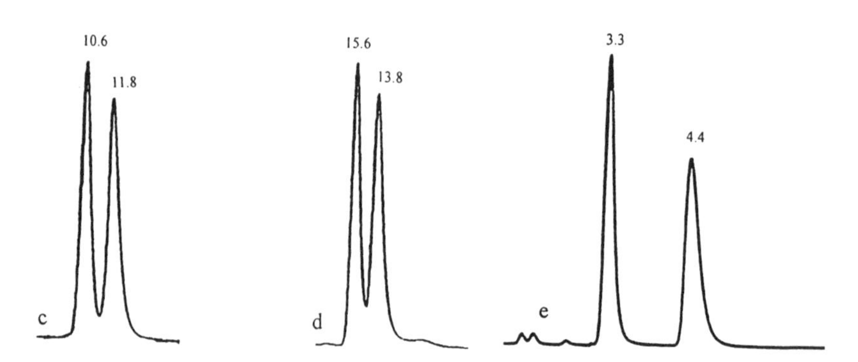

FIGURE 4 Chromatograms of (a) naphthyl derivatives of 1-phenyl-2-aminopropane on 3,5-dinitrobenzoyl phenylglycine CSP [29], (b) 4-fluoro-7-nitro-2,1,3-benzoxadiazole derivatives of leucine (I) and penylalanine (II) on Sumichiral OA-2500(S) [18], (c) 3-(4-chlorobenzyl)-chroman-4-one on Whelk-O1 [24], (d) 6-chloro-3-(4-chlorobenzyl)chroman-4-one on Whelk-O1 [24], and (e) benzylleucine on Ulmo column [28].

TABLE 2 Enantiomeric Resolution of Some Racemic Compounds on Pirkle-Type CSPs

Racemic compounds	CSPs	Refs.
π-Acidic CSPs		
α- and β-Blockers	DNBGP	13, 30–35
Amines	DNBGP	5–7, 29, 32, 36–51
	DNBLeu	7, 29, 48–51
Arylacetamide	DNBGP	13, 30
Alcohols	DNBGP	4, 13, 14, 30, 32, 52–66
	DNBLeu	51–66
Hydantoins	DNBGP, DNBLeu	13, 30, 50, 52, 53, 67, 68
2-Carboalkoxyindolines	DNBGP	50
Barbiturates	DNBGP, DNBLeu	68
Naphthols	DNBGP	13, 30, 32, 37, 52, 68, 69
Benzodiazepinones	DNBLeu	32, 37, 70, 71
Carboxylic acids	DNBGP	43, 44, 52, 72–85
Imides	DNBGP, DNBLeu	13, 30, 52, 67
Lactams	DNBGP	13, 30, 52, 86
	DNBLeu	87
Aryllactones	DNBGP, DNBLeu	88, 89
Phthalides	DNBGP, DNBLeu	13, 52, 53, 71, 88
Phosphorus compounds	DNBGP	13, 30, 90–95
	DNBLeu	91–95
Selenoxides	DNBGP	96, 97
Sulfoxides	DNBGP, DNBLeu	13, 30, 53, 88, 98
Dipeptides	DNBGP	99
Rotenoids	DNBGP	100, 101
π-Basic CSPs		
Alcohols	1*R*,3*R*-*trans* Chrysanthemic 4 acid chloride	4
	N-Hydroxysuccinamide esters of *N*-(1*R*,3*R*)-*trans*-Chrysanthemoyl-D-phenylglycine	102
	1-(α-Naphthyl)ethylamine and 2-(4-chlorophenyl)-isovaleric acid	103
	(*R*)-*N*-(2-Naphthyl)alanine and (*R*)-*N*-acylated α-aryl-α-aminoalkane	104
	N-(*S*)-2-(4-Chlorophenyl) isovaleroyl-D-phenylglycine	105, 106

(continued)

TABLE 2 Continued

Racemic compounds	CSPs	Refs.
	N-3,5-Dinitrobenzamide of (*S*)-leucine and (*R*)-phenylglycine	107
Amino acids	Amino acids, amines, phosphine, oxide, alkaloids based	108–118
Phenylthiohydantoin amino acids	Supelcosil LC(R)–urea	119
Amines	Dinitrobenzoic acid based	108
	Amino acids and amines based	47, 107, 109–111
Aminophosphonates	Amino acid based	107, 111
Naphthols	Alkaloid based	120
Carboxylic acids	Amino acids and amine based	87, 102, 111
Hydroxy acids	1*R*,3*R*,-*trans*-Chrysanthemic acid chloride	4
Lactones and lactams	1*R*,3*R*,-*trans*-Chrysanthemic acid chloride	4
	Amino acid based	87
Sulfoxides	1*R*,3*R*,-*trans*-Chrysanthemic acid chloride	4
Thiols	1*R*,3*R*,-*trans*-Chrysanthemic acid chloride	4
π-Acidic–Basic CSPs		
Alcohols	Sumipax	121
	Pirkle urea	122
	Chiris series	123
	Chirex	124
	Chirex	125
	Kromasil phases	126
	Pirkle-1J	127
	α-Burke 1	127
	Sumipax	119
Amino acids	Naphthyl urea	128
	Whelkosil-II 3C18 RS	129
	Chirex	126
	Kromasil phases	127
	Sumipax	119
Amines, amides, and imides	Amino acids based	22
	Chirex	124, 126
	Kromasil phases	127
	Pirkle-1J	128
	α-Burke 1	128
	Sumipax	18–22, 119

(continued)

Table 2 Continued

Racemic compounds	CSPs	Refs.
Naphthols	Kromasil phases	127
Carboxylic acids and esters	Whelk-O1	26
	Ulmo	27
	Chirex	124, 126
	Kromasil phases	127
	Sumipax	119
Ketones	Whelk-O1	24, 25
	Chiris series	123
	Chirex	126
	Kromasil phases	127
	α-Burke 1	128
Lactones and lactams	Kromasil phases	127
Sulfoxides	Kromasil phases	127
Phosphorus compounds	Whelk-O1	23
	Chirex	126
Uridine analogs	Whelk-O1	130

Note: DNBGP=3,5-dinitrobenzoyl phenylglycine; DNBLeu=3,5-dinitrobenzoyl leucine.

Because of the development of the new types of CSP (π-acidic–basic types), the use of π-acidic and π-basic types of CSP are not common. In view of this, attempts are made to describe the art of the optimization of the chiral resolution using only π-acidic–basic-type CSPs. The optimization of the chiral resolution on Pirkle-type CSPs are presented in Scheme 1.

5.4.1 Composition of the Mobile Phase

Generally, the normal phase mode has been used frequently for the chiral resolution of racemic compounds on Pirkle-type CSPs. Hexane, heptane, and cyclohexane are the nonpolar solvents of choice on these phases. Aliphatic alcohols may be considered as hydrogen donors and acceptors and may thus interact at many points with the aromatic amide groups of CSPs generating the hydrogen bonds. Therefore, the addition of aliphatic alcohols improve the chiral resolution; hence, the alcohols are considered as organic modifiers. The most commonly used alcohol is 2-propanol. However, methanol, ethanol, 1-propanol, and *n*-butanol were also used. Some reports also indicated the use of dichloromethane and chloroform as the organic modifiers with hexane. In addition to this, acidic and basic additives improve the chromatographic resolution. A small amount of acetic acid, formic acid, or trifluoroacetic acid improve the peak shape

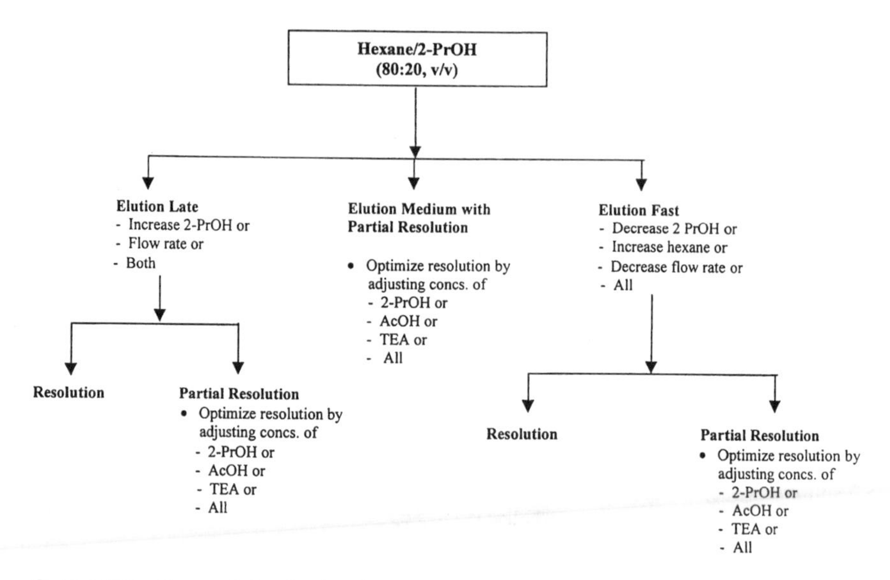

SCHEME 1 Protocol for the development and optimization of mobile phases on Pirkle-type CSPs under the normal phase mode. AcOH, acetic acid; TEA, triethylamine.

and enantioselectivity for acidic and basic solutes. Sometimes, there is a need to combine an acid and organic amine (e.g., triethylamine) for strong basic racemic compounds. We have carried out an extensive search of literature on these phases and proposed a protocol for the development and use of the mobile phases; it is presented in Scheme 1.

Fell and Days [137] studied the effect of the chain length of the alcohols on the chiral resolution of propranolol on 3,5-dinitrobenzoyl-α-phenylglycine CSP. In one study, the separation of the phenylurea derivatives of the enantiomers of propranolol using *n*-dodecane and *n*-pentane as the mobile phase was carried out and is shown in Figure 5 [11]. This figure indicates the greater resolution using *n*-pentane in comparison to using *n*-dodecane. The authors also reported an improved selectivity and column efficiency when using *n*-pentane as the mobile phase.

Uray and Kosjek [27] used different ratios of the hexane and 2-propanol for the chiral resolution of flobufen and its metabolites. The CSPs used were Whelk-O1 and Ulmo. In another study, the same authors [28] utilized the mixtures of heptane, 2-propanol, and trifluoroacetic acid for the chiral resolution of a variety of racemic compounds using Ulmo and related CSPs. Caccamese et al. [23] studied the effect of the concentration of 2-propanol on the chiral resolution of

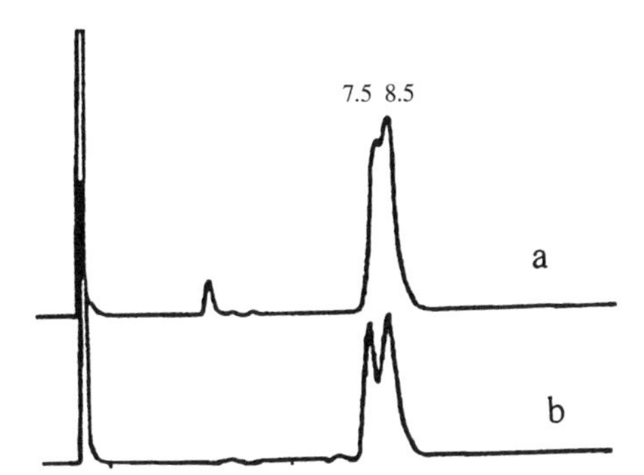

FIGURE 5 Chromatograms of the chiral resolution of propranolol enantiomers on 3,5-dinitrobenzoyl-α-phenylglycine CSP using (a) *n*-dodecane and (b) *n*-pentane as the major component of the mobile phases. (From Ref. 11.)

some α-hydroxybenzoyl phosphonate diethyl esters. The authors reported a different pattern of chiral resolution for different esters. Baeyens et al. [26] reported the best resolution of several nonsteroidal anti-inflammatory drugs and benzoin using hexane-2-propanol-acetic acid (70 : 30 : 0.5, v/v/v) as the mobile phase on Whelk-O1 CSP. Imai et al. [19] carried out the chiral resolution of derivatized amino acids on a series of Sumipax CSPs using methanol containing citric acid as the mobile phase. The authors studied the effect of the concentration of citric acid on the chiral resolution of amino acid derivatives. They found the 10 mM concentration of citric acid to be the best one (Fig. 6). Stringham and Blackwell [138] studied the effect of organic modifiers (i.e., methanol, ethanol, 1-propanol, 2-propanol, *n*-butanol, trifluoroethanol and tetrahydrofuran) on the chiral resolution of aromatic alcohols using the Whelk-O1 column at different temperatures. These findings are given in Figure 7, which shows different types of chiral resolution pattern. Magora and Abu-Lafi [130] used a variety of mobile phases for the chiral resolution of uridine analogs on the Whelk-O1 column. Macaudiere et al. [139] observed the reversal order of elution of some amino acid derivatives just by changing ethanol as an organic modifier to chloroform or dichloromethane in hexane as a mobile phase.

A few reports are available on the use of reversed-phase eluents on the ionically bonded CSPs. However, with the development of new and more stable CSPs, the use of reversed-phase eluents could be possible. The reversed-phase mode is usually chosen to minimize nonstereoselective polar adsorption of the enantiomers on the stationary phase. The reversed-phase mode is preferred when solubility problems occur in the mobile phase or when injection without analyte

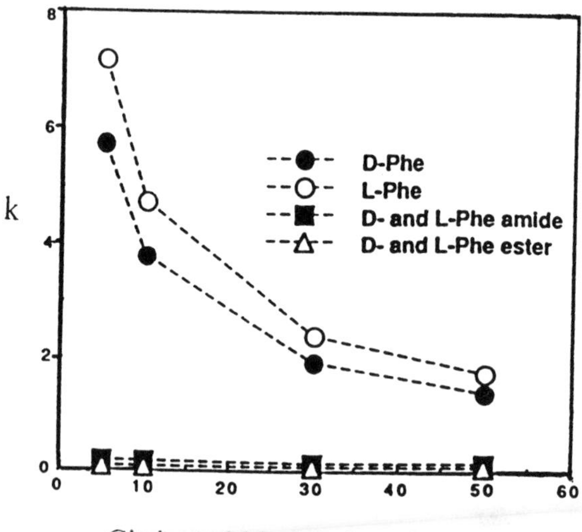

FIGURE 6 Effect of the concentration of citric acid in methanol on the chiral resolution of phenylalanine and its derivatives on Sumichiral OA-2500(S) CSP: (●) D-phenylalanine, (○) L-phenylalanine, (■): D- and L-phenylalanine amide and (△) D- and L-phenylalanine ester. (From Ref. 20.)

prederivatization is needed. Iwaki et al. [140] used acetonitrile–0.15 M sodium acetate (pH 5.0) (70 : 30, v/v) as the mobile phase for the chiral resolution of *p*-bromophenylcarbamyl threonine and *p*-bromophenylcarbamyl phenylalanine on the naphthyl urea Pirkle phase. The same CSP was used for the chiral resolution of other amino acids derivatives using different ratios of acetonitrile and phosphate buffer (pH 6.0) [140].

5.4.2 Temperature

The effect of the temperature is also important for the chiral resolution on these CSPs. Pirkle and Murray [141] reported the inversion of the elution order of 3,5-dinitrobenzoyl-α-phenylethylamine when the temperature was successively changed. The observed effect of temperature was dependent on the content and polarity of the organic modifier in the mobile phase. Stringham and Blackwell [138] studied the effect of the temperature on the chiral resolution of aromatic alcohol on the Whelk-O1 column. The results are given in Figure 7 and it is clear from this figure that the resolution decreased by increasing the temperature. Recently, Magora et al. [130] studied the effect of temperature on the chiral

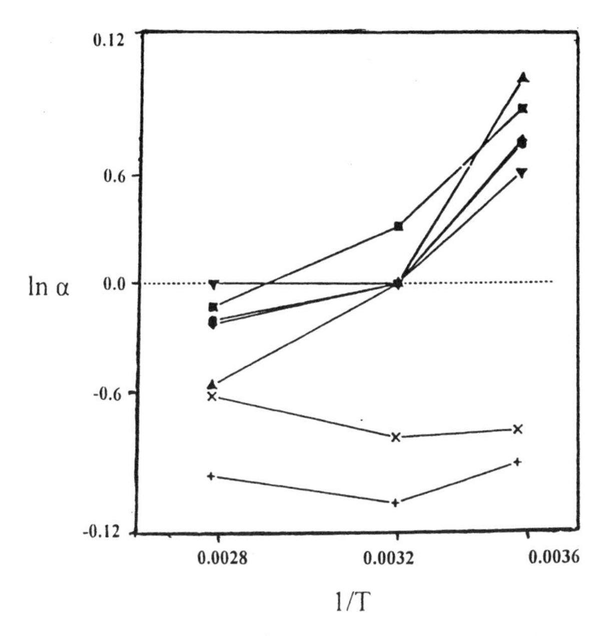

FIGURE 7 Effect of the organic modifiers on the chiral resolution of aromatic alcohol using Whelk-O1 CSP at different temperatures with negative values of ln *k* indicating the reversal order of elution: (■) ethanol, (●) 1-butanol, (▲) 2-propanol, (◆) 1-propanol, (+) trifluoroethanol, and (×) tetrahydrofuran. (From Ref. 138.)

resolution of uridine analogs on the Whelk-O1 column. Generally, the increase in temperature results into a decrease in the differences of the binding energies of the two enantiomers and, hence, a decrease in the enantioselectivity occurs.

5.4.3 Structures of the Solutes

As in the case of other CSPs, the chiral resolution is effected by the structure of the solute. The chiral resolution of amino acids may be considered as the best example for this study. The work of Fukushima et al. [20] (i.e., the chiral resolution of amino acids) indicated the different behavior of the chiral resolution on (*S*)-*N*-3,5-dinitrobenzoyl-1-naphthylglycine CSP. Altomare et al. [142] studied the chiral resolution of a series of 3-phenyl-4-(1-adamantyl)-5-*X*-phenyl-Δ^2-1,2,4-oxadiazolines on *N*,*N*′-(3,5-dinitrobenzoyl)-1(*R*),2(*R*)-diaminocyclohexane CSP. The effect of the influence of aromatic ring substituents on enantioselectivity was studied by traditional linear free-energy-related equations and comparative molecular field analysis methods. The authors reported that an increase in retention was favored by the π-basicity and the hydrophilicity of the solutes. In one of the studies, Pirkle and Spence [143] developed a relationship between the structures of aryl-substituted heterocycles, focusing on lactones and analogs. The authors obtained the enantioselectivity and elution order of heterocycles practically via a mechanistic hypothesis developed by them.

5.4.4 Structures of the CSPs

The chiral resolution on these CSPs varied from one CSP to another; therefore, it can be optimized by selecting the suitable chiral stationary phases. In one of the studies, the effect of the structures of the CSPs on the chiral resolution of certain nonsteroidal anti-inflammatory drugs was carried out by Pirkle and Welch [144]. The studied CSP 1 had the long aliphatic chain; this chain was removed in CSP 2 (Fig. 8a). The results of this study are given in Table 3. The CSP 2 was found to have significantly greater chiral selectivity than the long-chain CSP 1. For a better selectivity of CSP 2, the authors claimed the stronger and different interactions between CSP 2 and the enantiomers. However, the elimination of the long aliphatic chain would also greatly reduce the dispersive interactions of the solute with the stationary phase and thus relatively augment the polar interactions. Imai et al. [19] carried out an interesting study of the chiral resolution of 4-fluoro-7-nitro-2,1,3-benzoxadiazole (DNB) derivatives of amino acids on several Sumichiral OA series CSPs. The various studied CSPs are Sumichiral OA 2000(S), OA-2500(S), OA-2500(R), OA-3000(S), and OA-3200(S) and their structures are given in Figure 8b. The chiral centers of these CSPs are derived from *N*-acylamino-(*S*)-amino acid amides. The results of the chiral resolution of amino

TABLE 3 Effect of the Structures of Pirkle-Type CSPs on the Chiral Resolution of Profens

	CSP 1		CSP 2	
Profens	k	α	k	α
Naproxen	3.96	2.26	1.71	2.93
Ibuprofen	0.94	1.12	0.19	1.47
Ketoprofen	4.53	1.11	1.39	1.29
Flurbiprofen	1.63	1.19	0.37	1.59
Pirprofen	2.53	1.38	0.85	1.81
Fenoprofen	1.48	1.22	0.38	1.61
Cicloprofen	3.03	1.71	1.16	2.15
Tioprofenic acid	6.15	1.09	2.02	1.23

Note: CSP 1 contains larger aliphatic alkyl chain than CSP 2.
Source: Ref. 144.

a

NO_2 NO_2
H
N
O
CH_3
$(CH_2)_n$
Si — O
CH_3
Silica Gel
Matrix

CSP 1: n = 9
CSP 2: n = 1

b

OA-2000 (S)
O
Si
O
N
H
O
H
N
O
NO_2
NO_2

OA-2500 (S)
O
Si
O
N
H
O
H
N
O
NO_2
NO_2

OA-2500 (R)
O
Si
O
N
H
O
H
N
O
NO_2
NO_2

OA-3000 (S)
O
Si
O
N
H
O
H
N
O
H
N

OA-3100 (S)
O
Si
O
N
H
O
H
N
O
H
N
NO_2
NO_2

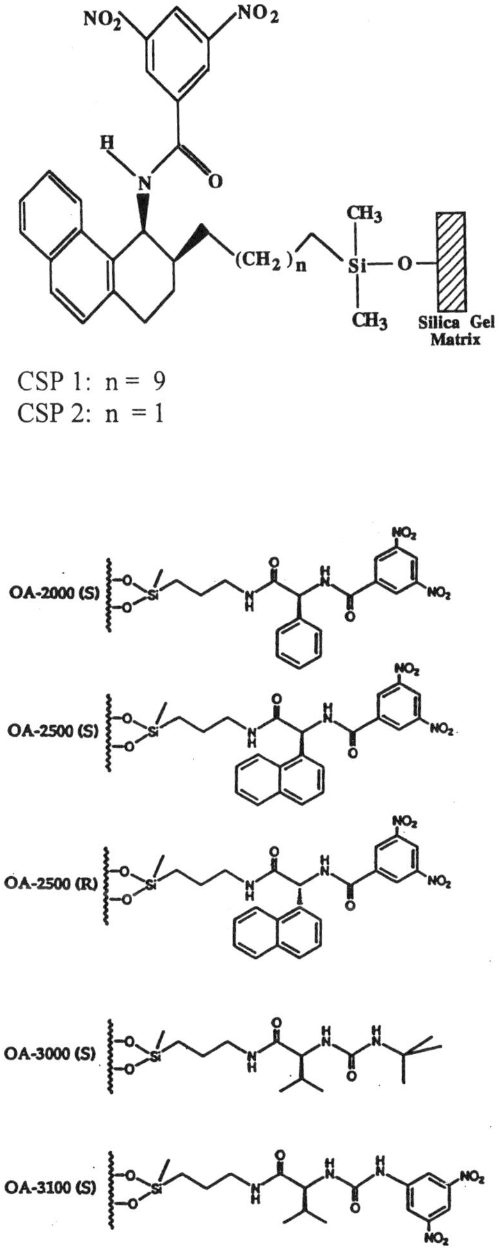

FIGURE 8 The chemical structures of (a) Pirkle-type CSPs with small and large alkyl chains (from Ref. 144) and (b) different Sumichiral OA CSPs. (From Ref. 20.)

TABLE 4 Effect of the Structures of Sumichiral OA CSPs on the Chiral Resolution of DNB Amino Acids

	OA-2000(S)		OA-2500(S)		OA-2500(R)		OA-3000(S)		OA-3100(S)	
DNB-amino acids	*k*	α	*k*	α	*k*	α	*k*	α	*k*	α
Leu	0.91	1.10	2.73	1.12	2.19	1.17	1.48	1.08	1.01	1.52
Ile	1.09	1.07	3.28	1.15	2.72	1.19	1.58	1.00	1.18	1.25
Val	1.31	1.00	3.53	1.14	2.93	1.20	1.71	1.00	1.29	1.22
Ala	2.58	1.06	4.66	1.14	4.29	1.19	2.34	1.00	2.05	1.31
Pro	5.47	1.00	7.31	1.09	7.87	1.44	3.30	1.00	5.48	1.06
Thr	2.36	1.06	5.43	1.12	4.62	1.16	2.77	1.00	2.13	1.35
Ser	4.68	1.00	7.54	1.11	6.96	1.15	3.96	1.00	3.10	1.33
Phe	2.62	1.10	6.37	1.21	6.18	1.28	2.65	1.04	2.16	1.36
Met	2.55	1.10	4.91	1.17	5.18	1.24	2.55	1.06	2.08	1.40
Gln	3.90	1.00	6.76	1.10	6.55	1.12	2.97	1.00	3.00	1.27
Lys	8.16	1.12	12.04	1.18	15.04	1.24	2.65	1.00	4.96	1.57

Mobile phase: 5 mM citric acid in methanol; DNB=4-fluoro-7-nitro-2,1,3-benzoxadiazole.
Source: Ref. 20.

acids derivatives on these CSPs are given in Table 4. It is clear from this table that the chiral resolution varies from one CSP to another. The larger values of separation factors were obtained on OA-3100(S), which bears the 3,5-dinitrophenyl group instead of the *tert*-butyl group in OA-3000(S). The extent of the separation factors of DNB-amino acids on OA-2000(S), bearing a phenyl group, were lower than OA-3100(S). The retention of DNB-amino acids on OA-2500(S), which has a naphthyl group, was greater than OA-2000(S). The authors explained the different patterns of the chiral resolution as being caused by the interactions of different magnitudes on these CSPs.

5.4.5 Other Parameters

In addition to the above-discussed chromatographic parameters, other factors are also important for controlling the chiral resolution on these phases. These include injection amount, particle size of silica gel, spacer between the chiral selector and the silica gel, and the dimension of the column. Recently, Baeyens et al. [26] carried out the chiral resolution of pirprofen on Whelk-O1 columns of diameters of 2.1 and 4.6 mm. The authors reported the best resolution on the column having 4.6 mm as the internal diameter (Fig. 9).

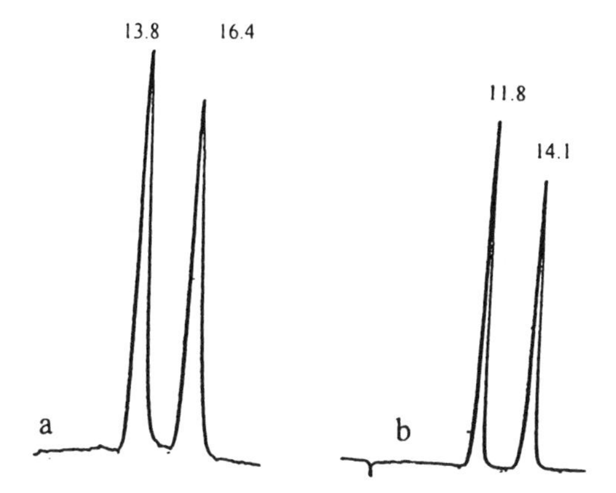

FIGURE 9 A comparison of the chiral resolution of pirprofen on Whelk-O1 CSP of (a) 2.1 and (b) 4.6 mm internal diameters. (From Ref. 26.)

5.5 CHIRAL RECOGNITION MECHANISM

Pirkle type CSPs contain a chiral moiety having phenyl ring and, therefore, the formation of a $\pi-\pi$ charge-transfer diastereoisomeric complex of the enantiomers (with the phenyl group) with CSP is supposed to be essential. In view of this, the π-acidic CSPs are suitable for the chiral resolution of π-donor solutes and vice versa. However, the newly developed CSPs containing both π-acidic and π-basic groups are suitable for the chiral resolution of both types of solute (i.e., π-donor and π-acceptor analytes).

Different studies were carried out to predict the chiral recognition mechanisms on these CSPs. The most important methodologies include the use of chromatography [9,72,109,145,146], X-ray [142], NMR [9,20,147,148], molecular modeling [9,25,142], and computer-aided chemistry [9,149–151]. Initially, Pirkle et al. [72,109] obtained the chiral resolution of a series of homologous compounds on different CSPs. The authors advocated two antinomic mechanisms involving either hydrogen-bondings or the dipole stacking phenomenon. In another study, Pirkle and Däppen [7], starting from the reciprocality concept, established the presence of hydrogen-bondings and dipole stacking mechanisms. Wainer and Alembic [39] observed the reverse order of elution of some amides by changing the dipole group and the authors reported the existence of hydrogen-bondings and dipole stacking processes. Vinkovic et al. [145] studied the chiral resolution of 2-arylpropionic acid enantiomers on different CSPs. The authors reported that hydrogen-bondings and $\pi-\pi$ interaction forces are responsible for the chiral resolution. Even though chromatographic studies provide adequate evidence for the chiral recognition mechanisms, they are also limited because they give an overall view of the mechanisms; hence, more models were developed with the help of X-ray, nuclear magnetic resonance (NMR), and molecular modeling studies.

Altomare et al. [142] carried out the X-ray and circular dichroism measurements of 3-phenyl-4-(1-adamantyl)-5-*X*-phenyl-Δ^2-1,2,4-oxadiazolines on *N*,*N*′-(3,5-dinitrobenzoyl)-1(*R*),2(*R*)-diaminocyclohexane CSP. The authors claimed the chiral recognition is the result of π-basicity and hydrophobicity of the analytes. Pirkle and Pochapsky [147] studied the chiral binding of 3,5-dinitrobenzoyl leucine propylamide on (*S*)-methyl(*N*)-(2-naphthyl)alaninate CSP by NMR. The chiral recognition model of this study is shown in Figure 10, which shows the presence of hydrogen-bondings and $\pi-\pi$ interactions. Similarly, Fukushima et al. [20] postulated the presence of $\pi-\pi$ interactions and hydrogen-bondings in the diastereoisomeric complexes of the enantiomers and CSP. In addition to these studies, computation aided chemistry and molecular modeling calculations were carried out and applied to the chromatographic experiments. The combination of these calculations and the results obtained from the experimental methodologies (X-ray, NMR, chromatography) indicated that $\pi-\pi$ interactions, hydrogen-bondings, steric effects and dipole–dipole forces are responsible for the chiral resolution on Pirkle-type CSPs.

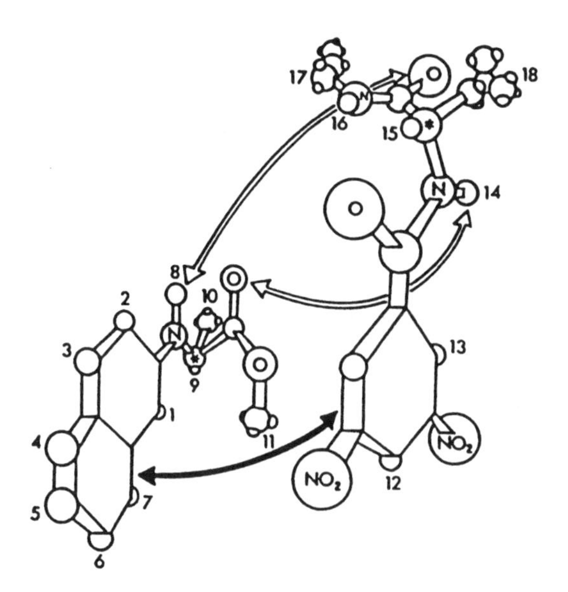

FIGURE 10 Chiral recognition model showing three simultaneous bondings between (*S*)-methyl-*N*-(2-naphthyl)alaninato and (*S*)-*N*-(3,5-dinitrobenzoyl)leucine *n*-propylamine. (From Ref. 9.)

Briefly, Pirkle-type CSPs contain a chiral moiety which provides the chiral environment to the enantiomers. Therefore, the enantiomers interact with this chiral moiety with different fittings (due to different spatial configuration of the enantiomers). In this way, the two enantiomers form the diastereoisomeric complexes having different binding energies. The different binding energies of the diastereoisomeric complexes are the result of the various interactions, as mentioned earlier, of different magnitude. Therefore, the two enantiomers eluted at different retention times with the flow of the mobile phase and hence the chiral resolution occurred. A general graphical representation of the chiral resolution of naproxen enantiomers on (*S*,*S*)-Whelk-O1 CSP is shown in Figure 11, which represents the above-mentioned concept.

Naproxen

(S,S)-Whelk-O1

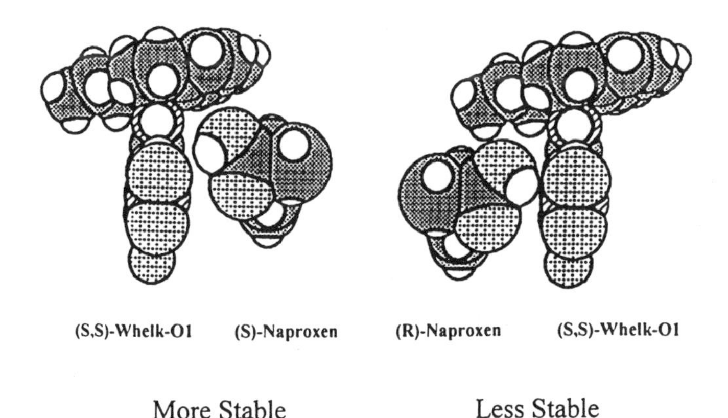

FIGURE 11 Graphical representation of the chiral recognition mechanisms of naproxen enantiomers on (*S*,*S*)-Whelk-O1 CSP.

5.6 APPLICATIONS IN SUB- AND SUPERCRITICAL FLUID CHROMATOGRAPHY

Pirkle type CSPs have been used in sub-FC and SFC models of chromatography since 1985. Mourier et al. [95,152] reported the chiral resolution of phosphine oxides on DNBPG CSP with carbon dioxide mixed with various modifiers. Macaudiere et al. [38,153,154] compared the chiral resolution of amides and phosphine oxides using SFC and high-performance liquid chromatography (HPLC). Later, several reports were published on the chiral resolution of a variety of racemic compounds using sub-FC and SFC techniques [155–158]. Blum et al. [159] reported the chiral resolution of warfarin, *trans*-stilbene oxide, benzoin, verapamil, ketoprofen, abscissic acid, *N*-mesitylsulfonyl-D-alanine, and phenylmethyl sulfoxide on the Whelk-O1 column at analytical and preparative scales. Furthermore, the authors separated 200 mg of warfarin enantiomers within 10 min on the analytical Whelk-O1 column (preparative separations). Terfloth et al. [160] resolved the enantiomers of warfarin, flubiprofen, and benzoin on Whelk-O1 CSP using sub-FC and SFC models of chromatography. The enantioseparations of β-blockers was achieved on the ChyRoSine-A CSP with a modified carbon dioxide eluent, but the same compound could not be resolved by HPLC [161]. According to one of the reviews published by Terfloth [162], the enantiomers of propranolol were resolved at the preparative scale on ChyRoSine A CSP using carbon dioxide as the mobile phase in SFC. The chromatograms of the chiral resolution of ibuprofen and ketoprofen on Whelk-O1 column are shown in Figure 12. The applications of the chiral resolution on Pirkle-type CSPs using SFC are summarized in Table 5.

The effects of the subcritical and supercritical chromatographic parameters on the chiral resolution of racemic compounds on these phases were also investigated. Macaudiere et al. [38,153,154] compared the chiral resolution of amides and phosphine on SFC and HPLC. The authors reported greater values of resolution factors with fivefold smaller values of retention times in SFC in comparison to HPLC. The comparative results for their studies are shown in Figure 13. Furthermore, they also reported the effect of various parameters such as composition of the mobile phase, use of organic modifiers, and temperature on chiral resolution using SFC. According to them, the use of a small amount of water in the mobile phase resulted in an improved chiral resolution (Fig. 13). They concluded that the effects of these parameters on the chiral resolution by SFC were comparable with HPLC. Blum et al. [159] compared the chiral resolution of some drugs on SFC and HPLC and the authors advocated the superiority of SFC in terms of speed, efficiency, and method development procedures. Stringham [163] presented a model which predicted the relationship between the chiral sub-FC and the resolution times. The author reported that the increased column efficiency of sub-FC at typical flow rates rescued separations

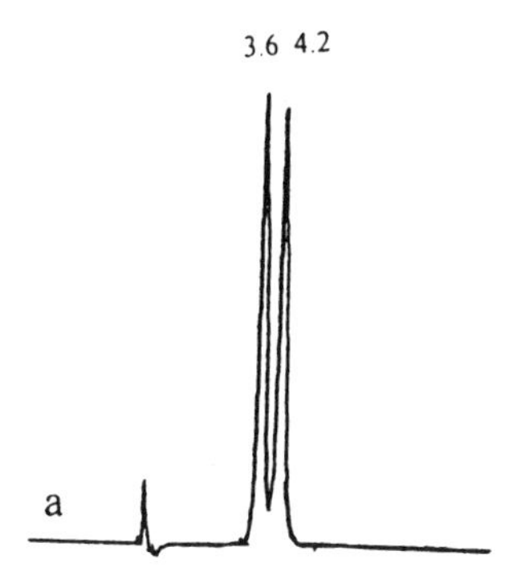

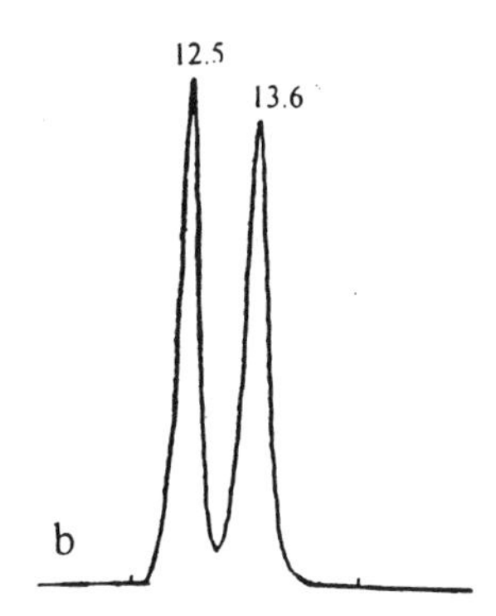

FIGURE 12 Chromatograms of the chiral resolution of (a) ibuprofen and (b) ketoprofen on Whelk-O1 CSP using SFC. (From Ref. 158.)

that failed in HPLC. The effect of the temperature [155,156] on the chiral resolution of some drugs was evaluated. Generally, it was observed that the enantioselectivities decreased at higher temperature. Blackwell et al. [158] presented an empirical relationship which related properties of the mobile phase modifiers to the chiral selectivities for a number of analytes. Furthermore, Stringham and Blackwell [164] studied the effect of the temperature on the chiral resolution of benzoins and *Z*-phenylalaninol and related compounds using SFC model of liquid chromatography. The authors reported a decrease in chiral resolution by increasing the temperature. The effect of the temperature on the chiral resolution of benzoins and *Z*-phenylalaninol is shown in Figure 14. In other studies, the same authors [165,166] achieved the chiral resolution of a set of

TABLE 5 Enantiomeric Resolution of Racemic Compounds on Pirkle-Type CSPs Using Sub-FC and SFC

Racemic compounds	CSPs	Refs.
Phosphine oxides	DNBPG	38, 95, 153, 154
Amidotetralin	Whelk-O1	156
Amines	Whelk-O1	162
Anticoagulants	Whelk-O1	162
Aryloxypropionic acids	Whelk-O1	162
Arylpropionic acids	Whelk-O1	159, 160, 162
Chlorohydrin	Whelk-O1	163
Dinitrobenzamide	Whelk-O1	162
Epoxides	Whelk-O1	162
Hydantoins	Whelk-O1	162
β-Blockers	ChyRoSine-A	161, 162

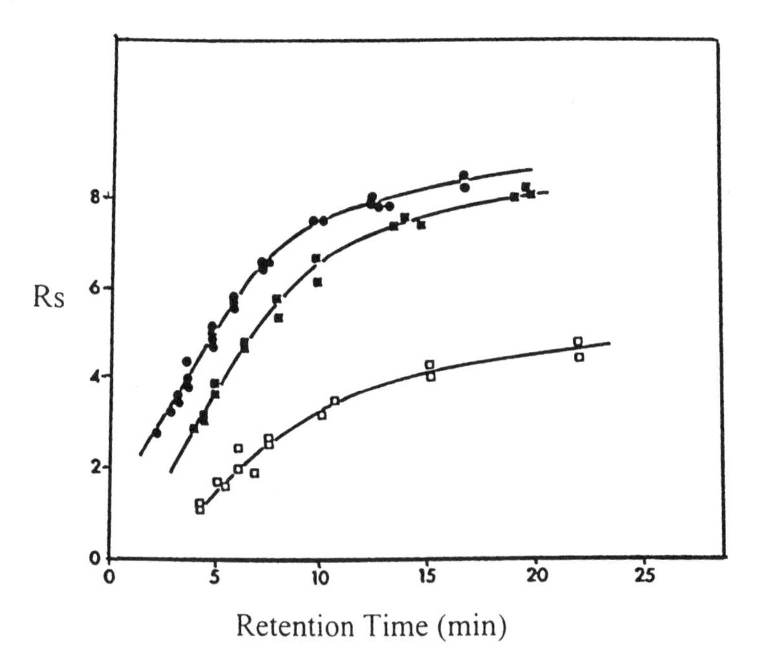

FIGURE 13 A comparison of the chiral resolution of 2-naphthyl amide derivatives of 2-aminooctane enantiomers on 3,5-dinitrobenzoyl phenylglycine CSP by HPLC and SFC using (□) hexane–2-propanol (95 : 5, v/v), (■) carbon dioxide–2-propanol (95 : 5, v/v), and (●) carbon dioxide–2-propanol–water (95 : 4.8 : 0.2, v/v/v) as the mobile phases. (From Ref. 9.)

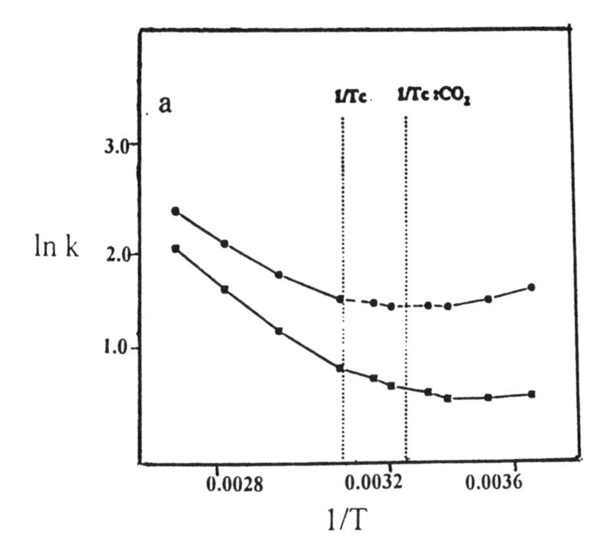

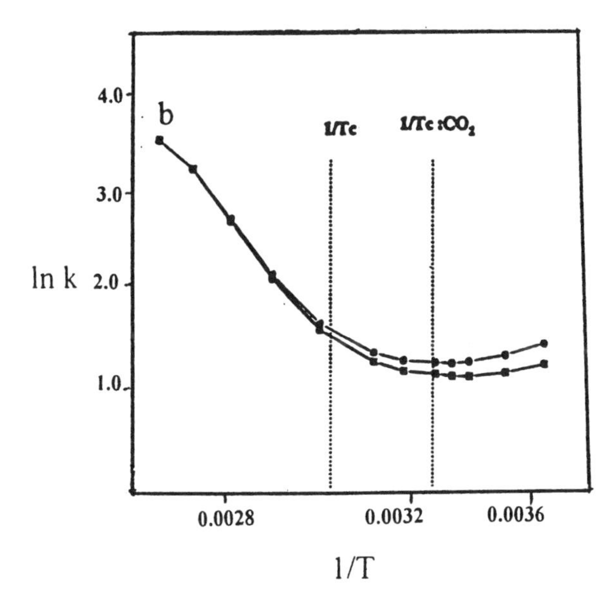

FIGURE 14 Effect temperature on the chiral resolution of (a) benzoin and (b) *Z*-phenylalaninol on Whelk-O1 CSP using SFC, $1/T_c$ and $1/T_c : CO_2$ are the critical temperatures of the mobile phase and CO_2, respectively, ■ and ● are the ln *k* values of the two enantiomers. (From Ref. 163.)

phenyl-substituted *N*-*t*-butoxycarbonyl phenylalanine analogs on the α-Bruke 2 column using various mobile phase additives. The acidic mobile phase additives used were acetic acid, chloroacetic acid, dichloroacetic acid, trifluoroacetic acid, methanesulfonic acid, and trifluoromethanesulfonic acid. It was reported that trifluoromethanesulfonic acid resulted in good separation. The authors modeled the selectivity factors and predicted the presence of hydrogen-bondings and $\pi-\pi$ interactions between the enantiomers and CSP. Briefly, the chiral recognition mechanisms on these CSPs using SFC are similar to that of HPLC chiral separations.

5.7 MISCELLANEOUS APPLICATIONS

In spite of the high speed and efficiency of capillary electrochromatography, the Pirkle-type CSPs could not be used frequently for chiral resolution purposes. Only a few reports are available on this issue. Cavender et al. [167] and Wolf et al. [168] utilized Pirkle-type CSPs, in the form of packed capillary columns, for the chiral resolution of *N*-[-(4-bromophenyl)]-2,2-dimethylpropionamide, 3,5-dinitrobenzoyl amino acids, hydroxy benzoin, and cyclopropane derivatives. Peters et al. [169] prepared several monolithic chiral stationary phases for reversed-phase electrochromatography within the confines of untreated fused silica capillaries by the direct copolymerization of the chiral monomer 2-hydroxyethyl methacrylate (*N*-L-valine-3,5-dimethylanilide) carbamate with ethylene dimethacrylate, 2-acrylamido-2-methyl-1-propanesulfonic acid, and butyl or glycidyl methacrylate in the presence of a protogenic solvent. The authors also reported that the hydrophilicity of the stationary phase, which may be enhanced further by the hydrolysis of the epoxide functionalities of the glycidyl methacrylate moieties within the monolith, was found to have a pronounced effect on the enantioseparation. Using the most hydrophilic monolithic capillary column and optimized elution conditions, the enantiomers of 3,5-dinitrobenzoyl leucine diallylamide were resolved successfully.

Lienne et al. [170] resolved the enantiomers of albendazole sulfoxides on a column derived from the (*S*)-*N*-(3,5-dinitrobenzoyl)tyrosine chiral selector. The developed method was applied for the enantiomeric resolution of albendazole sulfoxides in plasma samples. Witherow et al. [171] immersed a commercially available thin-layer plate (thin-layer chromatography) into a solution of *N*-(3,5-dinitrobenzoyl)-L-leucine solution. The developed plate was used for the chiral resolution of 2,2,2-trifluoro-(9-anthryl)ethanol and 1,1′-binaphthol enantiomers.

5.8 CONCLUSION

Pirkle-type CSPs have achieved a good status in the field of the chiral resolution by liquid chromatography. In these phases, the chiral moiety on the silica support can be bonded selectively and, therefore, these CSPs can be prepared as per the requirement of chiral resolution. There is no serious drawback associated with these phases. However, a search of literature indicates that the chiral resolution on these CSPs requires the prederivatization of some racemic compounds, which is, of course, a tedious and time-consuming job. Moreover, the derivatization of racemic compounds in natural samples (e.g., biological and environmental samples) is not possible; therefore, the application of these phases for the chiral resolution in natural samples is very limited. In addition, most of the chiral resolution reports on these CSPs deal with the use of the normal mobile phase mode which have, again, limited choice of optimization of the chiral resolution. However, with the development of the new and more stable CSPs, the reversed phase mode was used in some studies. In conclusion, it seems that these CSPs are not fully developed and are underway for their development. We hope that, in the future, these phases will achieve a good reputation, especially in natural samples, for chiral resolution.

REFERENCES

1. Mikeš F, Boshart G, Gil-Av E, J Chromatogr 122: 205 (1976).
2. Pirkle WH, Sikkenga DL, J Chromatogr 123: 400 (1976).
3. Pirkle WH, House DW, J Org Chem 44: 1957 (1979).
4. Pirkle WH, House DW, Fin JM, J Chromatogr 192: 143 (1980).
5. Pirkle WH, Finn JM, J Org Chem 46: 2935 (1981).
6. Pirkle WH, Welch CJ, J Org Chem 49: 138 (1984).
7. Pirkle WH, Däppen R, J Chromatogr 404: 107 (1987).
8. Finn JM, Rational design of Pirkle type chiral stationary phases, in Chromatographic Chiral Separations, Zief M, Crane LJ (Eds.), Chromatographic Science Series Vol. 40, Marcel Dekker, New York (1988).
9. Macaudiere P, Lienne M, Tambute A, Caude M, Pirkle type and related chiral stationary phases for enantiomeric resolution, in Chiral Separations by HPLC, Krstulovic AM (Ed.), Ellis Horwood, New York (1989).
10. Aboul-Enein HY, Wainer IW (Eds.), The Impact of Stereochemistry on Drug Development and Use, John Wiley & Sons, New York (1997).
11. Beesley TE, Scott RPW (Eds.), Chiral Chromatography, John Wiley & Sons, New York (1998).
12. Persson BA, Andersson S, J Chromatogr A 906: 195 (2001).
13. Pirkle WH, Fin JM, Schreiner JL, Hamper BC, J Am Chem Soc 103: 32,964 (1981).
14. Kip J, Van Haperen P, Karaak JC, J Chromatogr 356: 423 (1986).
15. Gargaro G, Gasparrini F, Misiti D, Palmieri G, La Torre F, Villani C, 16th International Symposium on Chromatography, Paris (1986).
16. Hyun MH, Min CS, Cho YJ, Na MS, J Liq Chromatogr 18: 2527 (1995).
17. Uray G, Niederreiter KS, Maier NM, Spitaler MM, Chirality 11: 404 (1999).
18. Imai K, Fukushima T, Uzu S, Biomed Chromatogr 7: 177 (1993).

19. Imai K, Fukushima T, Biomed Chromatogr 7: 275 (1993).
20. Fukushima T, Kato M, Santa T, Imai K, Biomed Chromatogr 9: 10 (1995).
21. Kato M, Fukushima T, Shimba N, Shimada I, Kawakami Y, Imai K, Biomed Chromatogr 15: 227 (2001).
22. Siddiqui MA, Jin H, Evans CA, Dimarco MP, Tse HLA, Mansour TS, Chirality 6: 156 (1994).
23. Caccamese S, Failla S, Finocchiaro P, Principato G, Chirality 10: 100 (1998).
24. Quaglia MG, Desideri N, Bossu E, Sgro R, Conti C, Chirality 11: 495 (1999).
25. Quaglia MG, Mai A, Artico SM, Ragno R, Massa S, Piano DD, Setzu G, Doratiotto S, Cotichini V, Chirality 13: 75 (2001).
26. Baeyens WRG, Van der Weken G, Aboul-Enein HY, Reygaerts S, Smet E, Biomed Chromatogr 14: 58 (2000).
27. Uray G, Kosjek B, Enantiomer 5: 329 (2000).
28. Kosjek B, Uray G, Chirality 13: 657 (2001).
29. Wainer IW, Doyle TD, J Chromatogr 259: 465 (1983).
30. Pirkle WH, Finn JM, Hamper BC, Schreiner J, Pribish JR, A useful and conveniently accessible chiral stationary phase for liquid chromatographic separations of enantiomers, in Asymmetric Reactions and Process in Chemistry, Eliels EL, Otsuka S (Eds.), ACS Symposium Series No. 185, American Chemical Society, Washington DC, p. 245 (1982).
31. Wainer IW, Doyle TD, Donn KH, Powell JR, J Chromatogr 306: 405 (1984).
32. Zief M, Crane LJ, Horvath J, J Liq Chromatogr 7: 709 (1984).
33. Wainer IW, Doyle TD, Hamidzadeh Z, Aldridge M, J Chromatogr 261: 123 (1983).
34. Wainer IW, Doyle TD, Fry FS, Hamidzadeh Z, J Chromatogr 355: 149 (1986).
35. Wainer IW, Doyle TD, Hamidzadeh Z, Aldridge M, J Chromatogr 268: 107 (1983).
36. Doyle TD, Wainer IW, J High Resolut Chromatogr 7: 38 (1984).
37. Smith DF, Pirkle WH, Psychopharmacology 89: 392 (1986).
38. Macaudiere P, Tambute A, Caude M, Rosset R, Alembik MC, Wainer IW, J Chromatogr 371: 177 (1986).
39. Wainer IW, Alembic MC, J Chromatogr 367: 59 (1986).
40. McErlane KM, Igwemezioe L, Kerr CR, J Chromatogr 415: 335 (1987).
41. Wainer IW, Doyle TD, Adams WM, J Pharm Sci 73: 1162 (1984).
42. Meyers AI, Bailey TR, J Org Chem 51: 872 (1986).
43. Wainer IW, Doyle TD, LC, HPLC Mag 2: 88 (1984).
44. Crowther JB, Covey TR, Dewey EA, Henion JD, Anal Chem 56: 2921 (1984).
45. Lee ED, Henion JD, Brunner CA, Wainer IW, Doyle TD, Gal J, Anal Chem 58: 1349 (1986).
46. Doyle TD, Adams WM, Fry FS, Wainer IW, J Liq Chromatogr 9: 455 (1986).
47. Pirkle WH, Hyun MH, J Chromatogr 322: 295 (1985).
48. Baker JK, Clark AM, Hufford CD, J Liq Chromatogr 9: 493 (1986).
49. Pirkle WH, Welch CJ, Mahler GS, Meyers AI, Fuentes LM, Boes M, J Org Chem 49: 2504 (1984).
50. Pirkle WH, Pochapsky TC, Mahler GS, Field RE, J Chromatogr 348: 89 (1985).
51. Pirkle WH, Pochapsky TC, J Org Chem 51: 102 (1986).
52. Kasai M, Froussios C, Ziffer H, J Org Chem 48: 459 (1983).

53. Pirkle WH, Finn JM, J Org Chem 47: 4037 (1982).
54. Perry JA, Rateike JD, Szczerba TJ, J Chromatogr 389: 57 (1987).
55. Perry JA, Rateike JD, Szczerba TJ, J Liq Chromatogr 9: 3297 (1986).
56. Pettit GR, Sing SB, Cragg GM, J Org Chem 50: 3404 (1985).
57. Yang SK, Weems HB, Anal Chem 56: 2658 (1984).
58. Weems HB, Yang SK, Anal Biochem 125: 156 (1982).
59. Weems HB, Fu PP, Yang SK, Carcinogenesis 7: 1221 (1986).
60. Yang SK, Mushtaq M, Fu PP, J Chromatogr 371: 195 (1986).
61. Weems HB, Mushtaq M, Fu PP, Yang SK, J Chromatogr 371: 211 (1986).
62. Fu PP, Yang SK, Biochem Biophys Res Commun 109: 927 (1982).
63. Mushtaq M, Weems HB, Yang SK, Biochem Biophys Res Commun 125: 539 (1984).
64. Yang SK, Li XC, J Chromatogr 291: 265 (1984).
65. Yang SK, Weems HB, Mushtaq M, Fu PP, J Chromatogr 316: 569 (1984).
66. Weems HB, Mushtaq M, Yang SK, Anal Chem 59: 2679 (1987).
67. Yang ZY, Barknan S, Brunner C, Weber JD, Doyle TD, Wainer IW, J Chromatogr 324: 444 (1985).
68. Pirkle WH, Schreiner JL, J Org Chem 46: 4988 (1981).
69. Pirkle WH, Finn J, Separation of enantiomers by liquid chromatographic methods, in Asymmetric Synthesis, Volume 1, Analytical Methods, (Morrison JD (Ed.), Academic Press, New York, p. 87 (1983).
70. Pirkle WH, Tsipouras A, J Chromatogr 291: 291 (1984).
71. Pirkle WH, Tsipouras A, Sowin TJ, J Chromatogr 319: 392 (1985).
72. Pirkle WH, Hyun MH, Tsipouras A, Hamper BC, Bank B, J Pharm Biomed Anal 2: 173 (1984).
73. Sioufi A, Colussi D, Marfil F, Dubois JP, J Chromatogr 414: 131 (1987).
74. Oi N, Matsumoto Y, Kitahara H, Miyazaki H, Bunseki Kagaku 35: 312 (1986).
75. Wainer IW, Doyle TD, J Chromatogr 284: 117 (1984).
76. McDaniel DM, Snider BG, J Chromatogr 404: 123 (1987).
77. Nicoll-Griffith DA, J Chromatogr 402: 179 (1987).
78. Demerson CA, Humber LG, Abraham NA, Schilling G, Martel RR, Pace-Asciak C, J Med Chem 26: 1778 (1983).
79. Chapman RA, J Chromatogr 258: 175 (1983).
80. Papadopoulou-Mourkidou E, Chromatographia 20: 376 (1985).
81. Cayley GR, Simpson BW, J Chromatogr 356: 123 (1986).
82. Dernoncour R, Azerad R, J Chromatogr 410: 355 (1987).
83. Blessington B, Crabb N, O'Sullivan J, J Chromatogr 396: 177 (1987).
84. Wainer IW, Doyle TD, Breder CD, J Liq Chromatogr 7: 731 (1984).
85. Mazzo DJ, Lindemann CJ, Brenner GS, Anal Chem 58: 636 (1986).
86. Lienne M, Caude M, Rosset R, Tambute A, J Chromatogr 448: 55 (1988).
87. Pirkle WH, Tsipouras A, Hyun MH, Hart DJ, Lee CS, J Chromatogr 358: 377 (1986).
88. Pirkle WH, Sowin TJ, J Chromatogr 387: 313 (1987).
89. Bonnaud B, Calmel F, Pastoiseau JF, N'Guyen NT, Cousse H, J Chromatogr 318: 398 (1985).

90. Brown TM, Grothusen JR, J Chromatogr 294: 390 (1984).
91. Oliveros L, Cazau M, J Chromatogr 409: 183 (1987).
92. Pescher P, Caude M, Rosset R, Tambute A, Oliveros L, Nouv J Chem 9: 621 (1985).
93. Pescher P, Caude M, Rosset R, Tambute A, J Chromatogr 371: 159 (1986).
94. Tambute A, Gareil P, Caude M, Rosset R, J Chromatogr 363: 81 (1986).
95. Mourier P, Eliot E, Caude M, Rosset R, Tambute A, Anal Chem, 57: 2819 (1985).
96. Shimizu T, Kobayashi M, Bull Chem Soc Jpn 59: 2654 (1986).
97. Shimizu T, Kobayashi M, J Org Chem 52: 3399 (1987).
98. Allenmark S, Nielson L, Pirkle WH, Acta Chem Scand B37: 325 (1983).
99. Pirkle WH, Däppen R, Reno DS, J Chromatogr 407: 211 (1987).
100. Abidi SL, J Liq Chromatogr 10: 1085 (1987).
101. Abidi SL, J Chromatogr 404: 133 (1987).
102. Oi N, Nagase M, Inda Y, Doi T, J Chromatogr 259: 487 (1983).
103. Oi N, Kitahara H, J Chromatogr 265: 117 (1983).
104. Pirkle WH, Malher GS, Pochapsky TC, Hyun MH, J Chromatogr 388: 307 (1987).
105. Takagi T, Itabashi Y, J Chromatogr 366: 451 (1986).
106. Itabashi Y, Takagi T, J Chromatogr 402: 257 (1987).
107. Pirkle WH, Sowin TJ, J Chromatogr 396: 83 (1987).
108. Däppen R, Arm H, Meyer VR, J Chromatogr 373: 1 (1986).
109. Pirkle WH, Hyun MH, Bank B, J Chromatogr 316: 585 (1984).
110. Däppen R, Meyer VR, Arm H, J Chromatogr 361: 93 (1986).
111. Pirkle WH, Hyun MH, Bank B, J Chromatogr 328: 1 (1985).
112. Däppen R, Meyer VR, Arm H, J Chromatogr 295: 367 (1984).
113. Tambute A, Begos A, Lienne M, Caude M, Rosset R, J Chromatogr 396: 65 (1987).
114. Pettersson C, Gioeli C, J Chromatogr 398: 247 (1987).
115. Pirkle WH, Hyun MH, Bank B, J Chromatogr 322: 287 (1985).
116. Berndt H, Krüger G, J Chromatogr 348: 275 (1985).
117. Krüger G, Grötzinger J, Berndt H , J Chromatogr 397: 223 (1987).
118. Lloyd MJB, J Chromatogr 351: 219 (1986).
119. Griffith OW, Campbel EB, Pirkle WH, Tsipouras A, Hyun MH, J Chromatogr 362: 345 (1986).
120. Rosini C, Altemura P, Pini D, Bertucci C, Zullino G, Savadori P, J Chromatogr 348: 79 (1985).
121. Ikuno Y, Maoka T, Shimizu M, Komori T, Matsuno T, J Chromatogr 328: 387 (1985).
122. Aboul-Enein HY, Al-Duraibi IA, J Liq Chrom & Rel Technol 21: 1817 (1998).
123. Stationary HPLC Phases for Chiral Separations, IRIS Technologies, Lawrence, KS (2001).
124. Clevland T, J Liq Chromatogr 18: 649 (1995).
125. The Innovative Direction in Chiral Separations, Phenomenex, Torrance, USA (1992).
126. Kromasil Chiral Phases for HPLC, SFC and SMB, Eka Chemicals AB, Marietta, GA (2000).
127. Pirkle-1J Chiral HPLC Column, Regis Technologies, Morton Grove, IL (1995).
128. Toyo'oka T, Jin D, Nagakawa K, Murofushi S, Biomed Chromatogr 13: 103 (1999).

129. Iwaki K, Yoshida S, Nimura N, Kinoshita T, J Chromatogr A 617: 279 (1993).
130. Magora A, Abu-Lafi S, Levin S, J Chromatogr 866: 183 (2000).
131. Welch CJ, Chem New Zealand 9 (1993).
132. Francotte E, Junker-Buchheit A,. J Chromatogr 576: 1 (1992).
133. Möller P, Sanchez D, Persson B, Andersson S, Allenmark S, 11th International Symposium on Preparative and Industrial Chromatography (1994).
134. Jageland PT, Bryntesson LM, Möller P, Chromatographic Workshop (1995).
135. Chiral Columns for Enantiomeric Separations by HPLC, Sumika Chemical Analysis Service, Osaka, Japan (1995).
136. Pirkle WH, Koscho ME, J Chromatogr A 840: 151 (1999).
137. Fell AF, Days AM, Analyst 116: 1343 (1991).
138. Stringham RW, Blackwell JA, Anal Chem 69: 1414 (1997).
139. Macaudiere P, Lienne M, Caude M, Rosset P, Tambute A, J Chromatogr 467: 357 (1989).
140. Iwaki K, Yoshida S, Nimura N, Kinoshita T, J Chromatogr 404: 117 (1987).
141. Pirkle WH, Murray PG, J High Resolut Chromatogr 16: 285 (1993).
142. Altomare C, Cellamare S, Carotti A, Barrreca ML, Chimirri A, Monforte AM, Gasparrini F, Villani C, Cirilli M, Mazza F, Chirality 8: 556 (1996).
143. Pirkle WH, Spence PL, J Chromatogr A 775: 81 (1997).
144. Pirkle WH and Welch CJ, J Liq Chromatogr 15: 1947 (1992).
145. Vinkovic V, Kontrec D, Sunjic V, Navarini L, Zanetti F, Azzolina O, Chirality 13: 581 (2001).
146. Hyun MH, Kang MH, Han SC, J Chromatogr 868: 31 (2000).
147. Pirkle WH, Pochapsky TC, J Am Chem Soc 109: 597 (1987).
148. Shah P, Hsu TB, Rogers LB, J Chromatogr 396: 31 (1987).
149. Lipkowitz KB, Demeter DA, Parish CA, Darden T, Anal Chem 59: 1731 (1987).
150. Norinder U, Sundholm EG, J Liq Chromatogr 10: 2825 (1987).
151. Suzuki T, Timofei S, Iuras BE, Uray G, Verdino P, Walter MF, J Chromatogr A 922: 13 (2001).
152. Mourier P, Sassiat P, Caude M, Rosset R, J Chromatogr 353: 61 (1986).
153. Macaudiere P, Caude M, Rosset R, Tambute A, J Chromatogr 405: 135 (1987).
154. Macaudiere P, Caude M, Rosset R, Tambute A, Proceedings of the International Symposium on Chiral Separations (1987).
155. Wolf C, Pirkle WH, J Chromatogr A 785: 173 (1997).
156. Selditz U, Copinga S, Grol CJ, Franke JP, De-Zeeuw RA, Wikstrom H, Gyllenhaal O, Pharmazie 54: 183 (1999).
157. Phinney KW, Sub- and supercritical fluid chromatography for enantiomer separations, in Chiral Separation Techniques: A Practical Approach, Subramanian G (Ed.), Weinham, Germany, p. 299 (2001).
158. Blackwell JA, Stringham RW, Xiang D, Waltermire RE, J Chromatogr 852: 383 (1999).
159. Blum AM, Lynam KG, Nicolas EC, Chirality 6: 302 (1994).
160. Terfloth GJ, Pirkle WH, Lynam KG, Nicolas EC, J Chromatogr A 705: 185 (1995).
161. Bargmann-Leyder N, Thiebaut D, Vergne F, Begos A, Tambute A, Caude M, Chromatographia 39: 673 (1994).

162. Terfloth G, J Chromatogr A 906: 301 (2001).
163. Stringham RW, Chirality 8: 249 (1996).
164. Stringham RW, Blackwell JA, Anal Chem 68: 2179 (1996).
165. Blackwell JA, Chirality 11: 91 (1999).
166. Blackwell JA, Stringham RW, Chirality 11: 98 (1999).
167. Cavender DM, Wolf C, Spence PL, Pirkle WH, Derrico EM, Rozing GP, J Chromatogr A 782: 175 (1997).
168. Wolf C, Spence PL, Pirkle WH, Cavender DM, Derrico EM, Electrophoresis 21: 917 (2000).
169. Peters EC, Lewandowski K, Petro M, Svec F, Frechet JM, J Anal Commun 35: 83 (1998).
170. Lienne M, Caude M, Rosset R, Tambute A, Delatour P, Chirality 1: 142 (1989).
171. Witherow L, Spurway TD, Raune RJ, Wilson ID, London K, J Chromatogr 553: 497 (1991).

6

Protein-Based Chiral Stationary Phases

Proteins are natural polymers and are made of amino acids, which are chiral molecules, with the exception of glycine, through amide bonds. However, some glycoproteins also contain sugar moieties. The protein polymer remains in the twisted form because of the different intramolecular bondings. These bonding are responsible for different types of loop/groove present in the protein molecule. This sort of twisted three-dimensional structure of protein makes it enantioselective in nature. Enantioselective interactions between small molecules and proteins in biological systems are well known [1]. For the first time in 1954, Karush [2] reported that the two enantiomers of *N*-aroylated phenylglycine gave nonidentical scatchard plots when the concentration dependence of their binding to human serum albumin (HSA) in a buffer solution was investigated. Later, in 1958, McMenamy and Oncley [3] also observed the enantioselective nature of protein. They observed that L-tryptophan binds more strongly with serum albumin than D-tryptophan in the isotopic labeling experiment. Müller and Wollert [4,5] reported in gel filtration studies, that the hemisuccinate of the drug oxazepam showed a highly enantioselective behavior to human serum albumin. *S*(+)-enantiomer was found to bind 30 times stronger than the *R*-(−)-enantiomer. Similarly, other studies were carried out and it was reported that proteins are enantiospecific in nature (e.g., the binding difference of the two antipodes of phenprocoumon, warfarin [6], ketoprofen [7], and oxazepam derivatives [8] was reported on protein molecules).

Because of the enantioselective nature of proteins, Stewart and Doherty [9] packed a column with bovine serum albumin (BSA)–succinoylaminoethyl–agarose and resolved the enantiomers of DL-tryptophan; the technique is called affinity chromatography (protein racemate complex formation). The same method was used for the resolution of other drugs [10,11]. Later, BSA was bonded on sepharose and some racemates were resolved using low-pressure liquid chromatography [12–14]. These types of column resulted into a good chiral resolution of certain racemates, but the column efficiency was poor and, therefore, further development in protein-based CSPs was realized [12–14]. Attempts were made to immobilize proteins on solid supports. The silica-gel-immobilzed proteins were packed into the columns and were used successfully for the chiral resolution of a variety of racemates by high-performance liquid chromatography (HPLC). Although all of the protein molecules are complex in structure and enantio-specific, they could not be used as successful chiral selectors because the enantiomeric separation varies from one protein to another. Albumin proteins used as chiral selectors in liquid chromatography are BSA, HSA, rat serum albumin (RSA), and guinea pig serum albumin (GPSA); BSA and HSA were found to be successful chiral selectors. However, other proteins molecules were explored for their chiral resolution [i.e., glycoproteins such as α_1-acid glycoprotein (AGP), ovomucoid (OVM), ovotransferin, avidin, trypsin (CT), and certain enzymes like chymotrypsin, riboflavin, lysozyme, pepsin, amyloglucosidase and lactoglobulin]. Cellobiohydrase-I (CBH-I), a protein obtained from fungus, was also used as chiral selector in HPLC [15]. Because of the importance of the protein molecules in chiral resolution by liquid chromatography, attempts were made to describe their structures, properties, applications, the effect of various chromatographic factors, and their chiral recognition mechanisms. Furthermore, the use of the protein molecules as chiral selectors in other modes of liquid chromatography has also been addressed.

6.1 STRUCTURES AND PROPERTIES

Although all proteins are complex in structure and chiral in nature, some of them could achieve the status of a chiral selector in liquid chromatography. The complex structures of proteins are the result of the different intramolecular hydrogen-bonding, disulfide bridges, and other types of bonding. All of the proteins used for chiral resolution in liquid chromatography are obtained from animals except for cellobiohydrolase-I. The structures and properties of some of the most commonly used proteins as chiral selectors are discussed herein.

6.1.1 Bovine Serum Albumin

Bovine serum albumin is a globular protein of molecular mass 66,210 consisting of 581 amino acids in a single chain and 17 intramolecular disulfide bridges connecting the 34 half-cysteines with the formation of nine double loops. The protein is relatively acidic and soluble in water, with an iosoelectric point of 4.7. At pH 7.0, the net charge is -18. The 17 disulfide bridges form 9 loops grouped into 3 similar domains. It is cigar shaped with molecular dimension 141×42 Å. BSA has the capacity to bind many of the organic compounds containing a hydrophobic contribution of at least five to six methylene groups [16]. However, BSA has the ability to bind many inorganic anions which are very hydrophobic in nature [17]. Fatty acids anions are also known to be very tightly bound and to exert a stabilizing effect. Therefore, BSA is stabilized with caprylic acid.

6.1.2 Human Serum Albumin

Human serum albumin is structurally related to BSA and shows similar properties when used as a chiral selector in liquid chromatography. The protein contains a single polypeptide chain containing 580 amino acids with a molecular mass of 69,000. The isoelectric point of this protein is 4.8. The 17 disulfide bridges give rise to 9 loops grouped into 3 homologous domains, each containing 3 loops. The slight difference between BSA and HSA lies in the third loop. The binding areas of this protein are benzadiazapine–indole and azapropazone–warfarin sites.

6.1.3 α_1-Acid Glycoprotein

α_1-Acid glycoprotein or orosomucoid is a human plasma protein and present in 55–140 mg/mL of human plasma. It is a glycoprotein with 181 amino acids and has a molecular mass of 41,000. It is acidic in nature, with an isoelectric point of 2.7. It is a stable protein with two disulfide bridges. It also contains 40 sialic acid residues. The sialic acid residues bind with ammonium-type compounds at neutral pH and are essential for the enantioselective process.

6.1.4 Ovomucoid

Ovomucoid protein is obtained from the white part of a chicken egg and it is an acidic glycoprotein. It contains 186 amino acids with a molecular mass of 55,000. The single chain of the protein forms three homologous domains and incorporates nine disulfide bridges. Four to five asparagine residues are glycosylated, and sialic acid constitutes 0.5–1% of the total weight. The isoelectric point is 4.5. It is capable of binding amines and acids.

6.1.5 Avidin

This is another protein obtained from chicken egg white and it is basic in nature, having an isoelectric point between 9.5 and 10.0. The protein is composed of four identical subunits, each having a molecular mass of 16,400. A glycosidic chain is attached to asparagine 17 of each subunit, which also can bind 1 mol of biotin. The advin biotin binding affinity is the strongest known and its pK is around 15, which approaches the strength of a covalent bond. Avidin has good enantioselective capacity particularly for anionic analytes (profen group) because of the net positive charge on the protein at neutral pH.

6.1.6 Ovotransferrin

Ovotransferrin is also obtained from the white portion of a chicken egg and has been used as a chiral selector in liquid chromatography. This protein is also called conalbumin. It is a metal ion (iron, copper, manganese, and zinc) binding protein of molecular mass 70,000–78,000 and with an isoelectric point of 6.1–6.6. This protein is sensitive to acids and heat.

6.1.7 Chymotrypsin

This protein is extracted from pancreatic tissues. This protein occurs in α-form and β-form, but the α-form is used as chiral selector in liquid chromatography. The molecular mass is 25,000, with an isoelectric point of 8.1–8.6. It is inhibited by metal ions. The protein is useful for chiral resolution of amino acids and amino esters.

6.1.8 Cellobiohydrolase-I

Most of the protein used as chiral selectors in liquid chromatography are obtained from animals, but cellobiohydrolase-I is the first plant protein. It is obtained from *Trichderma reesi* fungus. Basically, it is one of the cellulose-degrading enzymes (cellulases). It is an acid glycoprotein having an isoelectric point of 3.6. The molecular weight of this protein is 60,000. This protein contains two parts (viz. cellulose binding and glycosylated regions).

6.2 PREPARATION AND COMMERCIALIZATION

As discussed earlier, the proteins used as chiral selectors in affinity chromatography cannot be used under the high-pressure HPLC with a variety of mobile phases; therefore, these proteins were immobilized with some solid support such as hydroxyethylmethacrylate, polystyrene–divinylbenzene, polyethylene fibers, and silica gel [14,15]. A variety of techniques have been used for the immobilization of proteins on a solid support. Some of these make use of reactions causing cross-linking of proteins, whereas others leave the protein in its monomeric state. In the early stage, BSA was adsorbed onto a silica gel surface [18] and anion-exchange resin [19] and was used for chiral resolution. However, the limited choice of mobile phases makes this technique less attractive.

Later, proteins were bonded covalently to a spherical silica gel surface (most commonly used solid in liquid chromatography). There are a number of synthetic procedures available to achieve the goal. The protein can be linked to a silica surface through amino or carboxylic groups, but the most common methods involve binding through the amino group. The amino group is protected in many ways and the most common reagent used is *tert*-butyloxycarbonyl azide; the procedure is shown in Figure 1. Similarly, silica gel is also reacted to bind the appropriate organic moiety (by silane reagent) to which the peptide or protein can be attached. A commonly employed reagent for this purpose is 3-glycidoxypropyl triethoxysilane and the reaction path is presented in Figure 1. The protected

Step I

$$\underset{t\text{-Butyloxycarbonylazide}}{(CH_3)_3COCON_3} + \underset{\text{Protein}}{NH_2CRHCOOH} \rightarrow \underset{\text{Protected Protein}}{(CH_3)_3COCONHCRHCOOH}$$

Step II

$$\underset{\text{3-Aminopropyldimethylethoxysilane}}{NH_2(CH_2)_3(CH_3)_2SiOC_2H_5} + \underset{\text{Silica Gel}}{HO\text{-}Si\sim(\text{Silica Gel}} \rightarrow \underset{\text{Modified Silica Gel}}{NH_2(CH_2)_3(CH_3)_2SiO\sim(\text{Silica Gel}}$$

Step III

$$(CH_3)_3COCONHCRHCOOH + NH_2(CH_2)_3(CH_3)_2SiO\sim(\text{Silica Gel}$$

$$\downarrow$$

$$(CH_3)_3COCONHCRHCONH(CH_2)_3(CH_3)_2SiO\sim(\text{Silica Gel}$$

Step IV

$$\underset{\text{Protected Immobilized Protein}}{(CH_3)_3COCONHCRHCONH(CH_2)_3(CH_3)_2SiO\sim(\text{Silica Gel}} + HCl$$

$$\downarrow$$

$$\underset{\text{Unprotected Immobilized Protein}}{NH_2CRHCONH(CH_2)_3(CH_3)_2SiO\sim(\text{Silica Gel}}$$

FIGURE 1 The chemical pathway of protein immobilization on silica gel.

protein is allowed to react with modified silica gel in dichloromethane. If the protein is large and insoluble in dichloromethane, then a more polar solvent should be used. The protecting reagent is removed at the end of the reaction using hydrochloric acid to free the amino group, as shown in Figure 1. The influence of the cross-linking reagent on the enantioselectivity of the BSA column was studied in Ref. 20. Glutaraldehyde, formaldehyde, and (*N*,*N*-disuccinimidyl)carbonate were used as bifunctional reagents for the immobilization of BSA. The chiral stationary phases (CSPs) prepared this way showed different chromatographic enantioselectivities [20]. When there are number of amino and carboxylic acid groups available on the same protein, care must be taken not to allow these groups to react with themselves because these types of phase have a limited choice of mobile phases. In this way, the proteins can be bonded covalently to silica gel and are packed in stainless-steel columns of different dimensions for use in HPLC. Different companies have commercialized these columns by different trade names. The commonly used chiral columns supplied by various companies are given in Table 1.

TABLE 1 Various Protein-Based Commercial Chiral Columns

Trade names (protein names)	Company
Chiral AGP (α_1-acid glycoprotein), Chiral HSA (HSA protein), Chiral CBH (cellobiohydrolase-I protein)	Advance Separation Tech., Whippany, NJ, USA
Chiral AGP (α_1-acid glycoprotein), Chiral CBH (cellobiohydrolase-I protein), Chiral HSA (HSA protein)	Chrom Tech, Ltd., Cheshire, UK
Resolvosil BSA-7, Resolvosil BSA-7PX (BSA proteins)	Macherey-Nagel, Duren, Germany
Chiral AGP (α_1-acid glycoprotein), Chiral CBH (cellobiohydrolase-I protein), Chiral HSA (HSA protein)	Regis Technologies, Morton Grove, IL, USA
AFpak ABA-894 (BSA protein)	Showa Denko, Kanagawa, Japan
Keystone HSA (HSA protein), Keystone BAS (BSA proteins)	Thermo Hypersil, Bellefonte, PA, USA
TSKgel Enantio L1, TSKgel Enantio-OVM (ovamucoid protein)	Tosoh, Tokyo, Japan
EnantioPac (α_1-acid glycoprotein)	LKB Pharmacia, Bromma, Sweden
Bioptic AV-1 (advin protein)	GL Sciences, Tokyo, Japan
Ultron ES-BSA (BSA protein), Ultron ES-OVM Column (ovomucoid protein), Ultron ES-OGP Column (ovoglycoprotein), Ultron ES-Pepsin (pepsin)	Shinwa Chemical Industries, Kyoto, Japan

6.3 APPLICATIONS

6.3.1 Analytical Separations

In 1973, Stewart and Doherty [9] resolved enantiomers of tryptophan on a column packed with BSA–succinoylaminoethyl–agarose in a discontinuous elution procedure. The mobile phase used was 0.1 M borate buffer (pH 9.2). The chromatograms of this classical research are shown in Figure 2. Several years later, this technique was applied for the chiral resolution of warfarin enantiomers [10]. In 1981, the enantiomers of tryptophan and warfarin racemates were resolved on various serum albumin CSPs [11,21,22]. The same method was used for the resolution of other drugs [12–14]. Allenmark et al. [23] studied the resolution of a series of active racemic sulfoxides on a BSA column using 0.08 M phosphate buffer (pH 5.8) as the eluting solvent.

The protein phases were the first to be developed and are still in use for the chiral resolution of different racemates. Because both BSA and HSA are similar in structure, they have similar enantiorecognition capabilities. Warfarin and

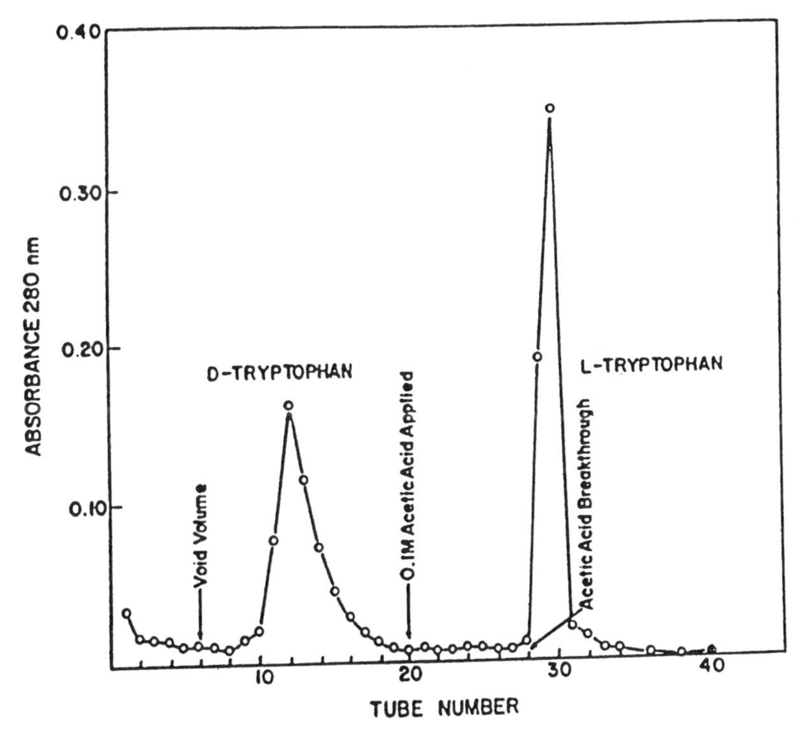

FIGURE 2 Enantiomeric resolution of DL-tryptophan on defated BSA–succinoyl-aminoethyl–agrose column (0.9×25 cm) using borate buffer (0.1 M). (From Ref. 9.)

indole are the main sites for chiral recognition for racemates (Sec. 6.1.2). It has been demonstrated that compounds belonging to benzodiazepines bind to the indole site [24–26]. Basic and neutral racemates were resolved on these CSPs. However, some acidic compounds were also separated [14]. Allenmark carried out an extensive and remarkable work on the chiral resolution of various racemates using BSA-based CSPs. The enantiomers of glutamic acid, warfarin, *N*-benzoylamino acids [27,28], benzonals, *N*-acetyl tryptophan, *N*-benzoyl phenylglycine [29], chlorthalidone (a diuretic and closely related to the benzothiadiazine group), tryptophan derivatives [30], benzoin, *N*-2,4-dinitrophenyl-aspartic (DNP) acid [31,32], some sedative, diuretics, muscle-relaxant drugs [33], *N*-acetyl tryptophan ethyl ester, *N*-(4-nitrobenzoyl) [34], enantiomers of degradation of certain drugs [35], phthalimido-threonine, *N*-benzenesulfonyl serine [36], anticonvulsant benzonals and a series of analogs [37], *N*-(2,4-dinitrophenyl)- and dansyl amino acids [38], and other derivatized amino acids [39] were resolved successfully on BSA-based CSPs. The mobile phases used for these studies were phosphate buffers of different concentrations and pHs. The organic modifiers used were 1-propanol, 2-propanol, octanoic acid, and hexanoic acid. A mixture of three peptides was obtained by enzymatic cleavage of BSA and were immobilized on silica gel. The developed CSPs were tested for the chiral resolution of tryptophan derivatives, oxazepam, and ibuprofen drugs [40]. Lienne et al. [41] resolved cytotoxic α-methylene-γ-butyrolactone enantiomers on Resolvosil CSP (BSA) using phosphate buffer (7.5 mM, pH 6.9)–2-propanol (97.5 : 1.5, v/v) as the mobile phase. The BSA fragments (FG75 and F2) were isolated and bonded to silica gel. The enantioselectivity was investigated for 2-arylpropionic acid derivatives, benzadiazepines, warfarin, and benzoin racemates [42]. Recently, the enantioselectivity of BSA coated to zirconia was also investigated [43]. To show the nature of chiral resolution on BSA-based CSP, the chromatograms of chiral resolution of DNP- and dansyl glutamic acid are shown in Figure 3.

The enantiomers of oxazepam hemisuccinate were resolved on HSA and acylated HSA proteins [44]. The native enantioselectivities in the binding of HSA toward 2-aryl propionic acid nonsteroidal anti-inflammatory drugs (fenoprofen, benoxaprofen, indoprofen, naproxen, ketoprofen, flurbiprofen, ibuprofen, suprofen, and pirprofen) were investigated [45]. The chiral resolution was achieved using phosphate buffer (50 mM, pH 6.9)–acetonitrile (85 : 15, v/v) as the mobile phase. The capacity factors were modified using octanoic acid as the mobile phase modifier. The chiral recognition of tryptophan enantiomers on the HSA column was studied [46]. The equilibrium constants for the two enantiomers were determined and were found to be different [46]. Derivatization of the free cysteine$_{34}$ in HSA anchored to silica was found to be enantioselective for the enantiomers of warfarin and phenylbutazone [47]. A series of 12 carboxylic acid racemates were resolved on the HSA column using potassium phosphate buffer (0.1 M, pH 7.0)–1-propanol (98 : 2, v/v) as the mobile phase [48]. The chiral

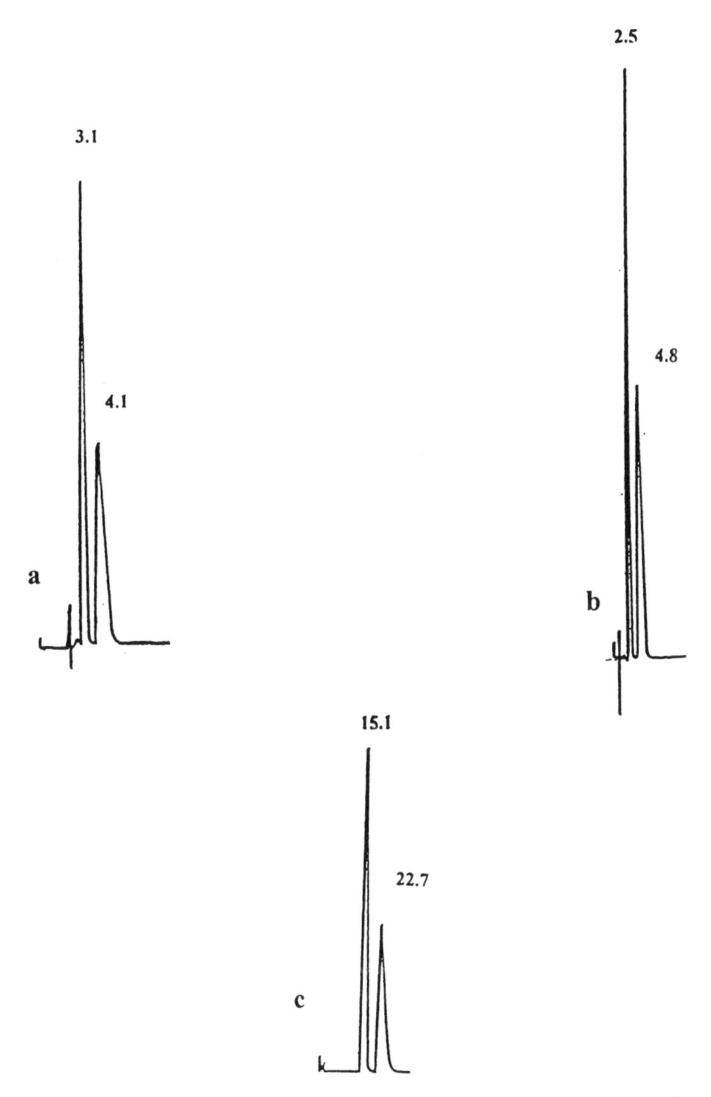

FIGURE 3 Chromatograms of (a) DNP-glutamic acid, (b) dansyl glutamic acid, and (c) ibuprofen on BSA (a, b) and HSA (c) CSPs using (a) phosphate buffer (50 mM)–1-propanol (96 : 4, v/v, pH 7.7), (b) phosphate buffer (50 mM)–1-propanol (98 : 2, v/v, pH 7.9), and (c) phosphate buffer (50 mM, pH 6.9)–acetonitrile (85 : 15, v/v) containing 4 mM octanoic acid. (From Refs. 38 and 45.)

discrimination of *N*-(dansyl)-amino acids was carried out on the HSA protein column. Thermodynamic parameters for the transfer of enantiomers from the mobile phase to the CSP were determined [49,50]. In another study, the effect of a polymer sublayer on the chiral recognition ability of HSA was determined [51]. The chromatograms of the enantiomeric resolution of ibuprofen on HSA column are given in Figure 3.

α_1-Acid glycoprotein was found to recognize basic compounds. However, acidic and neutral compounds have also been resolved. The protein was used very frequently for the chiral resolution of a wide variety of racemates. The most commonly used mobile phases were buffered aqueous systems containing various amounts of organic modifiers (1–10%). The most important racemates resolved are verapamil, epibatidine, leukotriene D_4 antagonist, vamicamide [52], amines, amino alcohols, barbiturates, carboxylic acids and their derivatives, lactams, lactones, hydantoins [12–14,53], idrapril [54], thalidomide [55], cholecystokinin antagonist 4(3*H*)-quinazolone derivatives [56], and methadone [57]. Suckow et al. [58] coupled the chiral AGP column with Supelcosil LC-15 (achiral column) and used this combination for the chiral resolution of phenylmorpholinol metabolites of bupropion in human plasm, as shown in Figure 4.

Ovomucoid (OVM) protein-based CSP was found suitable for the chiral resolution of both nonprotolytes as well as acidic and basic protolytes. The CSP was explored for the chiral resolution of the two categories of racemates. Phosphate buffers with ethanol or acetonitrile were found to be suitable mobile phases [12–14,62]. Lammerhofer and Lindner carried out the optical purity determination of econazole and other imidazoles on OVM CSP using phosphate buffer (0.01 M, pH 6.5)–acetonitrile (70 : 30, v/v) as the mobile phase [63]. CSPs obtained from ovoglycoprotein and crude ovomucoid were compared for the chiral recognition. The CSP obtained from crude ovomucoid showed better chiral recognition ability than the CSP obtained from ovoglycoprotein for some racemic compounds [64]. The enantioselectivity of ovoglycoproteins obtained from chicken and Japanese quail egg whites were compared. The chicken egg protein column was found suitable for the basic compound and the CSP obtained from the quail egg protein showed good chiral recognition ability for acidic racemic compounds [65]. The effect of immobilization of ovoglycoprotein on chiral resolution was studied. The CSP prepared from the amino group of protein was found suitable for basic racemates, whereas the CSP obtained by bonding the carboxylic group of protein showed good chiral resolution characteristics for acidic compounds [66]. In 1992, Mano et al. [67] introduced a CSP based on ovotransferin (conalbumin) protein and has been used for the chiral resolution of basic drugs having a p*K* value greater than 8.0. The enantioselectivity of this CSP is pH-specific and a baseline resolution for azelastine was achieved using phosphate buffer (50 mM, pH 5.0)–ethanol (80 : 20, v/v) [64] as the the mobile phase. A new column-switching method for chiral analysis of 4-(4-chlorobenzoyl)-

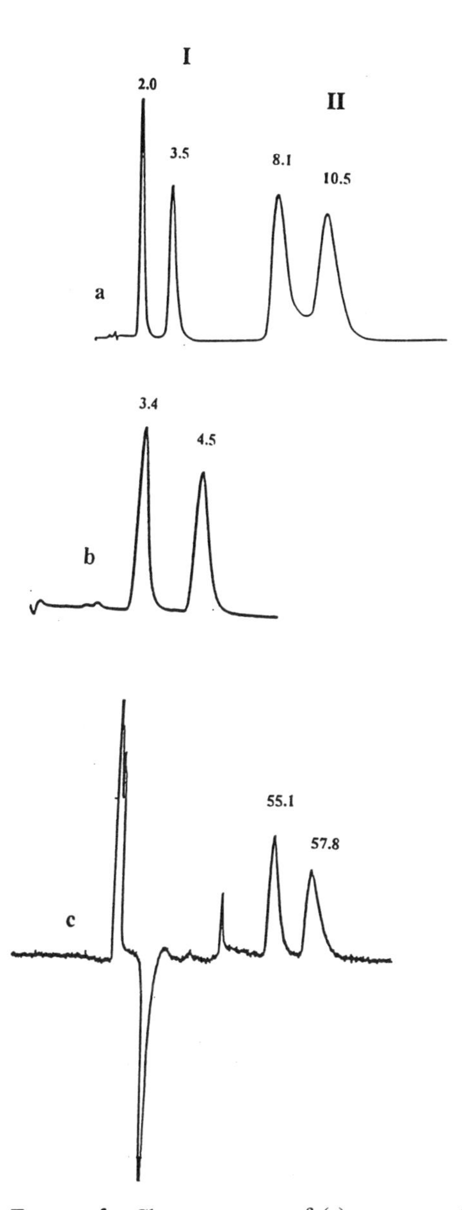

FIGURE 4 Chromatograms of (a) oxazepam (I) and oxazepam hemisuccinate (II) on Chiral AGP CSP, (b) 8-hydroxy-2-(di-*N*-propylamino)-tetraline hydrochloride on ovomucoid CSP and (c) 4-(4-chlorobenzyl)-2-(hexahydro-1-methyl-1*H*-azepin-4yl)-1-(2*H*)-phthalazinone on ovotransferin CSP using (a) phosphate buffer (0.01 M)–acetonitrile (95 : 5, v/v), (b) phosphate buffer (11 mM)–acetonitrile (83 : 17, v/v) and (c) phosphate buffer (1 mM)–methanol (90 : 10, v/v) as the mobile phases. (From Refs. 59–61.)

2-(hexahydro-1-methyl-1*H*-azepin-4-yl)-1-(2*H*)-phthalazinone in plasma was developed by coupling the chiral column with a mass spectrometer [59]. The chromatograms of the chiral resolution on ovomucoid and ovotransferin are given in Figure 4. Recently, Monaco et al. [68] immobilized chicken liver fatty acid binding proteins (FABPs) to silica gel and the developed stationary phase was used to examine the enantioselectivity of some chiral drugs. The enantiomers of basic and neutral drugs were poorly retained and not resolved by the developed CSP. Contrarily, the resolution of some acidic racemates was achieved.

Avidin, having a positive charge at neutral pH, was used for the chiral resolution of anionic analytes such as profens. An increase in the pH of the mobile phase has resulted in a drastic decrease in retention of racemates [69]. It has also been reported that avidin could not resolve the basic racemates even at partial level because of its basic character [69]. It was also shown that biotin completely eliminates the chiral recognition ability of this CSP, as observed from the experiments carried out by Miwa et al. [69]. Oda and co-workers [70] prepared avidin CSP and this was applied for the chiral resolution of ketoprofen in plasma. The mobile phase used was 0.1 M sodium hydrogen carbonate buffer [70]. In 1992, Oda et al. [71] carried out the chiral resolution of 1-benzyl-4-[(5,6-dimethoxy-1-indanon)-2-yl] methyl piperidine–hydrochloride and its metabolites in plasma on the avidin column. The authors reported that the hydrophobicity (alkyl chain length) of a spacer affected the enantioselectivity significantly. Later, Oda et al. [72] coupled avidin CSP with ovumucoid protein column and resolved the enantiomers of many drugs in plasma. This research group compared the chiral recognition capacities of avidin and ovomucoid proteins using flurbiprofen, warfarin, and propranolol racemates [73]. Both CSPs were found to have excellent stability and there was no significant decrease in retention or enantioselectivity even after the injection of 200 samples [73]. In another study, the same authors [74] studied the chiral recognition ability of avidin acylated protein at both amino and carboxylic groups separately. These CSPs were used for the successful chiral resolution of profen drugs. The chiral recognition capacities of these acylated avidin proteins were found to be different. Recently, the chiral resolution of some anti-inflammatory drugs was described on avidin CSP by Smet et al. [75]. The authors described the combination of the central composite design results in a multicriteria decision-making approach in order to obtain a set of optimal experimental conditions, leading to the most desirable compromise between resolution and analysis time. The chromatograms of the chiral resolution of piperofen on this CSP are presented in Figure 5.

Wainer et al. [77] presented a CSP based on α-chymotrypsin protein, and, initially, the chiral resolution of certain amino acids and amino ester was achieved on this protein CSP [14,77]. Later, this CSP was used for the chiral resolution of dipeptides and profens [78,79]. Recently, Felix and Descorps [80,81] used immobilized α-chymotrypsin for the chiral resolution of a variety of racemic compounds. Cellobiohydrolase-I (CBH-I) immobilized to silica gel was found to

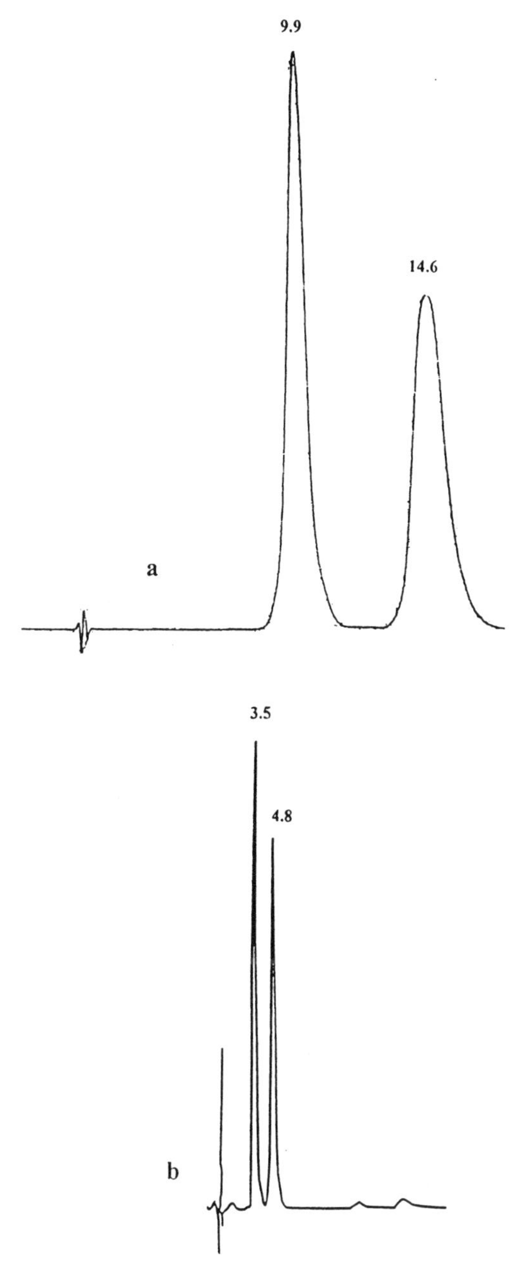

FIGURE 5 Chromatograms of (a) pirprofen on avidin CSP using phosphate buffer (0.1 M)–acetotirile (95 : 5, v/v) and (b) cathionine on Chiral CBH-I CSP using phosphate buffer (10 mM)–ethanol (95 : 5, v/v, pH 6.0 containing 50 μM Na_2EDTA) as mobile phases. (From Refs. 75 ad 76.)

be a highly interesting CSP for the chiral resolution of racemic drugs [82]. CBH-I was immobilized on silica gel using different methods, but it was found that the use of an epoxide silica gel gave the best chiral recognition. Trypsin-based CSP was presented by Thelohan et al. [83] and the authors resolved the enantiomers of O and N,O derivatives of amino acids. A successful chiral resolution of amino alcohols and barbiturates was achieved on CBH-I protein using phosphate buffer (pH 6.8) containing 2-propanol [84]. Allenmark and Andersson [85] achieved a baseline chiral resolution of β-blockers on the CBH-I protein column. Aboul-Enein and Serignese [76] described a direct isocratic HPLC method for the chiral resolution of cathinone and its metabolites on CBH-I CSP under reversed-phase conditions. The chromatograms of the chiral resolution of cathinone on CBH-I CSP are given in Figure 5. Recently, Gotmer et al. [86] studied the influence of solute hydrophobicity on the enantioselectivity of CBH-I CSP using racemic β-blockers. The authors fit the adsorption data to the bi-Langmuir model. The adsorption of the *S*-enantiomer was found exothermic at low pH, endothermic at high pH and athermal at a narrow pH range depending on the β-blocker under investigation. The order of increasing energy of the chiral interactions was the same as that of hydrophobicity (i.e., propranolol > alprenolol > metoprolol). Some applications of the protein-based CSPs are summarized in Table 2.

In addition to the use of the above-cited proteins as chiral selectors, other proteins were also explored as suitable tools for chiral resolution in liquid chromatography. Lysozyme was bonded to silica gel by Haginaka et al. [92]. The authors used this CSP for the chiral resolution of basic and neutral racemic compounds, whereas no resolution of acidic compounds was reported. The same authors [93] introduced a new pepsin-based CSP. The developed CSP was found useful for the chiral resolution of basic and neutral racemic compounds, whereas the chiral resolution of acidic compounds had not been reported. Furthermore, it has also been noted that the pepsin column lost chiral recognition capacity using an eluent of basic pH. In another study, Haginaka et al. [94] immobilized a pepsin and ovamucoid protein mixture on porous aminopropyl silica gel. The enantioselectivity of this CSP was compared and it was found that this column was superior to the pepsin-based column. On the other hand, the enantioselectivity of this mixed-protein CSP was comparable to the CSP obtained from the ovomucoid protein. In 1999, Karlsson et al. [95,96] immobilized amyloglucosidate on aldehyde silica gel and the developed CSP was tested for the chiral resolution of β-blockers. The authors observed a decrease in retention of some β-blockers by increasing the buffer ionic strength without any loss of enantioselectivity.

6.3.2 Preparative Separations

It is well known that a small fraction of protein is involved in the chiral resolution at the analytical scale and, therefore, the preparative-scale capabilities of protein

TABLE 2 Enantiomeric Resolution of Different Racemic Compounds Using Protein-Based CSPs

Racemic compounds	CSPs	Refs.
Amino acids (aromatic)	BSA	21
Glutamic acid, warfarin and *N*-benzoylamino acids	BSA	27, 28
Benzonals, *N*-acetyl tryptophan, *N*-benzoyl phenylglycine	BSA	29
Chlorthalidone and tryptophan derivatives	BSA	30
Benzoin and *N*-DNP-aspartic acid	BSA	31, 32
Sedative, diuretics and muscle relaxant drugs	BSA	33
N-Acetyl tryptophan ethyl ester and *N*-(4-nitrobenzoyl)	BSA	34
Enantiomers of degradation of certain drugs	BSA	35
Phthalimido-threonine and *N*-benzenesulfonyl serine	BSA	36
Anticonvulsant benzonals and a series of analogs	BSA	37
N-(2,4-Dinitrophenyl)- and dansyl amino acids	BSA	38
Derivatized amino acids	BSA	39
Tryptophan derivatives, oxazepam and ibuprofen drugs	BSA	40
Cytotoxic α-methylene-γ-butyrolactone	BSA	41
2-Arylpropionic acid derivatives, benzadiazepines, warfarin and benzoin racemates	BSA	42
Carboxylic acids	BSA	76
Sulfoxides and sulfoximines	BSA	22,23
2-Aryl propionic acid non-steroidal anti-inflammatory drugs (fenoprofen, benoxaprofen, indoprofen, naproxen, ketoprofen, flurbiprofen, ibuprofen, suprofen and pirprofen)	HSA	45
Tryptophan enantiomers	HSA	46
Carboxylic acid racemates	HSA	48
Derivatized amino acids, profens and other organic acids	HSA	53
Verapamil, epibatidine, leukotriene D_4 antagonist, and vamicamide	AGP	52
Amines, amino alcohols, barbiturates, carboxylic acids and their derivatives, lactums, lactones, and hydantoins	AGP	12–14, 53
Idrapril	AGP	54
Thalidomide	AGP	55
Chlecystokinin antagonist 4(3*H*)-quinazolone derivatives	AGP	56
Methadone	AGP	57
Amines, amino alcohols, barbiturates, benzodiazepinenones, benzothiadiazines, coumarin derivatives, and dihydropyridines	AGP	87

(continued)

TABLE 2 Continued

Racemic compounds	CSPs	Refs.
Carboxylic acids and derivatives, hydantoins, and lactams	AGP	88
Vinca alkaloids and other drugs	AGP	53
Econazole and other imidazoles	OVM	60
Amines and amino alcohols	OVM	89
Barbiturates	OVM	90
Verapamil	OVM	91
Profens, β-blockers, and other drugs	OVM	59
β-Blockers, halofantrine, and other drugs	OVM	60
Azelastine	Ovotransferin	67
4-(4-Chlorobenzoyl)-2-(hexahydro-1-methyl-1*H*-azepin-4-yl)-1	Ovotransferin	59
Ketoprofen	Avidin	70
1-Benzyl-4-[(5,6-dimethoxy-1-indanon)-2-yl] methyl piperidine–hydrochloride and its metabolites	Avidin	71
Anti-inflammatory drugs	Avidin	74
Amino acids and amino esters	α-Chymotrypsin	14
Cathinone and its metabolites	α-Chymotrypsin	79
Amino alcohols and barbiturates	CBH-I	85
β-Blockers	CBH-I	86
β-Blockers and other drugs	CBH	53

columns are not very good. However, Erlandsson et al. [18] immobilized BSA, an inexpensive and readily available protein, and packed a 500 × 22-mm column. This column was used to resolve the enantiomers of tryptophan at the preparative scale. Only 0.25 mg of tryptophan was resolved in a single run using this column. The conditions for the preparative chiral separation on BSA column is described by Jacobson et al. [97].

6.4 OPTIMIZATION OF HPLC CONDITIONS

One of the advantages of protein-based CSPs is that chiral chromatography is carried out under the reversed-phase mode; that is, aqueous mobile phases are used frequently and, therefore, there is a great chance to optimize the chiral resolution. The most important parameters to be optimized are the composition of the mobile phase, the pH of the mobile phase, ionic strength, and temperature. The optimization of these parameter are discussed herein.

6.4.1 Mobile Phase Compositions

Primarily, buffers of different concentrations and pHs are used for the chiral resolution on these CSPs because these phases are stable under the reversed-phase mode. The elution is carried out by the isocratic mode, with the exception of a few reports for which gradient elution was used. The most commonly used buffers are phosphate and borate. These buffers were used in the concentration range 20–100 mM with a 2.5–8.0 pH range. However, with all silica-based CSPs, the prolonged use of alkaline pH buffer (>8.5) is not suitable. Contrarily, at lower pH, irreversible changes are possible in cross-linked protein phases and, hence, the long-term of buffers of low pH for the best are not recommended. In view of this, buffers in the 3–7 pH range are of choice. An ammonium acetate buffer of pH 4.5 may be useful. For the mobile phase development, any buffer (50 mM, pH 7) can be used and the optimization is carried out by changing the concentration and pH. The use of organic solvents in the optimization of chiral resolution on these CSPs has been reported successfully because the hydrophobic interactions are effected by their use. The most commonly used organic solvents are methanol, ethanol, 1-propanol, 2-propanol, acetonitrile, tetrahydrofuran (THF), hexanoic acid, and octanoic acid. These organic modifiers were used in the 1–10% range. Care must be taken when using these organic modifiers, as they can denature the protein. However, the high concentrations of methanol and acetonitrile have been used on some of the cross-linked protein CSPs. The selection of these organic modifiers depends on the structure of racemic compounds and the CSP used. In some cases, charged modifiers such as octanoic acid and quaternary ammonium compounds have been used for the optimum chiral resolution [87,98]. Allenmark et al. [32] studied the effect of 1-propanol concentration on the chiral resolution of certain substituted amino acids on BSA (Fig. 6a). It was found that 1–10% 1-propanol was suitable for the best chiral resolution [32]. Furthermore, Andersson and Allenmark [29] have carried out a study on the effect of octanoic acid and octyl amine as the organic modifiers on the chiral resolution ibuprofen and ketoprofen. The results of these findings are presented in Figure 6b, which indicates 1–2 mM as suitable concentrations of these organic modifiers. A higher concentration of both modifiers resulted in poor enantioselectivity. Oda et al. [99] studied the effect of THF concentration on the chiral resolution of ketoprofen using ovomucoid CSP. These findings are shown in Figure 6c, which indicates 4–8% as the best concentration of THF under the reported chromatographic conditions. Furthermore, Oda et al. [73] studied the chiral resolution of flurbiprofen, propranolol, and warfarin on avidin and ovomucoid CSPs using methanol, ethanol, propanol, and acetonitrile organic

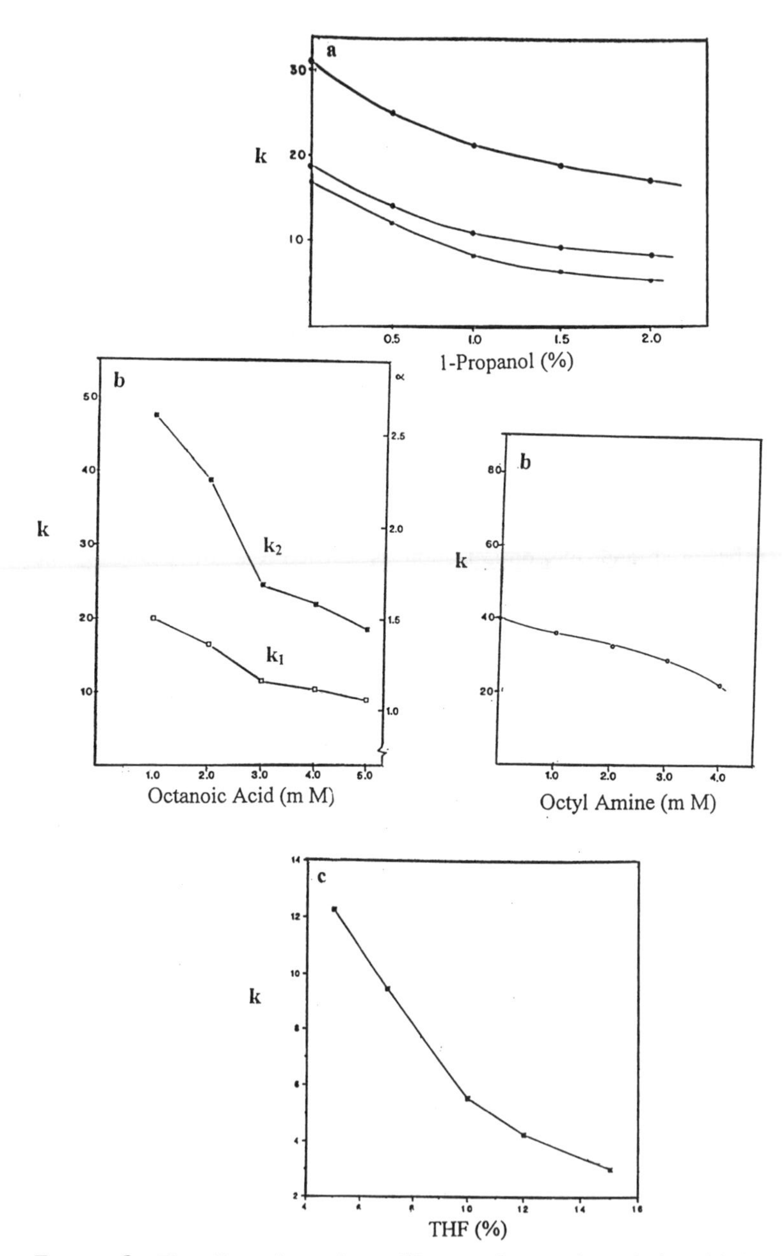

FIGURE 6 The effect of organic modifiers on the enantioresolution: (a) 1-propanol on enantiomers of $C_6H_5CONHCHRCO_2H$ where R is (▲) $-CH_2OH$, (■) $-CH_3$ and (●) $-CH_3$ on BSA CSP; (b) octanoic acid (on ibuprofen) and octyl amine (on ketoprofen) on BSA CSP; (c) THF (ketoprofen) on ovomucoid CSP with phosphate buffer as the main component of the mobile phases in all the studies. (From Refs. 29, 32, and 99.)

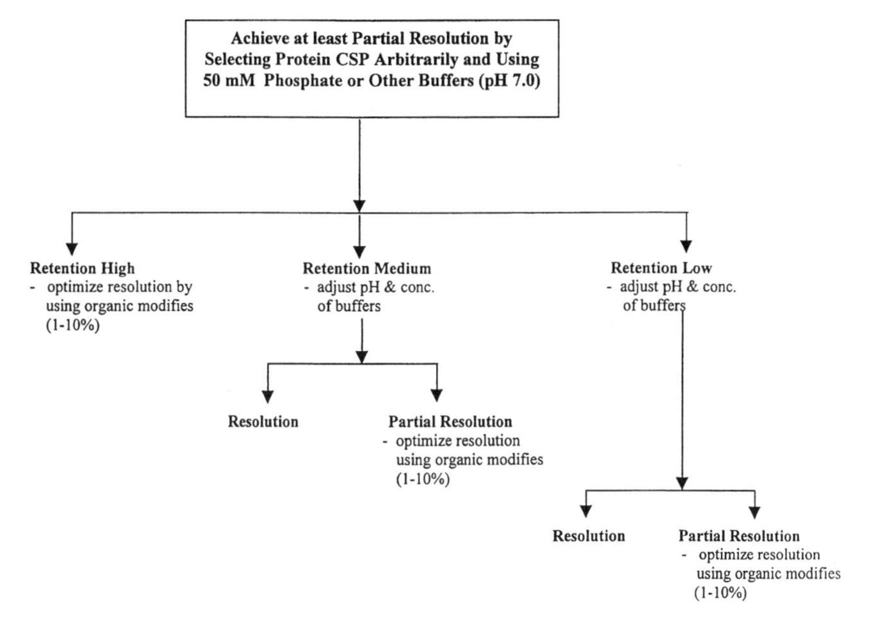

SCHEME 1 The protocol for the development and optimization of mobile phases on protein CSPs.

modifiers. Generally, the chiral recognition behavior of these modifiers on avidin and ovomucoid CSPs was in the order methanol > ethanol > propanol > acetonitrile. Similarly, various reports have been published on the optimization of chiral resolution using different types of mobile phase and the concentrations of organic modifiers [100–103]. Briefly, the selection of mobile phases varies from one solute to another and also depends on the structure of the protein CSP used. A protocol for the development and optimization of mobile phases for the chiral resolution on protein-based CSPs is presented in Scheme 1.

6.4.2 pH of the Mobile Phase

The influence of pH is more pronounced on charged analytes than the uncharged ones. The decrease in retention of carboxylic acids by increasing pH is due to the decreasing number of positive charges on the protein, leading to fewer charge interactions with ionic analytes [14]. Similarly, retention factors of amines and other basic racemates decrease with a decrease of pH due to the decreasing number of negative charges on the protein for charge interactions with the cationic analytes. Therefore, chiral resolution on protein-based CSPs can be controlled by adjusting the pH of the eluent. The effect of pH on the chiral resolution of different racemic compounds on protein-based CSPs was studied by various workers. The effect of pH on the chiral resolution of certain amino acids on the BSA-based CSP was carried out by Allenmark et al. [32] and is given in Figure 7a. It is clear from this figure that pH values of 4–6 are suitable for the chiral resolution under the reported chromatographic conditions. Fitos et al. [61] studied the effect of pH on the chiral resolution of oxazepam on the Chiral AGP CSP (Fig. 7b). It may be concluded from this Figure 7b that, again, values of pH between 5 and 6 are suitable for the chiral resolution of the reported drugs (in the presence of acetonitrile and isopropanol as organic modifiers separately). Similarly, the best chiral resolution of *N*-benzenesulfonyl alanine was achieved on BSA-based CSP using the mobile phase having pH values of 5–6 (Fig. 7c) [104]. Mano et al. [67] reported the best chiral resolution of azelastine on conalbumin CSP at 5.5 pH. Furthermore, Mano et al. [105] studied the effect of pH on the chiral resolution of profens, warfarin, and benzoin racemic compounds using flavoprotein-conjugated silica gel CSP and the best results were obtained at pH values of 4 – 4.8. Oda and co-workers [73] studied the chiral resloution of a variety of racemic compounds on ovomucoid and avidin protein CSPs. It was observed that higher pH values were suitable for chiral resolution with avidin CSP, whereas good resolution was observed at lower pHs when ovomucoid CSP was used. Recently, Munro and Walker [103] studied the effect of pH on the chiral resolution of buprion enantiomers on the ovomucoid column. Finally, there are several reports dealing with the optimization of chiral resolution by pH control and it was found that pHs ranging from 3 to 6 are the best for chiral resolution on most of the protein-based CSPs.

6.4.3 Ionic Strength of Mobile Phase

The chiral resolution on protein CSPs is affected greatly by the change of ionic strength. A decrease of ionic strength results in an increase of charge interactions between analytes and protein CSP and, accordingly, an increase in the retention of enantiomers results. At high ionic strength, the hydrophobic interactions are often favored and this may result to a more complex dependence, yielding a retention minimum at a particular salt (buffer) concentration. Allenmark et al. [32] studied the chiral resolution of certain amino acids over a range of buffer concentrations and the results are given in Figure 8a, indicating the optimum resolution at 10–50 mM concentration of the buffer. Furthermore, Allenmark et al. [28] also studied the effect of ionic strength on the chiral resolution of *N*-benzoyl alanine on the BSA-based CSP and the results of this are shown in Figure 8b, which shows that a 50 mM buffer concentration is suitable for the purpose. In another study, Allenmark and Andersson [33] studied the effect of increasing ionic strength on some simple structurally rigid uncharged analytes (diuretics) at constant pH. The authors reported that an increase in ionic strength resulted into an increase of retention without any affect on enantioselectivity. Oda et al.

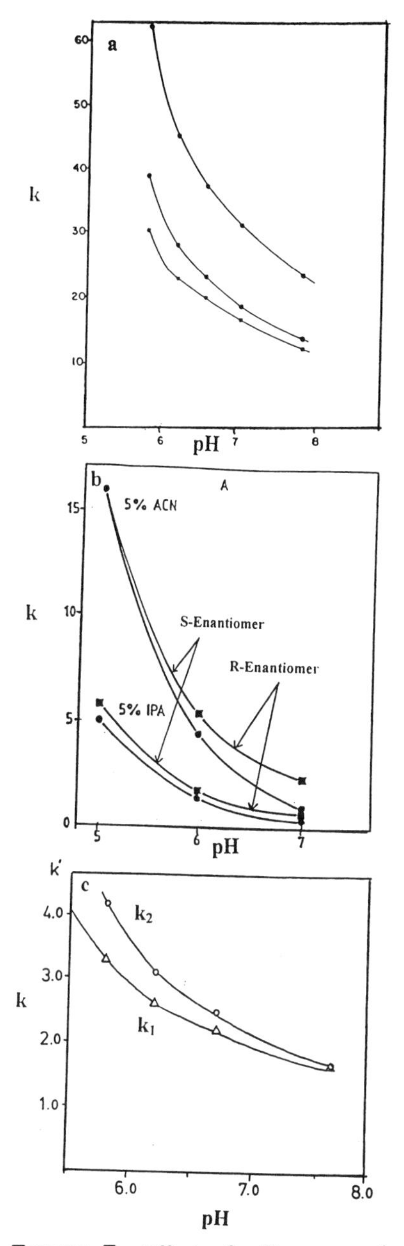

FIGURE 7 Effect of pH on enantioresolution of (a) enantiomers of C_6H_5CON $CHCHRCO_2H$, where R is (▲) $-CH_2OH$, (■) $-CH_3$ and (●) $-CH_3$ on the BSA-based CSP, (b) oxazepam on Chiral AGP CSP, and (c) *N*-benzenesulfonyl alanine on the BSA-based CSP with phosphate buffer as the main component of the mobile phases in all of the studies. (From Refs. 37, 61, 104.)

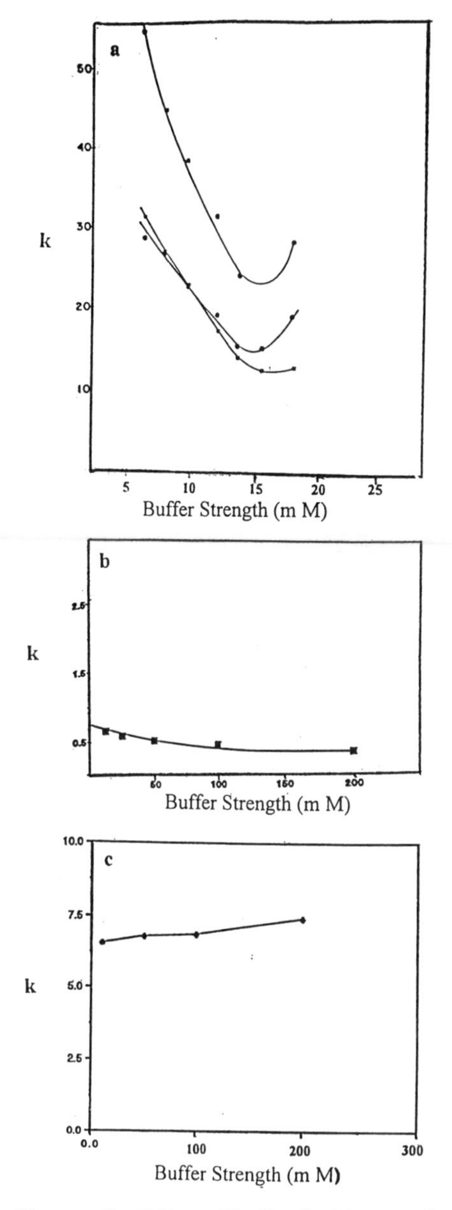

FIGURE 8 Effect of buffer (ionic) strength on enantioresolution of (a) enantiomers of $C_6H_5CONHCHRCO_2H$, where R is (▲) $-CH_2OH$, (■) $-CH_3$ and (●) $-CH_3$ on the BSA-based CSP, (b) *N*-benzoyl alanine on BSA-based CSP, and (c) ketoprofen on avidin CSP with phosphate buffer as the main component of the mobile phases in all of the studies. (From Refs. 28, 32, and 70.)

[70,99] studied the effect of ionic strength on the chiral resolution of ketoprofen, as shown in Figure 8c. It is obvious from this figure that 10 mM phosphate buffer is suitable for chiral resolution. In another study, the same authors [91] reported the best resolution of verapamil enantiomers using 50 mM as the salt concentration on ovomucoid CSP. Mano et al. [67] studied the effect of ionic strength on the enantiomeric resolution of azelastine on conalbumin CSP and the best results were obtained using the 20 mM concentration of buffer. Mano et al. [105] studied the effect of eluent concentration on the chiral resolution of profens, warfarin, and benzoin and the authors reported 10–50 mM as the best concentration of the buffers. A comparative chiral resolution study was carried out by Oda et al. [73] using flurbiprofen, warfarin, propranolol, and benzyl benzoate racemates on avidin and ovomucoid columns. The authors claimed that 50–100 mM concentrations of buffers were found suitable for good chiral resolution on both CSPs. In the same study, the authors reported the effect of phosphate, citrate, acetate, and borate anions on chiral resolution. Generally, the chiral resolution capacity was in the order phosphate > citrate > acetate > borate for all racemates and on both CSPs. Recently, Munro and Walker [103] optimized the chiral resolution of bupropion enantiomers by controlling the ionic strength of the mobile phase. Apart from these few reports discussed herein, much work was carried out by scientists on the optimization of the chiral resolution on protein CSPs by controlling the ionic concentration of the mobile phase. Therefore, the ionic strength is the key parameter for controlling the chiral resolution on protein phases.

6.4.4 Temperature

The effect of temperature on the chiral resolution of protein phases is very crucial; however, few reports are available dealing with this. Gilpin and coworkers [106] found that the retention of D-tryptophan on the BSA-based CSP increased by decreasing the temperature, whereas for L-tryptophan, there was first an increase in retention, followed by a slight decrease. The effect of temperature on the retention factors is shown in Figure 9a. Similarly, Jönsson et al. [107] observed that an increase in temperature resulted in a decrease in the retention of *R*-propranolol, whereas the *S*-enantiomer became more retained, resulting in a large value of the separation factor on CBH-I CSP. The chromatograms of propranolol enantiomers at 10°C and 45°C are shown in Figure 9b. Similarly, Fulde and Frahm [108] studied the effect of temperature on the chiral resolution of sotalol on CBH-I CSP. A reversal order of elution was observed in the 17–28°C range. A change in the enantiomeric resolution order of felodipine and mosapride and its metabolites at 20–30°C on Chiral AGP CSP was also observed by Karlsson and Aspegren [109]. Recently, Williams et al. [102] optimized the chiral resolution of roxifiban enantiomers on the AGP column by controlling the

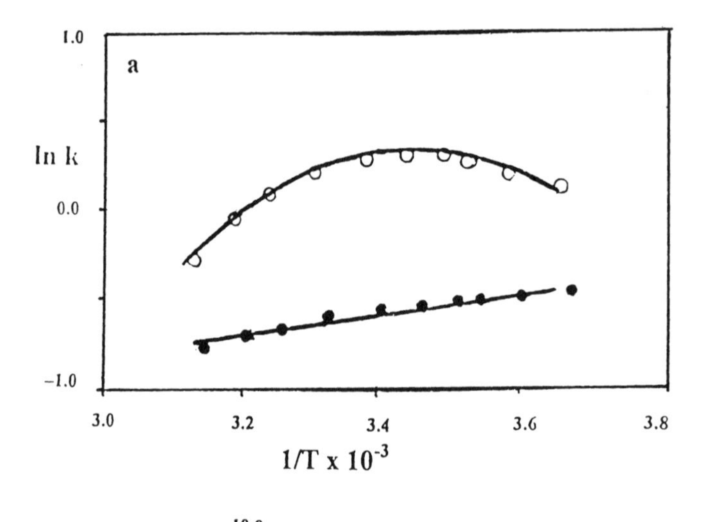

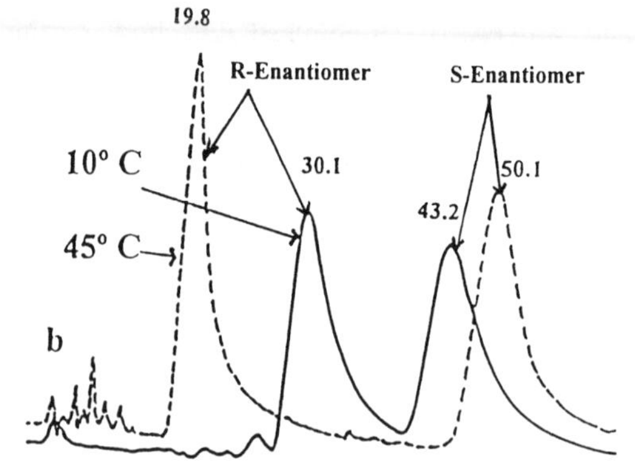

FIGURE 9 Effect of the temperature on the chiral resolution of (a) DL-tryptophan [(●) D-enantiomers and (○) L-enantiomers] on the BSA-based CSP using phosphate buffer (50 mM, pH 7.4) as the mobile phase and (b) propranolol at 10°C and 45°C on CBH-I CSP using acetate buffer (pH 5.5) as the mobile phase. (From Refs. 106 and 107).

temperatures. Munro et al. [103] also reported the optimization of the chiral resolution of bupropion enantiomers on ovomucoid CSP by temperature variations.

6.4.5 Flow Rate

The optimization of chiral resolution on protein phases by flow rate is seldom carried out. However, Kirkland et al. [60] optimized the chiral resolution of

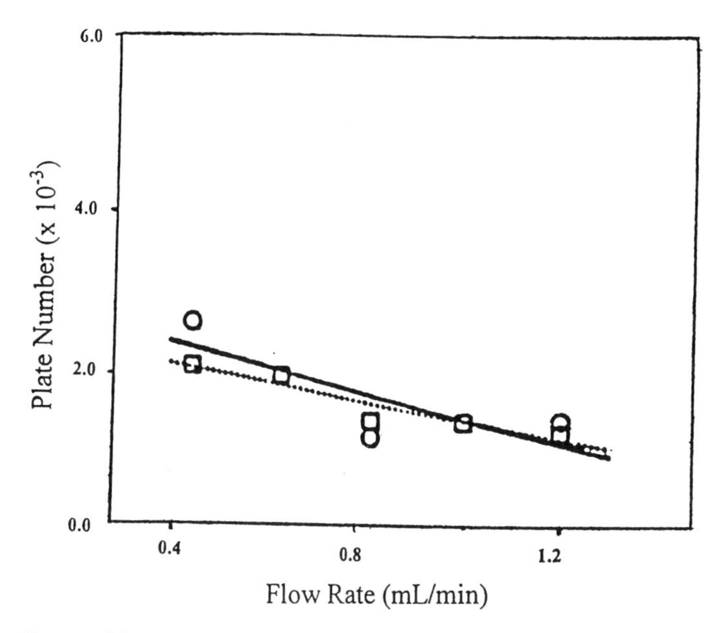

FIGURE 10 Effect of flow rate on the chiral resolution of lorglumide on ovomucoid CSP using phosphate buffer (13 mM)–acetonitrile (74 : 26, v/v, pH 6.3) as the mobile phase: (○) resolution and (□) selectivity factors. (From Ref. 60.)

lorglumide on ovomucoid protein CSP. The results of flow rate variation are given in Figure 10. The flow rate was varied from 0.2 to 1.2 mL/min and it was observed that the 1.0-mL/min flow rate was suitable for the chiral resolution of the reported drug. The authors reported that although plate numbers were somewhat smaller than those for the well-packed achiral reversed-phase column of the same dimension, the reduced plate number may be the result of an increase in kinetic interactions by the thick immobilized protein layer on the surface of CSP particles. Similar results were also reported for other drugs [110].

6.4.6 Structures of Solutes

As in case of other CSPs, the chiral resolution is also effected on protein phases by the structures of the racemic compounds. Allenmark and co-workers [23] resolved several sulfoxides on the BSA-based CSP. These sulfoxides contain different structures with various groups such as methyl, methoxy, and acetate. We analyzed the results of this study and, generally, it was found that the separation factor decreased by introducing larger groups in the sulfoxides. This behavior may be the result of a steric effect the author did not consider in the discussion

Compound No.	R_1	R_2	R_3	R_4
Ia	C_6H_5CO	H	CH_3	C_6H_5
Ib	C_6H_5CO	H	C_2H_5	C_6H_5
Ic	*p*-$O_2NC_6H_4CO$	H	C_2H_5	C_6H_5
Id	*p*-ClC_6H_4CO	H	C_2H_5	C_6H_5
Ie	C_6H_5CO	CH_3	C_2H_5	C_6H_5
If	C_6H_5	H	C_2H_5	C_6H_5
Ig	*p*-$O_2NC_6H_4$	CH_3	C_2H_5	C_6H_5

FIGURE 11 The chemical structures of anticonvulsants. (From Ref. 37.)

part [23]. Another study carried out by Allenmark [27] on the chiral resolution of 2-pyridinylmethyl-2-benzimidazolyl sulfoxides on the BSA-based CSP indicated the reverse results (i.e., the separation factors increased by introducing bulky groups in the sulfoxides). However, in the same study, no trend in the enantioselectivities of benzodiazepine derivatives was observed. Furthermore, Allenmark and co-workers [37] carried out a very interesting study of the chiral resolution of anticonvulsants on the BSA-based CSP. The structures of the studied anticonvulsants are shown in Figure 11 and the results are given in Table 3 which reflected a pronounced effect of the substituents, specially the introduction of *p*-chloro substituents in Ib caused a multiple increase in capacity factors. Also, a marked effect in elution behavior was found on the substitution of the *N*-benzoyl group for a phenyl group (If). Furthermore, Allenmark et al. [20,33] studied the chiral resolution of different racemates, having various substituents, on the BSA-based CSP and it was observed that the chiral resolution was affected by the substitution. Fitos et al. [61] studied the chiral recognition of benzodiazepines on Chiral AGP CSP. It was observed that, generally, retention increased by introducing the bulky alkyl groups. Recently, Abe and co-workers [111] described the chiral resolution of carbazole carbonyl amino acids with the linear alkyl side chain (C_1-C_4) on the BSA-based CSP. From these studies, it was concluded that the alkyl chains and groups of the racemates affect the chiral resolution on protein phases.

6.4.7 Injection Amount

The loading capacities of protein CSP may be understood from $\alpha = k_2/k_1$ and $k = k_s + k_n$, where k_s and k_n correspond to selective and nonselective contributions, respectively. When the amount of the analyte on CSP increases, the relative

TABLE 3 Effect of Substituents on the Chiral Resolution of Barbiturates (Fig. 11) on the BSA-Based CSP Using Phosphate Buffer (50 mM, pH 5.8)–1-Propanol (94 : 6, v/v) as the Mobile Phase

Barbiturates derivatives	k_1	k_2	α
Ia	14.6	23.2	1.6
Ib	34.9	44.8	1.3
Ic	55.1	70.6	1.3
Id	122.5	137.1	1.1
Ie	27.2	32.5	1.2
If	7.8	8.9	1.1
Ig	29.3	36.4	1.2

Source: Ref. 37.

contribution from k_n to the observed k values (k_1 and k_2) also increases. The critical selective sites become saturated, resulting in a decrease in separation factors because $k = a_s/a_m$, where a_s and a_m are the amounts of analytes in stationary and mobile phases, respectively. An increased analyte concentration means a decrease in the observed k values, because the relative amounts present in the stationary phase decrease due to the lower affinity of the nonselective sites. Kirkland et al. [60] described the loading of 1.5–3.0 nmol/g packing, producing a $\leq 15\%$ decrease in resolution (30% decrease in column plate number) on the 15×0.46-cm ovomucoid column. According to the authors, samples of micrograms per liter are typical for most of the analytical applications on the 15×0.46-cm column of ovomucoid protein. Andersson et al. [40] prepared BSA-based and BSA-fragment CSPs and the effect of loading amount of benzoin was carried out. At low concentration no effect on chiral resolution was observed on both CSPs. However, the chiral recognition capacity of the BSA-fragment CSP decreased at a higher solute concentration. The asymmetry factors were calculated for both the CSPs and are shown in Figure 12a. Furthermore, Andersson et al. [20] prepared three BSA-based CSPs (i.e., BSA–GLA (glutaraldehyde cross-linked BSA), BSA–FA (formaldehyde cross-linked BSA)), and BSA–DSC (*N*,*N*′-disuccinimidyl carbonate cross-linked BSA)). The authors studied the peak symmetry and calculated plate heights for all of the CSPs. These findings are shown in Figures 12b (asymmetry factors) and 12c (plate number). This figure indicates that BSA–DSC CSP has a larger column efficiency than the other two CSPs, whereas BSA–FA CSP showed the highest column-load threshold. For a small amount of solute injected (<10 nmol), there was no significant difference in the extent of deterioration of the peak symmetry and plate height on the three different CSPs for oxazepam enantiomers (Figs 12b and 12c). However, as the column

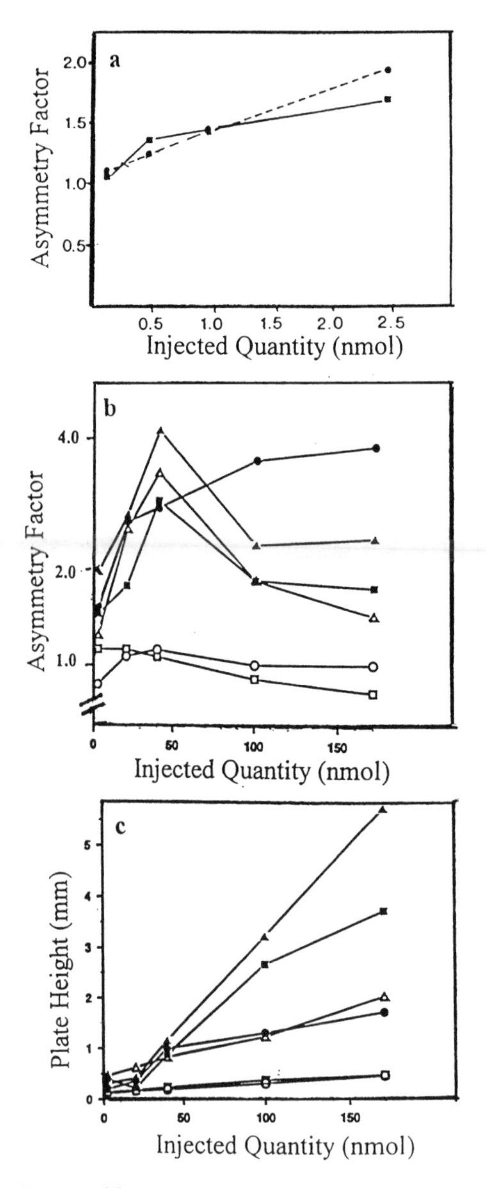

FIGURE 12 Effect of the loading amount on the enantioresolution of (a) benzoin on (■) the BSA-based CSP and (●) the BSA-fragment CSP using phosphate buffer (50 mM, pH 7.6)–1-propanol (96 : 4, v/v) as the mobile phase, (b, c) oxazepam on (△, ▲) BSA–DSC, (○, ●) BSA–FA, and (□, ■) BSA–GLA CSPs (for detail, see text) using phosphate buffer (50 mM, pH 7.1)–2-propanol (93 : 7, v/v) as the mobile phase. The open symbols represent the first eluted enantiomer. (From Refs. 20 and 40.)

load increased, the peak started to show tailing with poor resolution. Recently, Munro et al. [103] studied the loading capacity of ovomucoid CSP for the chiral resolution of bupropion enantiomers. In spite of the good chiral enantio-selectivities of different protein CSPs, the loading capacities are very poor.

6.4.8 Structures of Proteins

The complex structure of proteins having different types of group, loop, and bridge are responsible for the chiral resolution of racemic compounds. Therefore, a small change in the protein molecule resulted into a drastic change in enantioselectivity. A few reports are available which indicate that different structures of proteins are responsible for different chiral resolution capacities. Andersson et al. [20] prepared three types of BSA-based CSP [i.e. BSA–GLA, BSA–FA and BSA–DSC (described in Sec. 6.4.7)]. These columns were used for the chiral resolution of different racemic compounds and the enantioselectivities were found different for all three CSPs. The chiral resolution data of oxazepam, lorazepam, lormetazepam, and lopirazepam on these CSPs are given in Table 4. This table indicates that BSA–FA is the best CSP because all four drugs resolved on this CSP, whereas lormetazepam did not resolve on the other two CSPs. Moreover, the resolution factors on these CSPs were in the order BSA–FA > BSA–GLA > BSA–DSC. Oda et al. [71] compared the chiral resolution of 1-benzyl-4-[(5,6-dimethoxy-1-indanon)2-yl] methylpiperidine hydrochloride on avidin and avidin DSC (avidin disuccinimidyl carbonate cross-linked) CSPs. The authors observed that the best chiral resolution was achieved on avidin-based CSP (Fig. 13). Oda et al. [74] investigated the chiral resolution of some profens and other drugs on avidin and modified avidin CSPs and the authors reported different selectivities on the two columns. In another study, Oda et al. [73] compared the chiral resolution of flurbiprofen, propranolol, and warfarin on avidin and ovomucoid CSPs. The authors reported different results on these two CSPs. Accordingly, different proteins have different enantioselectivities for a

TABLE 4 Different Enantioselectivities of BSA-Based CSPs

	BSA–DSC			BSA–FA			BSA–GLA		
Racemic drugs	k_1	α	R_s	k_1	α	R_s	k_1	α	R_s
Oxazepam	4.38	5.78	5.69	5.87	5.16	9.10	5.70	3.53	7.16
Lorazepam	6.86	2.33	2.46	9.80	1.69	2.52	8.47	2.11	3.66
Lormetazepam	5.86	1.00	—	5.66	1.44	1.75	6.63	1.00	—
Lopirazepam	1.36	3.33	2.53	1.75	2.54	3.71	2.07	2.30	3.72

Source: Ref. 20.

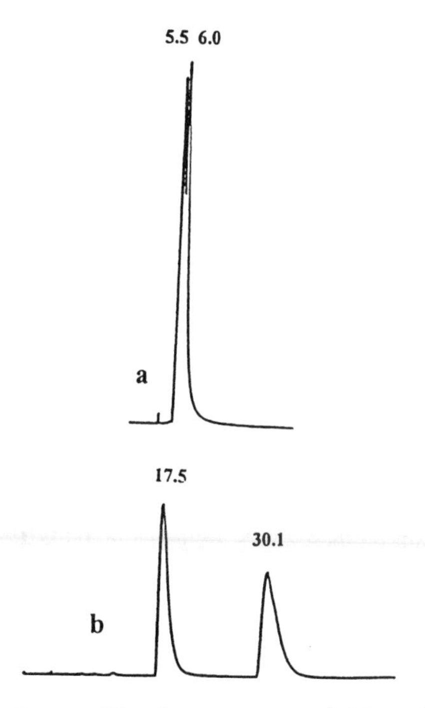

FIGURE 13 Chromatograms of 1-benzyl-4-[(5,6-dimethoxy-1-indanon)-2-yl] methylpiperidine hydrochloride enantiomers on (a) avidin–DCS and (b) avidin CSPs using ammonium acetate (0.1 M, pH 5.0) with (a) 5% and (b) 1% acetonitrile as the mobile phases. (From Ref. 71.)

particular racemic compound. Therefore, optimization of chiral resolution on suitable protein-based CSPs may be achieved by selecting the protein phase arbitrarily.

6.4.9 Other Parameters

In addition to the above-discussed parameters, other factors such as dimension of the column, particle size of the silica gel, and mobile phase additives may be considered for the optimization of chiral resolution on these CSPs. Metal ions often are important structural components in some proteins to organize conformation [112]. Therefore, the effect of metal ions as mobile phase additives may be useful. Oda et al. [74] used zinc ions as the mobile phase additive on avidin CSP to improve the chiral resolution of ibuprofen, chlormezanone, and other drugs. The results are summarized in Table 5, which indicates that the addition of zinc ions in the mobile phase has resulted into an improved resolution of some

TABLE 5 Effect of Zinc Metal Ions on the Enantiomeric Resolution of Some Drugs on Avidin CSP

	PP1		CM1		TP2		CP2		IP2	
$ZnCl_2$	k_1	α	k_1	α	k_1	α	k_1	α	k_1	α
0 mM	1.43	1.00	5.69	1.00	2.05	1.48	4.34	1.00	9.95	1.13
5 mM	2.51	1.00	5.16	1.17	1.14	1.61	2.20	1.15	10.90	1.00

Note: PP: 2-phenylpropionic acid; CM: chlormezanone; TP: trihexylphenidyl hydrochloride; CP: cloperastine; IP: ibuprofen. 1: 50 mM Tris buffer (pH 7)–methanol (90 : 10, v/v) mobile phase; 2: 50 mM Tris buffer (pH 7)–methanol (70 : 30, v/v) mobile phase.
Source: Ref. 74.

drugs. However, the authors reported that the use of zinc ions damaged the column and resulted in poor chiral recognition capacities. Recently, Haginaka and Takehira [113] studied the effect of silica pore size on the chiral resolution of ovoglycoprotein CSP. The authors reported the best chiral resolution on the CSP having the 12-nm pore size of silica gel. Furthermore, they have also studied the effect of the loading amount of ovoglycoprotein on silica gel and found that the best results were obtained on the CSP containing 80 mg ovoglycoprotein per 1 g silica gel of 12 nm pore size. Fitos et al. [114] studied the effect of ibuprofen enantiomers on the stereoselective binding of 3-acyloxy-1,4-benzodiazepines to human serum albumin (HAS) using native and Sepharose-immobilized proteins. This study indicated the different binding natures of the two types of protein in the presence of ibuprofen molecules. Thus, the above-cited parameters are also responsible for the different chiral resolutions on these CSPs.

6.5 CHIRAL RECOGNITION MECHANISM

The proteins are complex in structure and enantiospecific in nature with different functional groups, alkyl chains, bridges, and chiral loops. Therefore, the chiral recognition by proteins is a complex process. Recently, Hage [115] and Haginaks [15] reviewed the applications of protein phases in liquid chromatography with a discussion of certain studies carried out on chiral mechanisms. Generally, proteins have specific chiral sites for a pair of enantiomers with a few exceptions where the two antipodes bind to different sites [14]. The most common interactions involved in the chiral recognition on protein CSPs are nonpolar, dipole, or coulombic interactions, hydrogen-bondings, and steric effects [115,116]. Of course, the protein loops provide the chiral environment to the enantiomers if the enantiomers bind at the same site. Generally, if the enantiomers bind at different sites, the contribution of chiral loops may or may not be involved in the chiral recognition process. Ultimately, the combination of the above-cited forces resulted into enantiospecific binding of enantiomers and the chiral resolution is achieved. Important structural features with interaction forces, identified by a qualitative structure relationship, for the binding of basic drug (pheniramine) to AGP protein, are shown in Figure 14 [117].

As discussed earlier, the two binding sites of HSA [i.e. warfarin (I) and indole (II)] are responsible for the chiral resolution of a variety of racemates. In addition, other binding sites on this protein have been identified [15]. Recently, He and Carter [118] established the three-dimensional structure of HAS, which showed that I and II are the binding sites and are located in hydrophobic cavities. It has also been reported that the chemistry of binding site I is homologous to that of II. Yang and Hage [46,119,120] indicated that D- and L-tryptophan bind to single but distinct sites of HAS, whereas *R*- and *S*-enantiomers of ibuprofen and warfarin had one common binding site. Peyrin et al. [121,122] described II as the binding site for dansyl amino acids on HSA.

Kremer et al. [123] observed the hydrophobic pockets as the binding site on AGP protein. However, more than one binding site was reported. Haupt et al. [124] presented a retention model for the chiral resolution of uncharged solutes, felodipine, on AGP and the model has assumed the presence of two different stereoselective sites for different enantiomers. In another study, Waters et al. [125] carried out certain thermodynamic experiments for the determination of the mechanism of chiral resolution on AGP protein. The authors reported the two

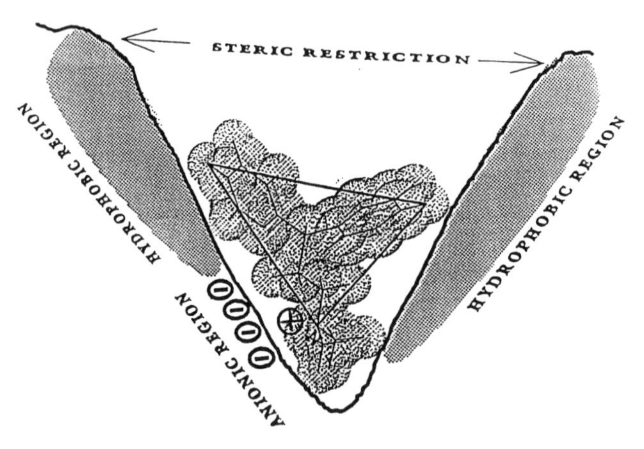

FIGURE 14 Structural features of the binding of pheniramine molecule on AGP protein with protonated aliphatic nitrogen guiding the drug toward the anionic region in the binding site. Hydrophobic moieties provide anchoring to the drug in the lipophilic region(s) of the site. (From Ref. 117.)

different binding sites for two enantiomers. Sometimes sialic acid residues are supposed to take part in the chiral resolution of enantiomers [126–128]. In spite of all these studies, the exact mechanism of chiral resolution on this protein is not known because of the lack of full knowledge of the tertiary structure of the protein.

In order to gain information on the chiral recognition mechanism on ovomucoid and ovoglycoproteins, various studies were carried out on these phases [64–66,127,129–132]. It has been observed that the three domains present on the protein took part in chiral resolution individually or in different combinations. It has also been concluded that sialic acid or sialic acid galactose groups sometimes play a role in the chiral recognition process on the reported proteins. Oda et al. [71] carried out certain studies on modified avidin protein and they have reported that avidin has multiple chiral recognition sites. Recently, Felix and Descorps [80] observed the enantioselectivity of α-chymotrypsin protein and the authors reported that enantioresolution was related to structures of enantiomers, hydrophobicity, and electrostatic interactions. These relationships determine the recognition mechanisms and the position of each enantiomer in the enzymatically active sites [133]. Cellobiohydrolase-I protein was found to have a binding site on both portions (Sec. 6.1.8) [134]. Recently, the three-dimensional structure of CBH-I was determined by X-ray studies [135,136]. It was found that the binding site is about a 40-Å long tunnel. Similarly, specific sites, responsible for chiral resolution on other proteins, were determined and discussed by Haginaka in his review [15]. Earlier, it was known that the three-point attachment of the enantiomers to the protein was sufficient for the chiral resolution, but, recently, Mesecar and Koshland [137] argued the importance of the four-point attachment model for the chiral discrimination on protein CSPs. In spite of various studies on the chiral recognition mechanisms on protein CSPs, the exact mechanism is not known because of the complex structure of proteins. In addition, the change in protein conformation in different solvents and environment may be another hindrance in the determination of chiral recognition mechanisms. However, further extensive physical and analytical studies are required to ascertain the chiral recognition mechanism.

6.6 MISCELLANEOUS APPLICATIONS

Much work has been done on chiral resolution on protein phases using HPLC, but only a few reports are available using the mode of liquid chromatography. In 1994, Ishihama et al. [138] utilized ovomucoid as a chiral pseudostationary phase in electrokinetic chromatography. The compounds separated were tolperisone, benzoin, eperisone, and chlorpheniramine. The chromatograms for the chiral resolution of eperisone are shown in Figure 15. Later, in 1999, Haddadian et al. [116] studied the enantioselectivities of two types of protein-based CSPs in

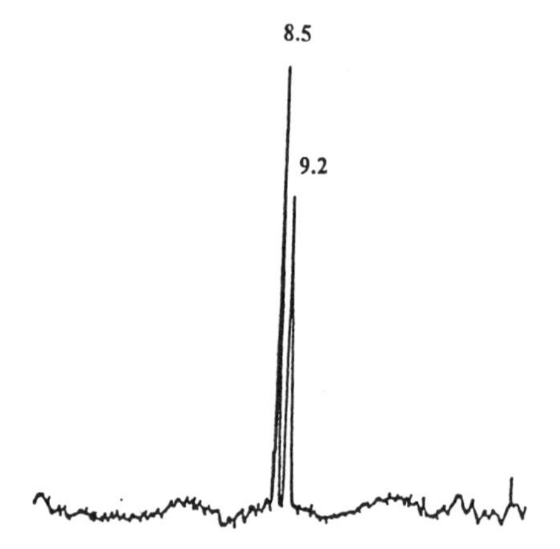

FIGURE 15 Chromatograms of eperisone enantiomers on ovomucoid CSP using phosphate buffer (10 mM, pH 5.0)–2-propanol (95 : 5, v/v) as the mobile phase in electrokinetic chromatography. (From Ref. 138.)

electrokinetic capillary chromatography. Two polymeric dipeptide chiral surfactants (PDCSs), poly (sodium *N*-undecanoyl isoleucyl)-valinate (SUILV) with three chiral centers and poly (sodium *N*-undecanoyl leucyl)-valinate (SULV), with two chiral centers, have been evaluated and compared as chiral pseudo-stationary phases. The enantioselectivity of these two CSPs was analyzed using anionic, cationic, and neutral analytes. Furthermore, the authors reported that the chiral resolution is controlled by the steric effect.

6.7 CONCLUSION

It is evident that proteins are successful chiral selectors in liquid chromatography due to the flexibility of the separation systems. The use of wide range of mobile phases makes them ideally suitable CSPs. In addition, the multiple binding interaction sites are responsible for good chiral recognition capacities of protein molecules. Various chromatographic parameters can be optimized easily to control the chiral resolution on these phases. However, there are certain limitations with these phases, as they are not suitable at the preparative scale. Also, the low loading capacity, lack of column ruggedness, and limited understanding of the chiral recognition mechanisms may be other drawbacks. Sometimes, after the prolonged use of organic modifiers, certain columns are denatured; however, these columns may be regenerated to some extent using a buffer of appropriate pH and concentration. In spite of this, the future of protein CSPs is bright. Other new protein fragments or mutant proteins, which may be more stable and enantioselective (better CSPs) than the native proteins, should be prepared by genetic engineering. Therefore, further research is required, especially on the chiral discrimination mechanisms, so that scientists can modulate future experiments in the exact and right directions.

REFERENCES

1. Lowe CR, Dean PGD, Affinity Chromatography, John Wiley & Sons, London (1974).
2. Karush F, J Am Chem Soc 76: 5536 (1954).
3. McMenamy RH, Oncley JL, J Biol Chem 233: 1436 (1958).
4. Müller WE, Wollert U, Res Commun Chem Pathol Pharmacol 9: 413 (1974).
5. Müller WE, Wollert U, Mol Pharmacol 11: 52 (1975).
6. Brown NA, Jähnchen E, Muller WE, Wollert U, Mol Pharmacol 13: 70 (1977).
7. Rendic S, Alebic-Kolbah T, Kajfez F, Sunjic V, Farmaco Sci 35: 51 (1980).
8. Gratton G, Rendic S, Sunjic V, Kajfez F, Acta Pharm Jugoslav 29: 119 (1979).
9. Stewart KK, Doherty RF, Proc Natl Acad Sci USA 70: 2850 (1973).
10. Lagercrantz C, Larsson T, Karlsson H, Anal Biochem 99: 352 (1979).
11. Lagercrantz C, Larsson T, Denfros I, Comp Biochem Physiol 69C: 375 (1981).
12. Allenmark S, Protein based phases, in Chiral Separations by HPLC: Applications to Pharmaceutical Compounds, Krstulovic AM (Ed.), Ellis Horwood, New York (1989).
13. Allenmark S, Chromatographic Enantioseparations: Methods and Applications, IInd (Ed.), Ellis Horwood, New York (1991).
14. Allenmark S, Separation of enantiomers by protein based chiral phases, in A Practical Approach to Chiral Separations by Liquid Chromatography, Subramanian G (Ed.), Wiley–VCH, Weinheim (1994).
15. Haginaka J, J Chromatogr A 906: 253 (2001).
16. McMenamy RH, Albumin binding sites, in Albumin Structure Function and Uses, Rosenoer VM, Oratz M, Rothschild MA (Eds.), Pergamon Press, Oxford, p. 143 (1977).
17. Lewin S (Ed.), Displacement of Water and its Control of Biochemical Reactions, Academic Press, New York (1974).
18. Erlandsson P, Hansson L, Isaksson R, J Chromatogr 370: 475 (1986).
19. Jacobson SC, Guiochon G, J Chromatogr 590: 119 (1992).
20. Andersson S, Thompson RA, Allenmark S, J Chromatogr 591: 65 (1992).
21. Allenmark S, Bomgren B, Boren H, J Chromatogr 237: 473 (1982).
22. Allenmark S, Bomgren B, J Chromatogr 252: 279 (1982).
23. Allenmark S, Bomgren B, Boren H, Lagerstrom PO, Anal Biochem 136: 293 (1984).
24. Sjöholm I, Ekman B, Kober A, Ljungstedt-Pahlman I, Mol Pharmacol 16: 767 (1979).
25. Kragh-Hansen U, Pharmacol Rev 33: 17 (1981).
26. Fehske KJ, Müller WE, Wollert U, Biochem Pharmacol 30: 687 (1981).
27. Allenmark S, J Liq Chromatogr 9: 425 (1986).

28. Allenmark S, Bomgren B, Boren H, J Chromatogr 264: 63 (1983).
29. Andersson S, Allenmark S, J Liq Chromatogr 12: 345 (1989).
30. Allenmark S, LC–GC 3: 348 (1985).
31. Thompson RA, Andersson S, Allenmark S, J Chromatog 465: 263 (1989).
32. Allenmark S, Bomgren B, Boren H, J Chromatogr 316: 617 (1984).
33. Allenmark S, Andersson S, Chirality 1: 154 (1989).
34. Allenmark S, Bomgren B, Andersson S, Prep Biochem 14: 139 (1984).
35. Allenmark S, Bomgren B, Boren H, Enzyme Microb Technol 8: 404 (1986).
36. Allenmark S, Bomgren B, J Liq Chromatogr 9: 667 (1986).
37. Allenmark S, Andersson S, Bojaraski J, Chromatogr 436: 479 (1988).
38. Allenmark S, Andersson S, J Chromatogr 351: 231 (1986).
39. Allenmark S, Andersson S, Chromatographia 31: 429 (1991).
40. Andersson S, Allenmark S, Erlandsson P, Nilsson S, J Chromatogr 498: 81 (1990).
41. Lienne M, Caude M, Rosset R, Tambutte A, J Chromatogr 448: 55 (1988).
42. Haginaka J, Kanagusi N, J Chromatogr A 769: 215 (1997).
43. Park JH, Ryu JK, Park JK, McNeff CV, Carr PW, Chromatographia 53: 405 (2001).
44. Noctor TAG, Wainer IW, Pharmacol Res 9: 480 (1992).
45. Noctor TAG, Felix G, Wainer IW, Chromatographia 31: 55 (1991).
46. Yang J, Hage DS, J Chromatogr A 766: 15 (1997).
47. Bertucci C, Wainer IW, Chirality 9: 335 (1997).
48. Andrisano V, Booth TD, Cavrini V, Wainer IW, Chirality 9: 178 (1997).
49. Peyrin E, Guillaume YC, Chromatographia 48: 431 (1998).
50. Peyrin E, Guillaume YC, Morin N, Guinchard C, J Chromatogr A 808: 113 (1998).
51. Millot MC, Loehman L, Hommel H, Chromatographia 50: 641 (1999).
52. Beesley TE, Scott RPW (Eds.), Chiral Chromatography, John Wiley & Sons, New York, p. 317 (1998).
53. User's Guide: Separation of Chiral Drugs and Other Chiral Compounds on Chiral–AGP, Chiral–CBH and Chiral–HSA, Chrom Tech, Hägersten, Sweden (1994).
54. Cirilli R, Di Bugno C, La Torre F, Chromatographia 49: 628 (1999).
55. Alvarez C, Sanchez-Brunete JA, Torrado-Santiago S, Cadorniga R, Torrado JJ, Chromatographia 52: 455 (2000).
56. Gyimesi-Forras K, Szasz G, Gergely A, Szabo M, Kokosi J, J Chromatogr Sci 38: 430 (2000).
57. Ortelli D, Rudaz S, Chevalley AF, Mino A, Deglon JJ, Balant L, Veuthey JL, J Chromatogr A 871: 163 (2000).
58. Suckow RF, Zhang MF, Cooper TB, Biomed Chromatogr 11: 174 (1997).
59. Mano N, Oda Y, Ohe H, Asakawa N, Yoshida Y, Sato T, J Pharm Biomed Anal 12: 557 (1994).
60. Kirkland KM, Neilson KL, McComb DA, DeStefano JJ, LC–GC 10: 322 (1992).
61. Fitos I, Visy J, Simonyi M, Hermansson J, J Chromatogr A 709: 265 (1995).
62. Ultron Technical Report for Ultron ES-OVM Column.
63. Lammerhofer M, Lindner W, Chirality 6: 261 (1994).
64. Haginaka J, Takehira H, J Chromatogr A 777: 241 (1997).
65. Haginaka J, Kagawa C, Matsunaga H, J Chromatogr A 858: 155 (1999).
66. Haginaka J, Okazaki Y, Matsunaga H, J Chromatogr A 840: 171 (1999).

67. Mano N, Oda Y, Miwa T, Asakawa N, Toshiba Y, Sato T, J Chromatogr 603: 105 (1992).
68. Monaco HL, Perduca M, Caccialanza, Wainer IW, J Chromatogr A 751: 117 (2001).
69. Miwa T, Miyakawa T, Miyake Y, J Chromatogr 457: 227 (1988).
70. Oda Y, Asakawa N, Abe S, Yoshida Y, Sato T, J Chromatogr 572: 133 (1991).
71. Oda Y, Ohe H, Asakawa N, Yoshida Y, Sato T, Nakagawa T, J Liq Chromatogr 15: 2997 (1992).
72. Oda Y, Asakawa N, Yoshida Y, Sato T, J Pharm Sci 1227 (1992).
73. Oda Y, Mano N, Asakawa N, Yoshida Y, Sato T, Nakagawa T, Anal Sci 9: 221 (1993).
74. Oda Y, Mano N, Asakawa N, Yoshida Y, Sato T, J Liq Chromatogr 17: 3393 (1994).
75. Smet E, Staelens L, Heyden YV, Baeyens WRG, Aboul-Enein HY, Van der Weken G, Garcia-Campana A, Chirality 13: 556 (2001).
76. Aboul-Enein HY, Serinese V, Biomed Chromatogr 11: 47 (1997).
77. Wainer IW, Jadaud P, Schombaum GR, Kadodkhar SV, Henry MP, Chromatographia 25: 903 (1988).
78. Marle I, Karlsson A, Pettersson C, J Chromatogr 604: 185 (1992).
79. Jadaud P, Wainer IW, J Chromatogr 476: 165 (1989).
80. Felix G, Descorps V, Chromatographia 49: 595 (1999).
81. Felix G, Descorps V, Chromatographia 49: 606 (1999).
82. Erlandsson P, Marle I, Hansson L, Isaksson R, Pettersson C, Pettersson G, J Am Chem Soc 112: 4573 (1990).
83. Theolohan S, Jadaud P, Wainer IW, Chromatographia 28: 551 (1989).
84. Marle I, Erlandsson P, Hansson L, Isaksson R, Pettersson C, Pettersson G, J Chromatogr 586: 233 (1991).
85. Allenmark S, Andersson S, Chirality 4: 24 (1992).
86. Gotmar G, Fornstedt T, Andersson M, Guiochon G, J Chromatogr 905: 3 (2001).
87. Hermasson J, Trends Anal Chem 8: 251 (1989).
88. Hermansson J, Eriksson M, J Liq Chromatogr 9: 621 (1986).
89. Miwa T, Kuroda H, Sakashita S, Asakawa N, Myake Y, J Chromatogr 511: 89 (1990).
90. Haginaka J, Seyama C, Yasuda H, Fujima H, Wada H, J Chromatogr 592: 301 (1992).
91. Oda Y, Asakawa N, Kajima T, Yoshida Y, Sato T, J Chromatogr 541: 411 (1991).
92. Haginaka J, Murashima T, Seyama C, J Chromatogr A 666: 203 (1994).
93. Haginaka J, Miyano Y, Saizen C, Seyama C, Murashima T, J Chromatogr A 708: 161 (1995).
94. Haginaka J, Miyano Y, Anal Sci 12: 727 (1996).
95. Nystrom A, Strandberg A, Aspergren A, Behr S, Karlsson A, Chromatographia 50: 209 (1999).
96. Strandberg A, Nystrom A, Behr S, Karlsson A, Chromatographia 50: 215 (1999).
97. Jacobson S, Felinger A, Guiochon G, Biotechnol Bioeng 40: 1210 (1992).
98. Hermansson J, J Chromatogr A 666: 181 (1994).
99. Oda Y, Asakawa N, Yoshida Y, Sato T, J Pharm Biomed Anal 10: 81 (1992).
100. Makamba H, Andrisano V, Gotti R, Carvini V, Felix G, J Chromatogr A 818: 43 (1998).

101. Cirilli R, Torre FL, J Chromatogr A 818: 53 (1998).
102. Williams RC, Myawa JH, Boucher RJ, Brockson RW, J Chromatogr A 844: 171 (1999).
103. Munro JS, Walker TA, J Chromatogr A 913: 275 (2001).
104. Bomgren B, Allenmark S, J Liq Chromatogr 9: 667 (1986).
105. Mano N, Oda Y, Asakawa N, Yoshida Y, Sato T, J Chromatogr 623: 221 (1992).
106. Gilpin R, Ehtesham SE, Gregory RB, Anal Chem 63: 2825 (1991).
107. Jönsson S, Schon A, Issakson R, Pettersson C, Pettersson G, Chirality 4: 505 (1992).
108. Fulde K, Frahm AW, J Chromatogr A 858: 33 (1999).
109. Karlsson A, Aspegren A, Chromatographia 47: 189 (1998).
110. Kirkland KM, Neilson KL, McComb DA, J Chromatogr 545: 43 (1991).
111. Abe Y, Shoji T, Matsubara M, Yoshida M, Sugata S, Iwata K, Suzuki H, Chirality 12: 565 (2000).
112. Berg JM, J Biol Chem 265: 6513 (1990).
113. Haginaka J, Takehira H, J Chromatogr A 773: 85 (1997).
114. Fitos I, Visy J, Simonyi M, Hermansson J, Chirality 11: 115 (1999).
115. Hage DS, J Chromatogr A 906: 459 (2001).
116. Haddadian F, Billiot EJ, Shami SA, Warner IM, J Chromatogr A 858: 219 (1999).
117. Kaliszan R, Nasal A, Turowski M, Biomed Chromatogr 9: 211 (1995).
118. He MX, Carter DC, Nature 358: 209 (1992).
119. Yang Y, Hage DS, J Chromatogr A 645: 241 (1993).
120. Yang Y, Hage DS, J Chromatogr A 725: 27 (1996).
121. Peyrin E, Guillaume YC, Guinchard C, J Chromatogr Sci 36: 97 (1998).
122. Peyrin E, Guillaume YC, Guinchard C, Anal Chem 69: 4253 (1998).
123. Kremer JMH, Wilting J, Janssen LHM, Pharmacology 40: 1 (1988).
124. Haupt D, Pettersson C, Westulund D, Chirality 7: 23 (1995).
125. Water MS, Sidler DR, Simon AJ, Middaugh CR, Thompson R, August LJ, Bicker G, Perpall HJ, Grinberg N, Chirality 11: 224 (1999).
126. Shiono H, Shibukawa A, Kuroda Y, Nakagawa T, Chirality 9: 291 (1997).
127. Kuroda Y, Shibukawa A, Nakagawa T, Anal Biochem 268: 9 (1999).
128. Haginaka J, Matsunaga H, Enantiomer 5: 37 (2001).
129. Pinkerton TC, Howe WJ, Urlich EL, Comiskey JP, Haginaka J, Murashima T, Walkenhorst WF, Westler WM, Markley JL, Anal Chem 67: 2354 (1995).
130. Haginaka J, Seyama C, Murashima T, J Chromatogr A 704: 279 (1995).
131. Haginaka J, Matsunaga H, Chirality 11: 426 (1999).
132. Haginaka J, Matsunaga H, Anal Commun 36: 39 (1999).
133. Norin M, Hult K, Biocatalysis 7: 131 (1993).
134. Marle I, Jonsson S, Isaksson R, Pettersson C, Pettersson G, J Chromatogr 648: 333 (1993).
135. Divne C, Stahlberg J, Reinikainen T, Ruohonen L, Pettersson G, Knowles J, Teeri T, Jones TA, Science 265: 524 (1994).
136. Divne C, Stahlberg J, Teeri T, Jones TA, J Mol Biol 275: 309 (1998).
137. Mesecar AD, Koshland DE Jr, Nature 403: 614 (2000).
138. Ishihama Y, Oda Y, Asakawa N, Yoshida Y, Sato T, J Chromatogr A 666: 193 (1994).

7

Ligand Exchange–Based Chiral Stationary Phases

In chiral ligand-exchange chromatography (CLEC), the separation occurs as a result of the exchange of the ligands and enantiomers on a metal ion. The ligand exchange involves the breaking and formation of coordinate bonds among the metal ions of the complex, ligands, and enantiomers. Therefore, ligand-exchange chromatography is useful for the chiral resolution of molecules containing electron-donating atoms such as oxygen, nitrogen, and sulfur. These types of molecule include amino acids and their derivatives, amines, hydroxy acids, and peptides, among other drugs. Some molecules containing π-electron-donating double bonds can form complexes with the metal ion and, therefore, can be separated by ligand exchange chromatography. In 1961, Helfferich [1] reported the ligand-exchange process by replacing two molecules of ammonia by a diamine molecule in a copper ammonia complex. It was found that coordination of the chiral selector and enantiomers occurred at the same metal ion and, hence, provided the best conditions for the chiral recognition of enantiomers. Sometimes the fast kinetics of ligand-exchange reactions in the metal ion coordination sphere makes this technique suitable for the chiral resolution of kinetically labile complexes. CLEC developed by Davankov [2,3] in 1969 was the first liquid-chromatographic technique successfully applied for the complete separation of enantiomers of amino acids. Subsequently, Davankov et al. [4–8] carried out extensive work on the chiral resolution of other racemic compounds using CLEC, which resulted in several publications. The scope of CLEC was also extended to the complexes formed by polydentate ligands and to outer-sphere complexation [3]. Chiral resolution using CLEC has also been reviewed by several authors [9–14]. Copper(II) was used as the ligand metal ion in most of the applications of CLEC. However, other metal ions such as nickel and zinc were also tested [15,16]. In spite of the development of more effective chiral selectors (Chaps. 2–6), CLEC still remains the best technique of choice from a theoretical point of view. Many theoretical concepts developed in CLEC are of general interest for the explanation and prediction of chiral recognition in all chiral chromatographic systems. Moreover, the simple and inexpensive nature of CLEC makes it a suitable and useful technique. This chapter describes chiral resolution by ligand-exchange chromatography.

7.1 STRUCTURES AND PROPERTIES

The main component of CLEC is the chiral selector (ligand). The different types of chiral molecule containing electron-donating atoms, such as nitrogen, oxygen, and sulfur, are considered the best ligands, as they can coordinate with metal ions. The most commonly used amino acids and their derivatives for the preparation of CSPs are proline, hydroxyproline, histidine, phenylalanine, aspartic acid, glutamic acid, methionine, thereonine, leucine, and valine [17–20]. However, cyclic amino acids, such as proline, hydroxyproline, allohydroxyproline, azetidene and some carboxylic acids, also proved to be suitable chiral ligands. The chiral ligand itself, or complexed with a metal ion, is bonded to a solid support. With regard to the use of metal ions, Davankov [12] tested copper(II), nickel(II), and zinc(II) metal ions; however, the best resolution was reported for the copper(II) ion. It has also been reported that tetra coordinate copper(II) ions produced the highest enantioselectivity effects when combined with bifunctional solute molecules and bidentate chiral matrix fixed ligands, whereas nickel(II) ion, having a coordination number of 6, was preferable in combination with tridentate ligands. These ligands provide the suitable chiral environment for the enantiomers because of their stereo-specific nature. Other functional groups, such as alkyl or phenyl groups, also take part in the interactions with the enantiomers. The presence of metal ions is essential, as it provides the site for the exchange process between the ligands and enantiomers. The chemical structures of some important chiral ligands used in CLEC are shown in Figure 1.

7.2 PREPARATION AND COMMERCIALIZATION

Cross-linked polystyrene resins incorporating chiral amino acids were utilized as the chiral stationary phases (CSPs) in CLEC [6,21]. Davankov et al. [12] reported the use of a resin incorporating L-proline in a polystyrene-type matrix. The resin was prepared by reacting L-proline with the chlormethylated styrene–divinyl-

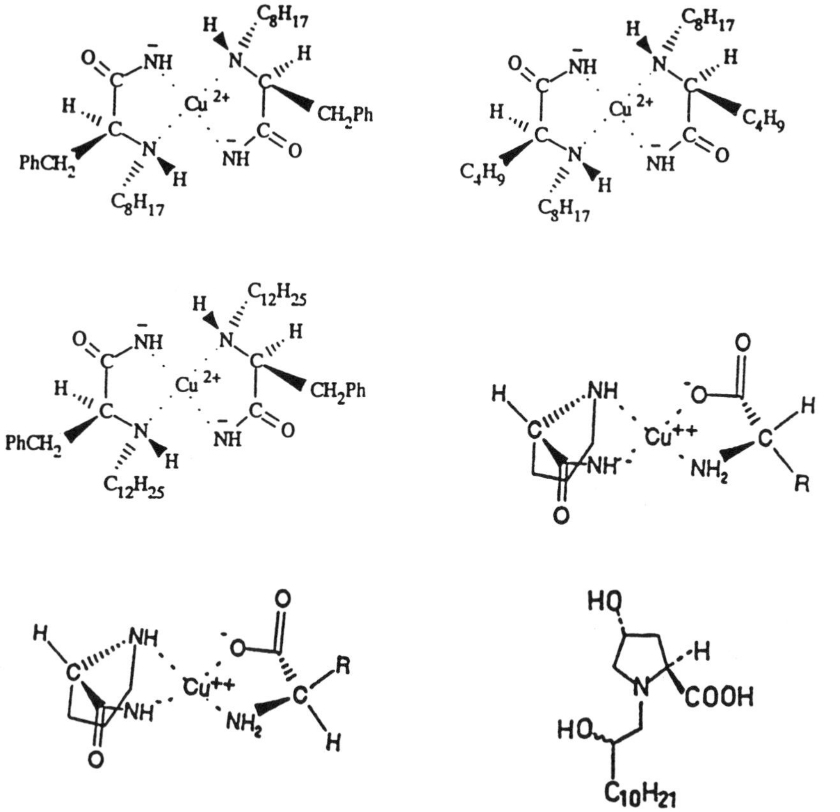

FIGURE 1 Chemical structures of ligand exchangers.

benzene copolymer [12]. A more general procedure involves the use of methyl esters or amides of amino acids and *p*-chloromethylated styrene-*co*-polymers [12]. Several other groups have synthesized chiral CSPs, starting from the conventional styrene–divinylbenzene copolymers containing amino acids [22–26]. A French research group later grafted the porous hydrophobic beads of cross-linked polyacrylamide and used it as the solid support for the chiral ligands [22,23]. The other hydrophobic polymeric support used were Separon H1000 [26] and TSK 2000 PW [27]. In spite of high enantioselectivity and exchange capacities, these CSPs suffer from high-pressure resistance and slow mass transfer as compared to other rigid solid supports such as silica gel. The swelling of these resins in the aqueous mobile phase is a major problem. Therefore, later, silica gel was used as the solid support for the preparation of ligand-exchange-based CSPs.

Initially, the chiral selectors were allowed to adsorb on the surface of silica gel and were packed into the columns [21,28–31]. Miyazawa et al. [32] deposited *N*,*S*-dioctyl-D-penicillamine onto the surface of RP-18 silica gel. The same chiral selector and (*R*,*R*)-tartaric acid mono(*R*)-1-(α-naphthyl)-ethylamide were applied by Oi et al. [33] for the modification of the Sumipax ODS column. Slivka et al. [34] synthesized *N*-alkylphenoxyglycinols and loaded them onto silica gel C_{18}. Kurganov et al. [35] prepared a CSP by loading Cu(II)[*N*-octadecyl-(*S*)-proline]$_2$ on the surface of Superspher-100 RP-18 silica gel. The coating of chiral ligands was achieved on both bare and chemically modified bonded silica gels. These types of CSP were prepared with low-molecular-mass ligands or complexes. Alternatively, the coated types of CSP were prepared by simply shaking the chiral ligand with silica gel at different time intervals [12]. Of course, these CSPs were used for the chiral separation of various racemates, but, again, they have certain drawbacks. These drawbacks include the limitations of the choice of mobile phases, especially with organic modifiers. Sometimes these CSPs deteriorated at a high column pressure. Therefore, more promising methods of the covalent binding of chiral ligands to the surface of silica gel, yielding highly stable and selective CSPs, were developed.

In 1979, three different groups [36–38] reported the preparation of covalently bonded chiral ligand-exchange CSPs and tested for the chiral resolution of amino acids. Gübitz et al. [39,40] prepared two CSPs containing L-proline covalently bonded to silica gel. These are popular CSPs in CLEC due to their excellent chiral recognition capabilities. A similar chiral stationary phase containing *S*-proline was prepared by Yang et al. [41]. The chiral selector was connected to silica gel through a single *S*-triazine spacer. Recently, Hyun et al. [42] prepared a CSP by covalently bonding (*S*)-*N*,*N*-carboxymethyl undecyl leucinol monosodium salt onto silica gel. About 40 different chiral ligand exchangers were developed and used for chiral resolution by CLEC. The structure of some of the important ligand-exchange CSPs are given Figure 2. There are many ways to prepare covalently bonded ligand-exchange CSPs [12,18–20,43–46]. As an example, one method for the preparation of chiral CSP is presented in Figure 3 [15].

Figure 3 describes the preparation of *N*-ω-undecenoyl-L-valine CSP bonded to silica gel. The carboxylic acid group of L-valine was protected by the reaction with isobutylene using the method of Roeske [47]. The formed *tert*-butyl ester of L-valine was precipitated from diethyl ether as the oxalate by the dropwise addition of a solution of 10% oxalic acid in absolute ethanol. The precipitate is dried and the oxalate group is removed by the reaction of sodium hydroxide. The *tert*-butyl ester of L-valine was treated with undecenoic acid in tetrahydrofuran (THF), which resulted in *N*-ω-undecenoyl-L-valine methyl ester. In another step, 10 mM of monochlorosilane was dissolved in 20 mL of dry pyridine and was allowed to react with *N*-ω-undecenoyl-L-valine methyl ester.

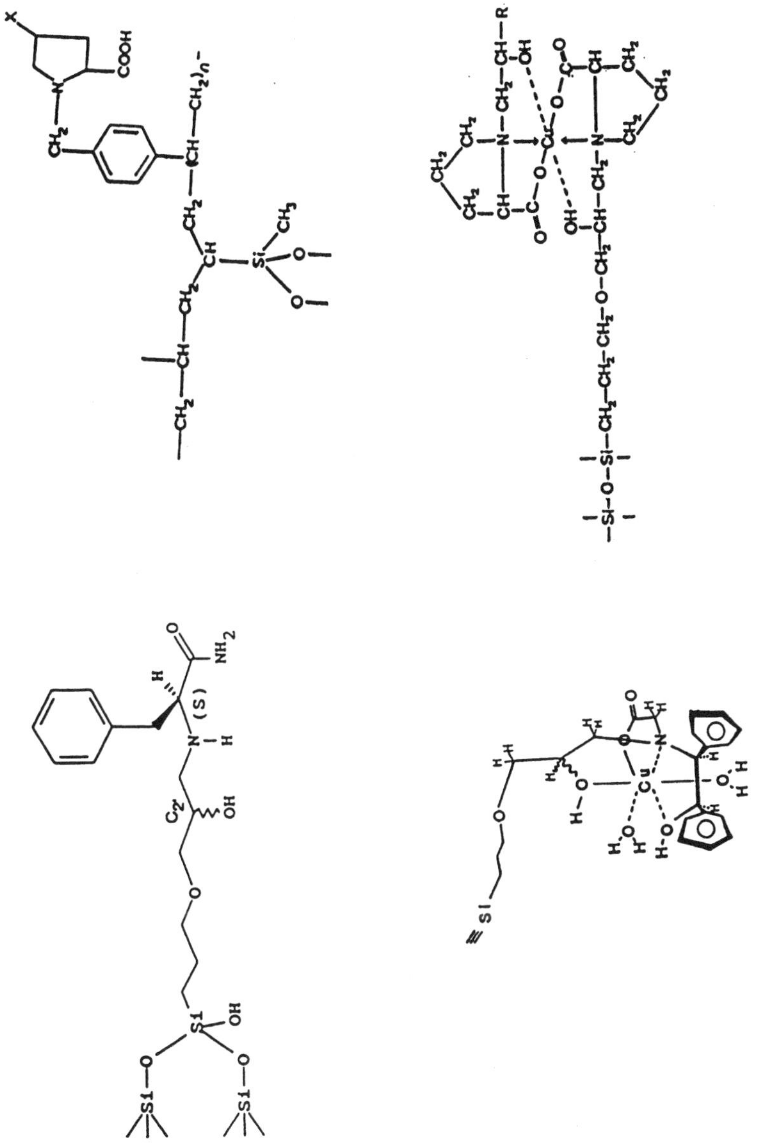

FIGURE 2 Chemical structures of ligand-exchange CSPs.

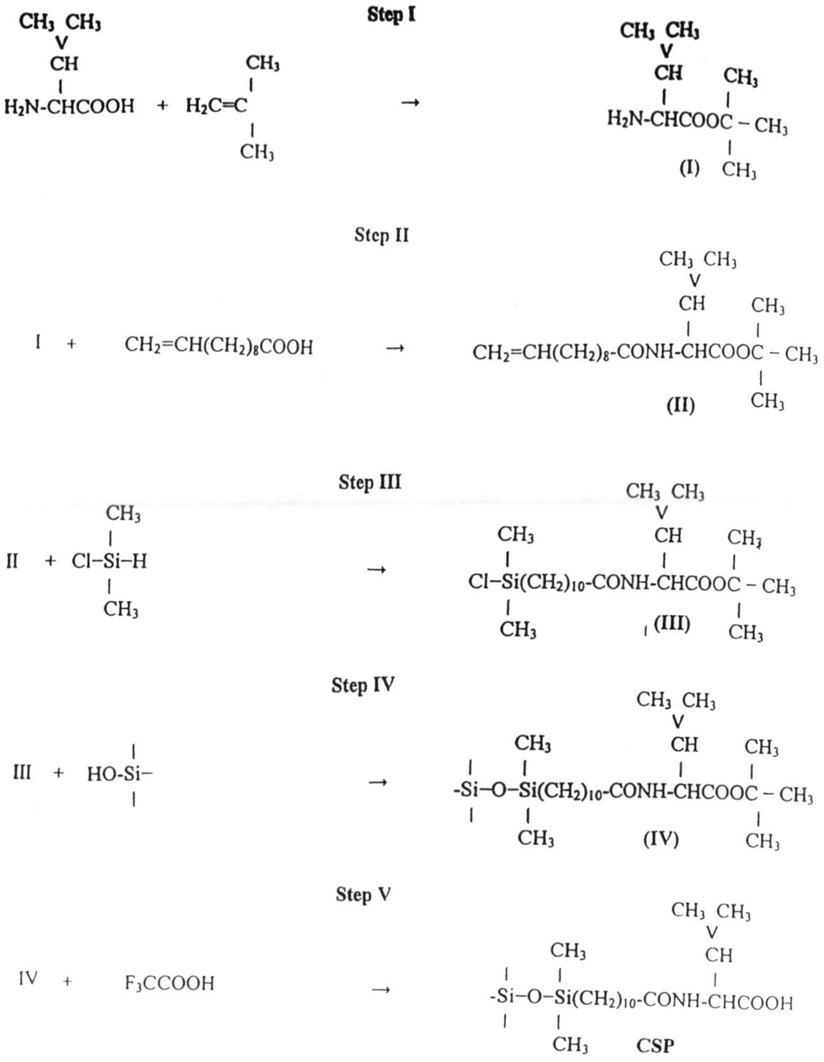

Step I

$$H_2N\text{-}CH(CH(CH_3)_2)COOH + H_2C{=}C(CH_3)_2 \rightarrow H_2N\text{-}CH(CH(CH_3)_2)COOC(CH_3)_3 \quad (I)$$

Step II

$$I + CH_2{=}CH(CH_2)_8COOH \rightarrow CH_2{=}CH(CH_2)_8\text{-}CONH\text{-}CH(CH(CH_3)_2)COOC(CH_3)_3 \quad (II)$$

Step III

$$II + Cl{-}Si(CH_3)_2{-}H \rightarrow Cl{-}Si(CH_3)_2(CH_2)_{10}\text{-}CONH\text{-}CH(CH(CH_3)_2)COOC(CH_3)_3 \quad (III)$$

Step IV

$$III + HO{-}Si{-} \rightarrow {-}Si{-}O{-}Si(CH_3)_2(CH_2)_{10}\text{-}CONH\text{-}CH(CH(CH_3)_2)COOC(CH_3)_3 \quad (IV)$$

Step V

$$IV + F_3CCOOH \rightarrow {-}Si{-}O{-}Si(CH_3)_2(CH_2)_{10}\text{-}CONH\text{-}CH(CH(CH_3)_2)COOH \quad \text{CSP}$$

FIGURE 3 Chemical pathway for covalent bonding of ligand exchanges with silica gel.

The obtained product was allowed to react with silica gel. The tertiary butyl protecting group was removed by the reaction of trifluoroacetic acid. The obtained product was washed thoroughly with 90% methanol followed by 100% methanol and dried at 60°C. The material was packed in appropriate stainless-steel columns. The ligand-exchange-based CSPs commercialized by different companies are given in Table 1.

TABLE 1 Various Ligand Exchange–Based Commercial CSPs

Commercialized CSP	Companies
Chirosolve	JPS Chemie, Switzerland
Chiralpak WH, Chiralpak WM, Chiralpak WE, and Chiralpak MA	Separations Kasunigaseki-Chrome, Tokoyo, Japan
Chiralpak WH and Chiralpak WM	Daicel Chemical Industries, Tokyo, Japan
Nucleosil Chiral-1	Macherey-Nagel, Duren, Germany
Phenylglycine and leucine types	Regis Technologies, Austin, TX, USA
Chirex types	Phenomene, Torrance, CA, USA
Orpak CRX-853	Showa Denko, Kanagawa, Japan

7.3 APPLICATIONS

It is well known that the chiral resolution of these CSPs occurred as a result of the exchange of ligands and enantiomers on the same metal ion. Therefore, these CSPs are suitable only for those racemates which can coordinate with the metal ion. Therefore, racemates like amino acids, amines, and hydroxy acids have been resolved successfully by the ligand-exchange process. As mentioned earlier, either the individual chiral ligand or one complexed with a metal ion is bonded onto silica gel support. Therefore, in the case of the first type of CSP, the metal ion is used in the mobile phase; no metal ion is required in the mobile phase in the latter case.

7.3.1 Analytical Separations

Davankov et al. [3,11,12] have carried out extensive and remarkable work on the chiral resolution using these CSPs. The chiral ligand-exchange resins based on macronet isoporous styrene copolymers containing L-hydroxyproline were used for the chiral resolution of amino acids. Because of the high resolving power of the resin, short (2 cm) columns were developed and used. L-Hydroxyproline-based resins were used for the chiral resolution of mandelic acid, β-phenyl-β-alanine, 2-aminopropanol-1, ^{1}N-benzyl-propane-diamine, and ephedrine [48,49]. Grafted L-phenylalanine has been used successfully for the chiral resolution of amino acids with values of a separation factor of more than 1.25 [50]. The chromatograms of the resolved amino acids on macronet isoporous polystyrene-type resin containing N^1-benzyl-(R)-diaminopropane-1,2 are shown in Figure 4a.

The coated ligand-exchange CSPs have been used with the mobile phases that did not cause bleeding effects. Otherwise, the CSP may lose its chiral ligand after a few runs [12]. It is very interesting to note that the values of separation factors were very high on the coated CSPs. The values of the separation factors for proline and valine amino acids were reported at 16.4 and 6.9, respectively

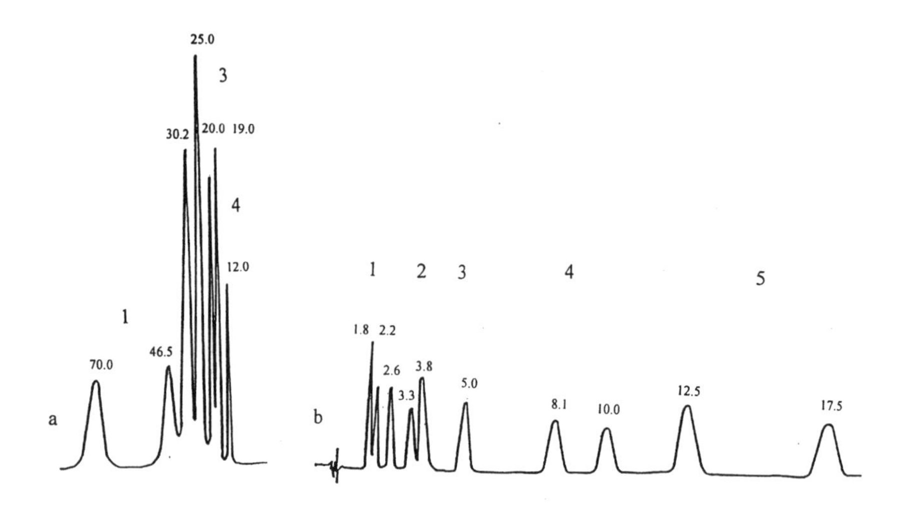

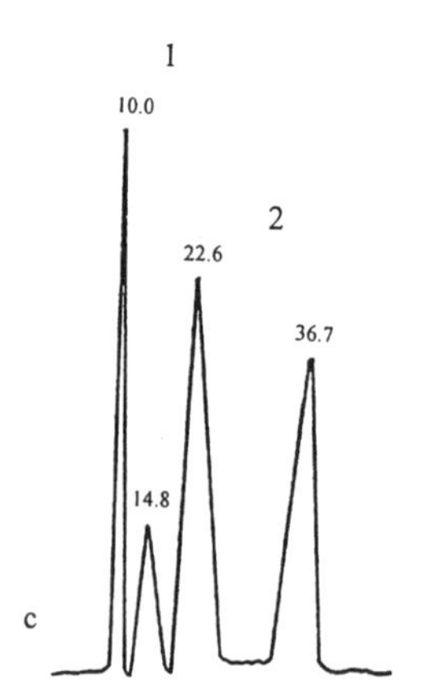

FIGURE 4 Chromatograms of the resolved amino acids and dipeptides on (a) leucine (1), serine (2), alanine (3), and lysine (4) on a glass microbore column with a macronet isoporous polystyrene-type resin containing groups of N^1-benzyl-(R)-diaminopropane-1,2 chiral selector using 0.25 M sodium acetate containg 1.5 M copper acetate; (b) alanine (1), proline (2), abrine (3), isovaline (4), and tyrosine (5) on Zorbax C_8 coated with C_{10}-L-histidine using 10^{-4} M aqueous copper acetate as the mobile phase; and (c) gly-ther (1) and gly-leu (2) on hydroxy proline-bonded CSP using aqueous 10^{-5} M copper(II) ions as the mobile phase. (From Refs. 12, 17, and 40.)

[12]. *N*-(2-Hydroxy-decyl)-L-hydroxyproline was used to determine the enantiomeric purity of D-penicillamine [51]. Zorbex ODS C_8 and LiChrosorb RP-18 columns were coated by *N*-decyl-L-histidine and charged with copper(II) ions. The developed CSPs were used for the chiral resolution of amino acids using aqueous copper acetate (0.0001 M) as the mobile phase [17]. Yuki et al. [52] developed a novel chiral phase consisting of (1*R*,2*S*)- and diastereoisomeric (1*S*,2*S*)-2-carboxymethylamino-1,2-diphenylethanol separately. These phases were used successfully for the chiral resolution of amino acids and their derivatives and hydroxy acids. Takeuchi et al. [53] coupled the Chiralpak WH column with a hydrophobic column (Hypersil ODS-3) in a series and achieved the chiral resolution of seven pairs of dansyl amino acids. The chromatograms of the chiral resolution of amino acids on a Zorbax C_{18} column coated with L-histidine are shown in Figure 4b.

Foucault et al. [46] bonded L-proline to 3-aminopropylsilated silica gel and the developed CSP was used for the chiral resolution of amino acids. Gübitz et al. [20] reported the successful separation of amino acids on silica gel covalently bonded with L-proline. Moreover, the authors reported the chiral separation of a mixture of four to five amino acids in less than 1 hr. The authors reported the high retention of L-enantiomers of all of the amino acids used in this study. The same authors [54,55] reported the resolution of biologically important molecules like thyroid hormones and 3,4-dihydroxyphenylalanine (DOPA). Furthermore, Gübitz [56] achieved the chiral separation of dipeptides on L-hydroxyproline-bonded CSP. Sinibaldi et al. [57,58] bonded the copper(II) complex of *trans*-cyclohexanediamine-1,2 to glycidoxypropyl silica gel and the developed CSPs were used for the chiral resolution of several amino acids and mandelic acid. In 1983, Watanabe [59] used L-histidine-based CSP for the chiral resolution of amino acids under acidic conditions using a phosphate buffer. A good resolution of dansyl amino acids was achieved on diamide-type phases prepared from L-proline [44]. The chiral separation of chlorquine and other amine derivatives were carried out on anion exchangers modified with heparin using a phosphate buffer with acetonitrile [60]. Recently, penicillamine bonded to silica gel was used as the chiral ligand for the chiral resolution of lactic acid in wine [61] and cyclic β-substituted α-amino acids [62]. Vidyasankar et al. [63] synthesized chiral ligands by molecular imprinting techniques and the developed CSPs were used for the chiral resolution of amino acids. Bulk polymers were prepared by allowing the functional polymer, Cu(II)-*N*-(4-vinylbenxzyl)iminodiacetic acid, to form complexes with the template amino acids in solution, followed by cross-linking with ethylene glycol dimethacrylate. The imprinted polymer was grafted onto derivatized silica gel. To show the nature of the chiral resolution on these phases, the chromatograms of the chiral resolution of dipeptides on the chemically bonded CSP (derivatized L-proline) are shown in Figure 4c. The most important racemic compounds resolved on CLEC are given in Table 2.

TABLE 2 Enantiomeric Resolution of Racemic Compounds on Ligand Exchange–Based CSPs

Racemic compound	CSPs	Refs.
Amino acids	L-Proline, L-hydroxyproline *N*-Octyl-(*S*)-phenylalaninamide, *N*-dodecyl-(*S*)-phenylalaninamide, *N*-Octyl-(*S*)-norleucinamide	18, 21, 40, 42, 53, 64, 65, 67
	(1*R*,2*S*)- and diastereoisomeric (1*S*,2*S*)-2-carboxymethylamino-1,2-diphenylethanol	52
	[Cu(II) 3-(L-propyl)propyl]$^+$, [Cu(II) 3-(L-hydroxypropyl)]$^+$	19
	L-Histidine	17, 59
	L-Tryptophan	30
Dansyl amino acids	*N*-ω-(Dimethylsiloxyl)undecanoyl-L-valine	15
	L-phenylalaninamide, L-alaninamide, L-prolinamide	72
Dipeptides	L-Proline, L-hydroxyproline	40
N-Benzyloxycarbonyl of amino acids	L-Valine	66
Penicillamine	(2*S*,4*R*,2′*RS*)-4-hydroxy-1-(2′-hydroxydodecyl) proline	51
Amino alcohols	L-Proline	71
Hydroxy acids	L-Hydroxyroline	69
	(1*R*,2*S*)- and diastereoisomeric (1*S*,2*S*)-2-carboxymethylamino-1,2-diphenylethanol	52
	L-Penicillamine	62
	L-phenylalaninamide, L-alaninamide, L-prolinamide	72
Thyroid hormones	L-Phenyalanine	55
Lactate	Shodex Orpack RX-453	68
Imidazole antifungal agents	Chiralpak WH	70

7.3.2 Preparative Separations

As a result of certain drawbacks attached to coated silica gel CSPs, they could not be used for preparative separations. However, the literature provides some reports on the preparative chiral separations on the bonded CSPs. The bonded CSPs are suitable for use at a preparative scale but the yield is very poor due to poor loading capacities. Davankov et al. [73] packed a column with 300 g of the L-hydroxyproline-type resin and used the chiral separation at a preparative scale.

The authors reported the chiral separation of proline and thereonine amino acid up to 20 and 6 g, respectively, in a single run. Micropreparative resolution of lecucine was presented. The resolution was discussed with respect to the degree of sorbent saturation with copper(II), elution rate, eluent concentration, temperature, and column loading condition [16]. Weinstein [74] reported the micropreparative separation of alkylated amino acids on a Chiral ProCu column. In another article, a preparative chiral resolution of 3-methylene-7-benzylidene-bicyclo[3.3.1]nonane was achieved on 7.5% silver(I)–*d*-camphor–10-sulfonate CSP [75]. Later, Shieh et al. [71] used L-proline-loaded silica gel for the chiral resolution of (*R*,*S*)-phenylethanolamine as the Schiff base of 2-hydroxy-4-methoxyacetophenone. Gris et al. [76] presented the preparative separations of amino acids on Chirosolve L-proline and Chirosolve L-pipecolic acid CSPs.

7.4 OPTIMIZATION OF HIGH-PERFORMANCE LIQUID CHROMATOGRAPHIC CONDITIONS

Ligand-exchange chromatography has been used for the chiral resolution of racemic compounds containing electron-donating atoms; therefore, their applications are confined. Mostly buffers, sometimes containing organic modifiers, were used as mobile phases. Therefore, the optimization was carried out by controlling the composition of the mobile phase. However, pH, effects of temperature, concentration of metal ions, and flow rate were also varied. The optimization of these parameters are discussed in detail.

7.4.1 Composition of the Mobile Phase

There are two strategies for the development and use of mobile phases on these CSPs. An aqueous mobile phase containing a suitable concentration of metal ions is used with the CSP having only a chiral ligand. In the case of the CSP containing a metal ion complex as the chiral ligand, the mobile phase is used without the metal ions. In most of the applications, aqueous solutions of metal ions or buffers were used as the mobile phases. The most commonly used buffers are ammonium acetate and phosphate. However, the use of a phosphate buffer is avoided if the metal ion is being used as a mobile phase additive to avoid the complex formation between the metal ion and the phosphate which may block the column. A review of literature indicates that these buffers (20–50 mM) were used frequently for successful chiral resolution, but in some instances, organic modifiers were also used to improve the resolution. Generally, acetonitrile was used as the organic modifier [15]. However, some reports deal with the use of methanol, ethanol, and THF [17,19,65,71]. The concentrations of these modifiers used vary from 10% to 30%. However, some reports indicated the use of these organic modifiers up to 75% [65,71]. There are only a few reports which indicated the use of these CSPs under normal phase modes. Perry et al. [77] used isopropanol–hexane (10 : 90, v/v) for the chiral resolution of trifluoro-1-(9-anthryl)-ethanol. Recently, Aboul-Enein and Ali [70] used hexane–2-propanol–diethylamine (400 : 99 : 1, v/v/v) as the mobile phase for the chiral resolution of the antifungal agents econazole, miconazole, and sulconazole on a Chiralpak WH column.

Generally, the chiral resolution of highly retained solutes is optimized using organic modifiers. The organic modifiers reduce the hydrophobic interactions resulting into improved resolution. Roumeliotis et al. [19,65] studied the effect of methanol, acetonitrile, and THF organic modifiers on the chiral resolution of amino acids and they reported methanol as the best solvent. Furthermore, the concentration of methanol was optimized and 50% was found to be a suitable concentration. These findings are shown in Tables 3 and 4. Kurganov et al. [21] carried out an extensive study on the variation of methanol and acetonitrile concentrations on the chiral resolution of amino acids. The authors reported 5–20% as the best concentrations of these modifiers. Similarly, Takeuchi et al. [53]

TABLE 3 Effect of Different Organic Solvents on the Chiral Resolution of Amino Acids on [Cu(II)–L-Hydroxypropyl)]methyl–Silica Gel

	Methanol		Acetonitrile		THF	
Amino acid	k_1	α	k_1	α	k_1	α
Aspartic acid	2.64	0.95	3.40	0.99	1.52	0.95
Glutamic acid	2.04	1.14	2.92	1.12	1.25	1.17
Histidine	4.42	0.46	5.86	0.59	2.65	0.54
Alanine	2.10	1.11	2.68	1.10	1.47	1.09
Asparagine	4.03	0.84	4.87	0.92	2.37	0.87
Serine	3.00	0.88	3.59	0.93	1.83	0.95
Proline	2.33	1.68	2.61	1.63	1.74	1.61
Threonine	3.41	1.03	4.00	1.11	2.08	1.12
Valine	2.49	1.01	3.01	1.02	1.71	1.03
Lysine	4.90	0.85	—	—	2.50	0.91
Tyrosine	3.13	0.78	3.12	0.85	2.04	0.85
Methionine	2.66	1.13	2.90	1.19	1.67	1.21
Arginine	4.42	1.05	—	—	2.60	1.08
Isoleucine	2.20	0.97	2.85	0.90	1.52	0.98
Leucine	2.56	0.90	2.92	0.90	1.75	0.89
Phenylalanine	3.04	0.87	2.87	0.94	1.96	0.95
Tryptophan	5.27	0.66	4.28	0.71	3.05	0.78

Source: Ref. 65.

TABLE 4 Effect of Methanol Concentration on the Chiral Resolution of Amino Acids on Cu(II)–L-Hydroxypropyl)methyl–Silica Gel

	Percentage of methanol							
	10		30		50		100	
Amino acid	k_1	α	k_1	α	k_1	α	k_1	α
Aspartic acid	1.00	1.00	1.13	1.00	2.64	0.95	0.44	1.00
Glutamic acid	0.88	1.09	0.74	1.14	2.04	1.14	0.39	1.08
Histidine	3.42	0.54	3.58	0.48	4.42	0.46	1.62	0.67
Alanine	1.18	1.07	0.79	1.10	2.10	1.11	0.47	1.08
Asparagine	1.94	0.93	1.74	0.93	4.03	0.84	0.81	0.96
Serine	1.52	0.93	1.17	0.95	3.00	0.88	0.57	0.96
Proline	1.46	1.42	0.97	1.26	2.33	1.68	0.79	1.14
Threonine	1.91	1.03	1.53	1.10	3.41	1.03	0.79	1.07
Valine	1.66	1.00	1.04	1.01	2.49	1.01	0.80	1.00
Lysine	2.59	0.91	1.47	0.90	4.90	0.85	0.99	0.93
Tyrosine	1.88	0.77	1.30	0.81	3.13	0.78	0.75	0.82
Methionine	1.87	1.09	1.18	1.18	2.66	1.13	0.91	1.07
Arginine	2.97	1.02	1.54	1.02	4.48	1.03	1.21	1.00
Isoleucine	1.66	0.98	1.12	0.96	2.20	0.97	0.88	0.96
Leucine	1.89	0.92	1.41	0.90	2.56	0.90	0.97	0.92
Phenylalanine	2.30	0.86	1.41	0.89	3.04	0.87	1.10	0.93
Tryptophan	3.77	0.64	2.42	0.65	5.27	0.66	1.58	0.70

Source: Ref. 65.

studied the effect of acetonitrile on the chiral resolution of amino acids on a Chiralpak WH column. The authors reported the best resolution at 5–30% acetonitrile concentrations. The results of this research is shown in Figure 5. Davankov et al. [17] used ethanol for the optimization of the chiral separation of amino acids and the authors reported the improved resolution at a higher concentration of ethanol. The maximum value of the separation factor was 1.81 using 100% ethanol. Okubo et al. [68] optimized the chiral resolution of lactate by varying the concentration of methanol on an Orpac CRX-453 B column. The authors reported 1.68, 1.77, and 1.60 as the values of separation factors without methanol and with 10% and 20% methanol, respectively. Schlauch et al. [62] observed the reversal order of elution of some amino acids when using different concentrations of acetonitrile. The protocol for the development and optimization of mobile phase is given in Scheme 1.

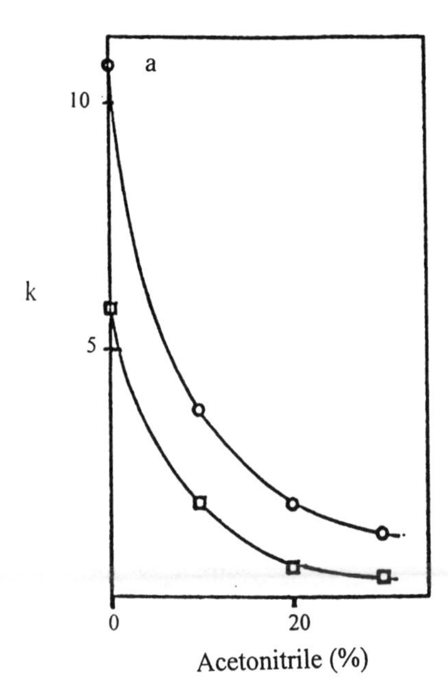

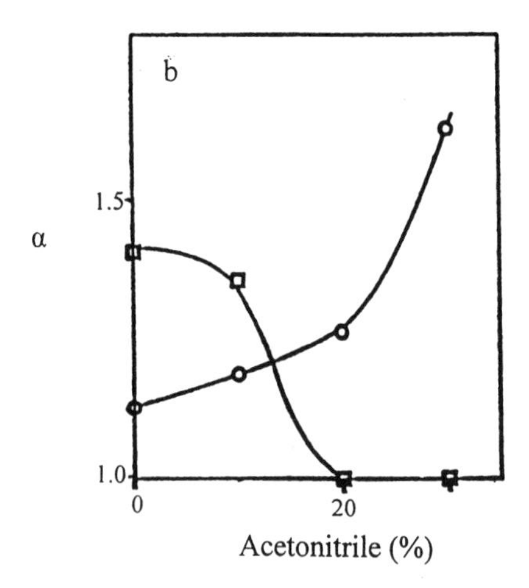

FIGURE 5 Effect of the concentration of acetonitrile on (a) the retention (k) and the (b) separation (α) factors for the chiral resolution of (○) dansyl asparagine and (□) dansyl valine on a Chiralpak WH column using 0.25 M ammonium acetate as the major component of the mobile phase. (From Ref. 53.)

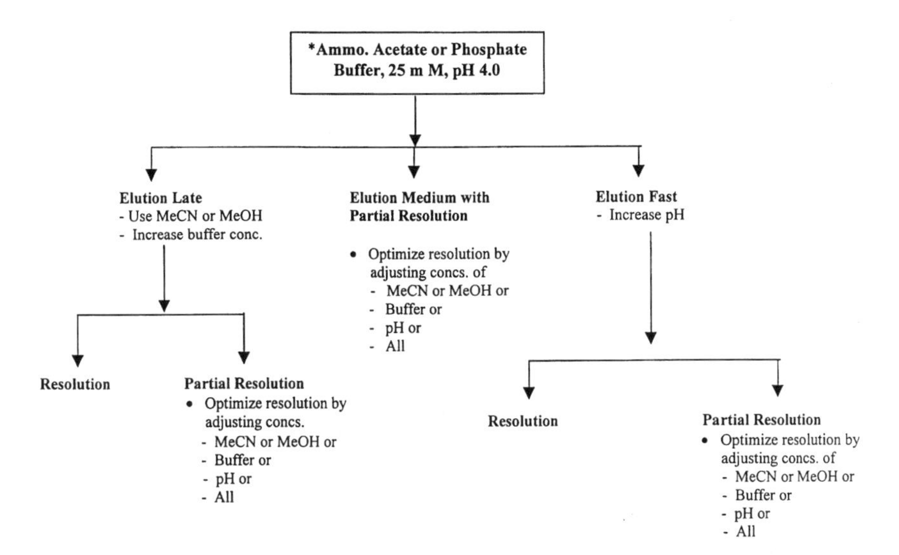

SCHEME 1 Protocol for the development and optimization of mobile phases on ligand-exchange-based CSPs. **Note*: Use phosphate buffer only with CSPs containing ligand metal complex as the chiral selector.

7.4.2 pH of the Mobile Phase

The pH of the mobile phase was recognized as one of the most important controlling factors of chiral resolution on ligand-exchange chiral phases [78]. The retention of all the racemates increased as the pH increased. The selectivity of separation was only affected moderately by pH changes, whereas the efficiency of the column showed a different trend depending on the relative retentions of racemates. Buffers having a pH from 3 to 5 were used. However, Gübitz and Juffmann [55] achieved the chiral resolution of thyroid enantiomers at pH 8.5. In another study, Gübitz and Jellenz [18] reported the maximum chiral resolution of certain amino acids at a pH higher than 8.0. Watanabe et al. [30] studied the effect of pH (4–5.5) on the chiral resolution of asparagine, thereonine, and histidine. The effect of pH on the chiral resolution of these amino acids is shown in Figure 6a, which indicates 5.5 as the best pH. Roumeliotis et al. [65] carried out an extensive study on the chiral resolution of amino acids with respect to pH variation. The pH varied from 4.5 to 6.0 and the best results were obtained at pH 4.5. Furthermore, Roumeliotis et al. [19] optimized the chiral resolution of amino acids on different CSPs using the mobile phase of different pHs and the best

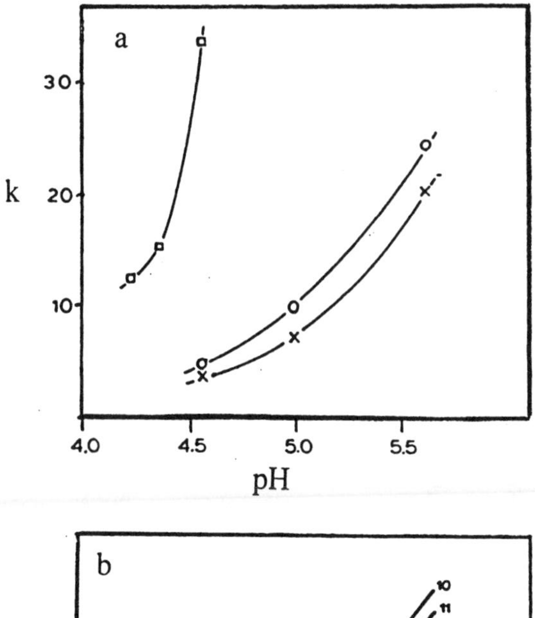

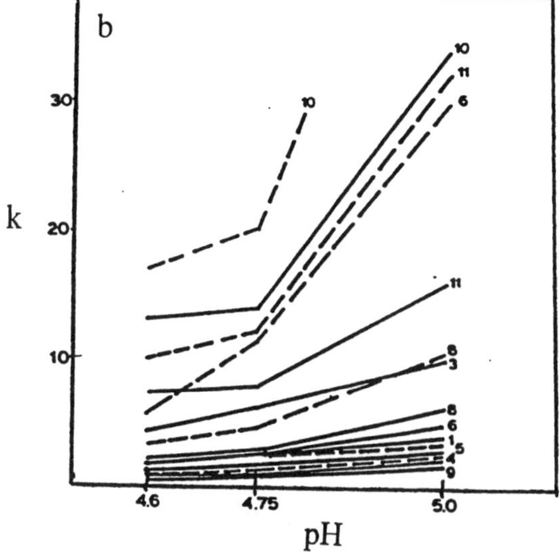

FIGURE 6 Effect of the pH on the chiral resolution of amino acids (a) (○) asparagine, (×) valine, and (□) histidine and (b) aspartic acid (1), glutamic acid (2), histidine (3), alanine (4), serine (5), proline (6), thereonine (7), valine (8), lysine (9), tyrosine (10), leucine (11), phenylalanine (12), and tryptophan (13) with L- (—) and D-enantiomers (- - -). (From Refs. 21 and 30.)

chiral resolution was reported at pH 4.75. Similarly, Kurganov et al. [21] varied the pH from 4.6 to 5.0 and observed the best resolution of amino acids at pH 5.0. The results of their findings are presented in Figure 6b. Schlauch et al. [62] resolved four pairs of amino acids in a single run by optimizing the pH conditions.

7.4.3 Ionic Strength of the Mobile Phase

The concentration of buffer is also a very important aspect in the optimization of the chiral resolution on these CSPs. It has been reported that an increase in buffer concentration caused a decrease in the retention and selectivity for all amino acids except for the basic amino acids. Therefore, the separation of basic amino acids is possible only with the most concentrated buffers. The buffers of concentrations in the 25–50-mM range were used for the chiral resolutions with some exceptions. In spite of this, few reports are available for the optimization of the chiral resolution by varying the ionic strength of the mobile phase. The effect of ionic strength of phosphate buffer on the chiral resolution of serine was carried out by Gübitz and Jellen [18] and the best resolution was achieved at 0.01 M concentration (Fig. 7). In another study, the concentration of ammonium acetate (0.001–0.01 M) was varied to optimize the chiral resolution of amino acids [19]. The effect of the concentration of ammonium acetate on the chiral resolution of amino

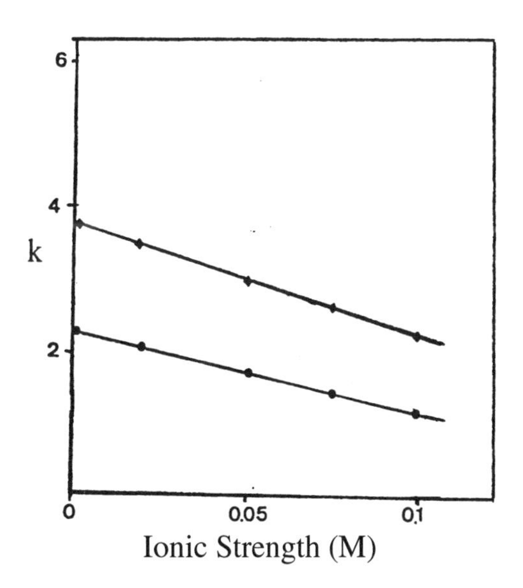

FIGURE 7 Effect of the ionic strength of the buffer on the chiral resolution of (●): D-serine and (●) L-serine. (From Ref. 18.)

TABLE 5 Effect of the Concentrations of Ammonium Acetate on the Chiral Resolution of Amino Acids

	Concentration of ammonium acetate					
	0.001 M		0.01 M		0.1 M	
Amino acid	k_1	α	k_1	α	k_1	α
Aspartic acid	2.19	0.97	0.58	1.00	0.20	1.00
Glutamic acid	2.51	1.16	0.49	1.14	0.06	1.00
Histidine	8.62	1.00	3.13	0.81	1.08	0.73
Alanine	4.37	1.05	0.83	1.09	0.11	1.27
Asparagine	4.27	0.97	1.08	1.00	0.32	1.00
Glutamine	4.29	0.98	0.84	0.58	0.17	0.35
Serine	4.21	0.98	1.06	0.83	0.23	1.04
Proline	5.39	1.27	1.06	1.78	0.16	1.75
Threonine	4.65	0.97	1.28	0.92	0.33	0.88
Valine	5.54	1.08	1.69	1.10	0.24	1.29
Lysine	14.4	1.13	2.95	1.04	0.31	1.00
Tyrosine	5.62	0.95	2.26	0.89	0.49	0.92
Methionine	5.70	1.04	1.94	1.05	0.38	1.08
Arginine	18.46	1.01	3.79	1.06	0.60	1.00
Isoleucine	5.34	6.11	1.76	1.20	0.29	1.32
Leucine	5.61	1.06	1.77	1.10	0.31	1.16
Phenylalanine	6.50	1.00	2.43	1.07	0.55	1.09
Tryptophan	10.67	0.97	5.12	0.96	1.30	0.98

Source: Ref. 19.

acids is given in Table 5, which indicates a decrease in the values of separation factors at a high concentration of ammonium acetate. Davankov et al. [17] studied the chiral resolution of some amino acids with and without ammonium chloride in the mobile phase. The authors reported a rapid decrease of chiral resolution by the addition of ammonium chloride in the mobile phase. It may have been the result of the combined effect of the increased ionic strength, partial protonation of both fixed and mobile amino acids ligands (due to a decrease in pH by adding ammonium chloride), and increased concentration of ligand (ammonia) competing for coordination with copper(II). Jin and He [79] observed the long retention of amino acids using chloride ions in the mobile phase.

7.4.4 Concentration of the Metal Ions

The retention and selectivity of enantiomeric resolution was also investigated with respect to metal ion concentrations on these CSPs. The chiral resolution on the CSPs containing only chiral ligands was carried out using metal ions in the mobile phase. It has been reported that different concentrations of metal ions were used for the chiral resolution of different racemic compounds under varied chromatographic conditions. The literature reveals few reports on the variation of metal ion concentrations. Watanabe et al. [30] studied the effect of the concentration of copper(II) ions on the chiral resolution of glutamine, thereonine, and asparagine and the best resolution was achieved at 0.0005 M concentration (Fig. 8a). The authors suggested avoiding the use of copper metal ions at higher concentrations, as the copper(II) ions absorb ultraviolet (UV) radiation, increas-

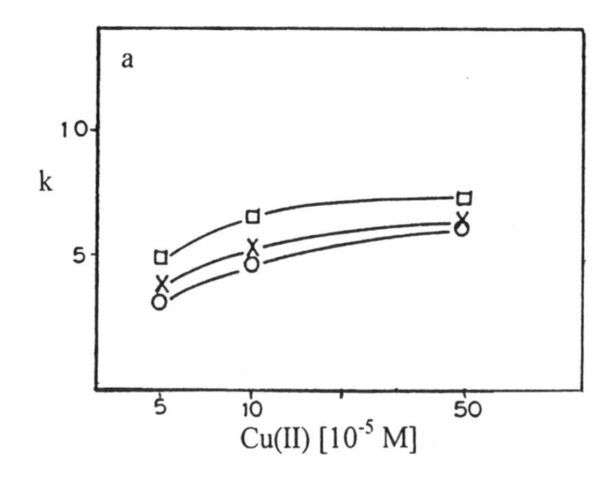

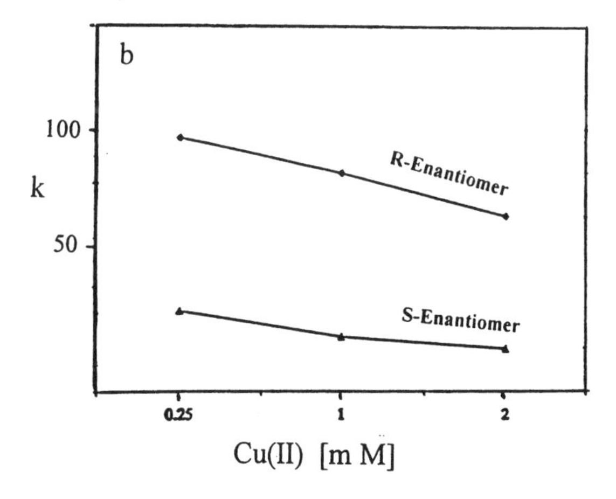

FIGURE 8 Effect of the concentration of copper(II) metal ion on the chiral resolution of (a) (○) glutamine, (×) threonine, and (□) asparagine; and (b) phenylalanine. (From Ref. 30.)

TABLE 6 Effect of the Concentrations of Copper Sulfate on the Chiral Resolution of Lactate Enantiomers

Copper sulfate (mM)	Retention times (min)		α	R_s
	L-Lactate	D-Lactate		
0.50	14.07	16.56	1.23	1.66
1.00	13.79	16.19	1.23	1.68
1.25	13.60	15.96	1.23	1.63
1.50	13.27	15.56	1.23	1.56
2.00	12.60	14.79	1.23	1.51

Source: Ref. 68.

ing the limit of detection. Davankov et al. [17] also carried out the chiral resolution of amino acids by varying the concentrations of copper acetate (0.5×10^{-4} to 2×10^{-4} M). It has been observed that an increase in the concentration of copper(II) ions produced a decrease in the values of retention factors; hence, 0.05×10^{-4} M was found as the suitable concentration. The effect of the copper(II) ions concentration on the retention of phenylalanine was also carried out, and higher values of retention factors were observed at 0.25 mM concentration [67]. This effect of copper(II) ions concentration on the retention of phenylalanine is shown in Figure 8b. Recently, Okubo et al. [68] varied the concentration of copper sulfate from 0.5 to 2.00 mM for the separation of lactate enantiomers. The values of the retention and separation factors are given in Table 6, which indicate the higher separation at 1 mM concentration.

7.4.5 Temperature

Temperature is also a very important parameter for controlling the chiral resolution of these CSPs. It has been observed that an increase in temperature has resulted in a decrease in the retention of all racemates, excluding basic amino acids [78]. The selectivity of resolution has been reduced by an increase in temperature. However, the column efficiency increased with a rise in temperature. The decrease in the retention times of the racemates at a high temperature indicated the exothermic nature of the ligand-exchange process. The effect of temperature (0–60°C) on the chiral resolution of arginine was presented by Gübitz and Jellen [18] and is shown in Figure 9a. This figure indicates no resolution at 0–20°C, partial resolution at 20–50°C, and complete resolution at 50°C or higher. The effect of temperature on the chiral resolution of arginine was also carried out by Watanabe et al. [30] (Fig. 9b) at 10–60°C. In another study [17], an increase in the temperature (i.e., 18–45°C) resulted in the decrease of

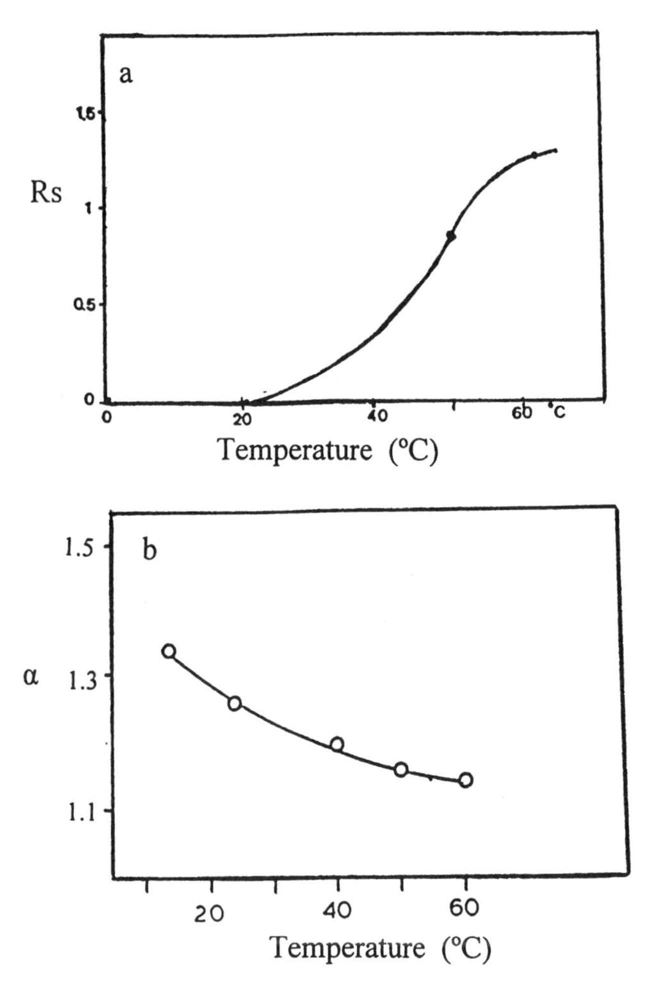

FIGURE 9 Effect of the temperature on the chiral resolution of (a) arginine and (b) asparagines. (From Refs. 18 and 30.).

retention of amino acids, which improved the chiral resolution. Roumeliotis et al. [19] observed a decrease in the retention of amino acids at higher temperature. The authors explained the decrease in the retention as the result of the weakness in the complexation and hydrophobic interactions at higher temperature. In another study [65], the same authors reported the reversal order of the elution of amino acids by varying the temperature. Jin and He [79] reported a decrease in the retention of amino acids at higher temperature. The authors explained the decrease in retention as the result of distortion of polymer conformation in the ligand-exchange process. Okubo et al. [68] also observed a decrease in the retention of lactate enantiomers at higher temperature, with little effect on enantioselectivity. Therefore, temperature may be used to control the enantio-separation on ligand-exchange CSPs.

7.4.6 Flow Rate of the Mobile Phase

In ligand-exchange high-performance liquid chromatography (HPLC), the efficiency of the chiral resolution may be controlled by adjusting the flow rate of the mobile phase. However, it has been reported that the chiral resolution at low flow rates did not follow the well-known Van Deemter's plot, which may be due to the slow ligand-exchange kinetics. There are very few studies dealing with the optimization of chiral resolution on these CSPs by adjusting the flow rates. However, Jin and He [79] studied the effect of flow rates (7.2–34.0 mL/hr) on the chiral resolution of amino acids. The authors reported an increase in the values of separation and resolution factors at low flow rates. However, on the other hand, the authors also reported the fast resolution of some amino acids at high flow rates. Aboul-Enein and Ali [70] reported a better resolution of econazole, miconazole, and sulconazole at a 0.5-mL/min flow rate. The authors rationalized the better resolution at the 0.5-mL/min flow rate as being the result of the greater residence time of enantiomers on the chiral stationary phase, which was sufficient for the ligand-exchange process to take place under the reported chromatographic conditions.

7.4.7 Structures of Solutes

As in the case of other CSPs, the chiral resolution on these CSPs is also affected by a change in the structures of the racemic compounds. The different selectivities of amino acids on these CSPs may be considerd as the best example. The effect of structures of the racemates on the chiral resolution may be understood from the work carried out by Shieh et al. [71]. The authors studied the chiral resolution of amino acids as their Schiff's bases. These racemates differ slightly in their structure and the substituent, such as alkyl groups, hence showed different values of enantioselectivities. The values of retention and separation factors decreased by introducing bulky groups in the racemates. Aboul-Enein and Ali [70] observed the lower values of retention factors of miconazole in comparison to econazole and sulconazole. The authors explained this sort of behavior on the basis of the steric effect exerted by the extra chlorine atom in miconazole molecule.

7.4.8 Structures of Ligand Exchangers

The chiral separations on these CSPs are based on the kinetics of ligand exchangers and the enantiomers. The exchange kinetics vary from one chiral selector to another. Therefore, the structures of the chiral selectors are very important from the selectivity point of view. Accordingly, the same racemate may behave in different ways on the different CSPs. This effect may be observed from the work carried out by Roumeliotis et al. [65] in which [Cu(L-hydroxypropyl)-methyl] silica and [Cu(L-hydroxypropyl)$_n$-octyl] silica gels were used as the chiral selectors for the enantiomeric resolution of amino acids (Table 7). The different values of retention and separation factors is presented in Table 7. Similar behavior for the chiral resolution of amino acids on different CSPs is shown in the research work of the same authors [19]. A very interesting study was carried out by Gübitz and Jellenz [18] using L- and D-enantiomers of proline as the chiral selectors separately for the chiral resolution amino acids. The two CSPs showed different chiral selectivities for certain amino acids. It has also been observed that L-proline is a better chiral ligand than D-proline. This could be because of the steric effects, as the two enantiomers differ only in their configurations. These

TABLE 7 Effect of the Structures of CSPs on the Chiral Resolution of Amino Acids

	[Cu(L-Hydroxy-propyl)methyl]$^+$ silica		[Cu(L-Hydroxy-propyl)$_n$-octyl]$^+$ silica	
Amino acid	k_1	α	k_1	α
Aspartic acid	0.13	1.08	2.01	1.06
Glutamic acid	0.10	2.50	1.57	1.00
Histidine	1.70	0.75	2.22	0.86
Alanine	0.48	1.21	0.40	1.45
Asparagine	0.52	1.54	0.89	0.92
Serine	0.65	1.00	0.82	0.90
Proline	0.63	1.43	1.07	1.93
Threonine	0.73	1.37	1.14	1.02
Valine	0.80	1.12	2.81	0.96
Lysine	1.42	1.04	0.07	0.29
Tyrosine	1.37	0.69	12.25	0.79
Methionine	1.23	1.14	4.05	1.13
Arginine	2.05	0.87	0.31	1.26
Isoleucine	1.23	1.00	5.86	1.01
Leucine	1.30	0.79	4.80	1.07
Phenylalanine	1.32	1.00	17.60	0.88

Source: Ref. 65.

TABLE 8 Effect of the Structures of L-Proline and D-Proline as Chiral Ligands on the Chiral Resolution of Amino Acids

Amino acid	L-Proline α	D-Proline α
Tryptophan	3.50	0.36
Phenylalanine	2.90	0.40
Tyrosine	3.10	0.38
Valine	1.50	0.66
Proline	0.60	2.30
Histidine	1.80	0.53
Thereonine	1.60	0.60

Source: Ref. 18.

results are presented in Table 8. In 1996, Galaverna et al. [67] carried out the chiral resolution of amino acids on three different CSPs: N^2-octyl-(*S*)-phenylalaninamide, N^2-octadecyl-(*S*)-phenylalaninamide, and N^2-octyl-(*S*)-norleucinamide. These three CSPs differ only in their functional groups and, therefore, showed different chiral recognitions. In 1997, Vidyasankar et al. [63] observed that adsorbents prepared from amino acids with larger aromatic side chains exhibited the highest selectivities for amino acids. Briefly, the different chiral resolution on different CSPs is due to the different physical, stereochemical properties of the CSPs.

7.4.9 Other Parameters

In addition to the above-discussed parameters, other factors are also important for controlling the chiral resolution of these CSPs. These parameters include particle size of the silica gel, amount of chiral selector on the solid support, and the injection volume. Roumeliotis et al. [19,65] used the silica support of different particle sizes and the authors reported that the best resolution was obtained on the silica gel of smaller particle size. It has also been reported that the packing of the chiral selectors affected the chiral resolution, which may be the result of the slowness of the ligand-exchange kinetics [5,27,64,80]. Boue et al. [64] studied the effect of the amount of sample loading on the chiral resolution of valine and the authors observed a shift from 3.0 to 3.5 in retention factor when the injected sample decreased from 1 to 0.15 mg. Jin and He [79] studied the chiral resolution of histidine using different loading amounts. They reported a decrease in the

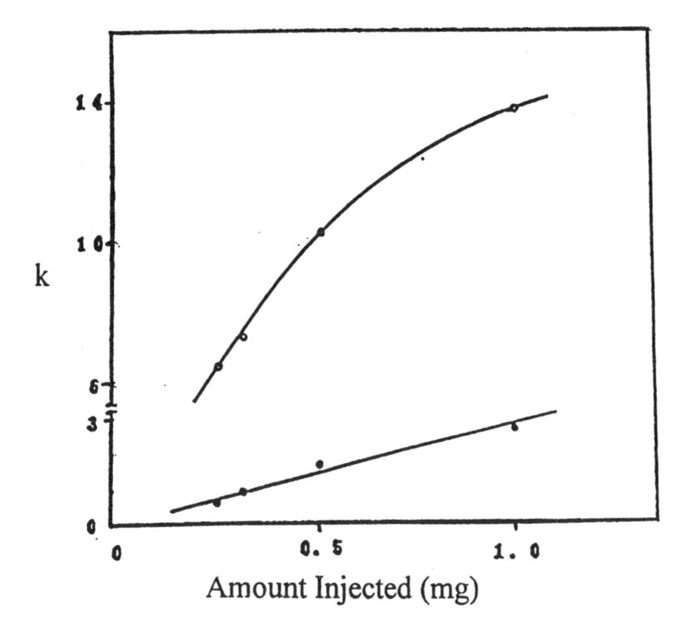

FIGURE 10 Effect of the injection amount on the chiral resolution of histidine. (From Ref. 79.)

retention at a higher concentration of histidine, which resulted in the improved separation of the enantiomers. The variation of the retention factors with respect to loading amounts is shown in Figure 10. The chiral resolution may be optimized by introducing a suitable spacer between the solid support and the chiral ligands.

7.5 CHIRAL RECOGNITION MECHANISM

The chiral resolution of these CSPs occur due to the exchange of the chiral ligands and the enantiomers of specific metal ions through coordinate bonds. The two enantiomers have different exchange capacities because of the stereospecific nature of the ligand-exchange process; hence, chiral resolution takes place. Davankov et al. [2,8,9] suggested a theoretical model for the mechanisms of the chiral resolution on these CSPs. In this model, the enantiomers were coordinated to the Cu(II) ions in different ways depending on their interactions with the ligands bonded to the stationary phase acting as chiral selectors. The authors explained that the chiral resolution is the result of different bondings

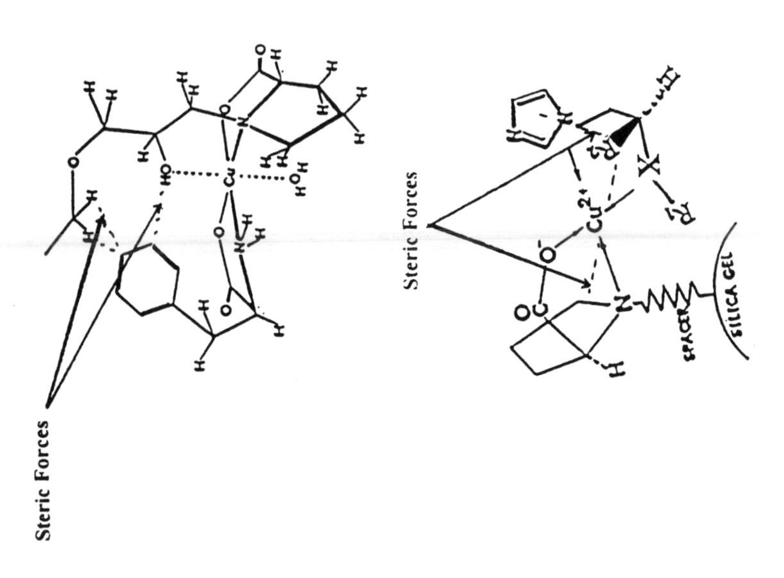
Steric Forces
Steric Forces
SPACER
SILICA GEL
a
b
SPACER
SILICA GEL

	R_1	R_2	X
Econazole	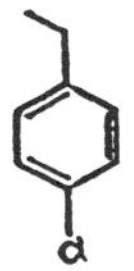	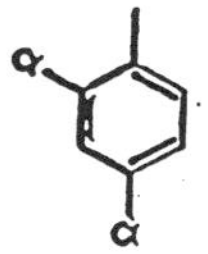	O
Miconazole	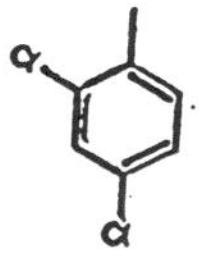	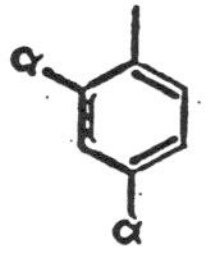	O
Sulconazole			S

FIGURE 11 Graphical representation of the chiral recognition mechanism of (a) phenylalanine and (b) antifungal agents on the chiral ligand-exchange CSPs. (From Refs. 18 and 70.)

along with the steric effects. In this way, the diastereoisomeric complexes are formed by the two enantiomers. These diastereoisomeric complexes are stabilized at different magnitudes by dipole–dipole interactions, hydrogen-bondings, van der Waal forces, and steric effects. The diastereoisomeric complexes of phenylalanine enantiomers and Cu(II)–L-proline are shown in Figure 11a. It is clear from this figure that the diastereoisomeric complex having the D-enantiomer is less stable, due to the presence of steric forecs, and hence eluted first followed by L-enantiomer.

To make the concept much clearer, an example of the work carried out by Aboul-Enein and Ali [70] is presented herein. The authors reported the chiral resolution of econazole, miconazole, and sulconazole (antifungal agents) on the Chiralpak WH column. The expected structures of the diastereoisomeric complexes of these antifungal agents with the CSP are shown in Figure 11b. All three antifungal agents contain electron-donating atoms such as nitrogen, oxygen, and sulfur (Fig. 11b). The Chiralpak WH CSP is a L-proline–Cu(II) complex attached to silica gel. The copper is a transition metal ion and contains empty *d*-orbitals, which are essential for coordinate bonding. Therefore, nitrogen, oxygen, and sulfur atoms of the antifungal agents are coordinated with a copper atom and hence the ternary complexes of L-proline–Cu(II)-enantiomers are formed. The enantiomeric resolution of these antifungal agents may be explained on the basis of the ligand-exchange mechanism through coordination bondings. Other interactions, such as hydrogen-bondings, dipole-induced dipole interactions, steric effect, and van der Waal forces, also participate in the chiral recognition phenomenon. Therefore, (+)- and (−)-enantiomers of the reported antifungal agents formed diastereoisomeric complexes of different stabilities with WH CSP in a stereo-specific way, which resulted in the complete resolution of the enantiomers.

Figure 11b shows that for (+)-isomers of antifungal agents, the substituents on the chiral carbon of antifungal agents are located far from the pyrrolidine group of L-proline, whereas the substituent on the chiral carbon of (−)-enantiomers are located near the pyrrolidine group of L-proline. Therefore, the L-proline–Cu(II)–(−)-enantiomer complexes feel some steric force, whereas the L-proline–Cu(II)–(+)-enantiomer complexes do not face any steric effect and, hence, the L-proline–Cu(II)–(−)-enantiomer complexes are less stable than L-proline–Cu(II)–(+)-enantiomer complexes. Because of this steric effect, the (−)-enantiomers of all three antifungal agents eluted first, followed by the (+)-enantiomers. It is also very interesting to observe that the coordination power of nitrogen, oxygen, and sulfur is in the order $S > N > O$. Therefore, it may be concluded that the stability of the L-proline–Cu(II)-enantiomer complex of sulcanazole is greater than the complexes of the enantiomers of econazole and miconazole. This explains the good resolution of the enantiomers as compared to the resolution of econazole and miconazole enantiomers.

7.6 MISCELLANEOUS APPLICATIONS

Other than HPLC, ligand-exchange chromatography has been used for chiral resolution using capillary electrochromatography and thin-layer chromatography. In 1998, Lämmerhofer and Lindner [81] prepared an anion-type chiral stationary phase by the reaction of *tert*-butyl carbamoyl quinine and the CSP was packed into fused silica capillaries of 25-cm lengths with diameters of 75 and 100 μm, respectively. These developed capillaries were tested for the chiral resolution of N-derivatized amino acids. The authors reported two to three times higher efficiency of chiral resolution in comparison to the efficiencies obtained from HPLC methods. Two years later, Tobler et al. [82] used the same CSP in nonaqueous capillary electrochromatography for the chiral resolution of derivatized amino acids and profens. Solvent composition of acetonitrile–methanol (80 : 20, v/v) and enhanced electrolyte concentrations up to 600 mM acetic acid at a constant acid–base (triethylamine and acetic acid) ratio of 100 : 1 with high applied voltage of 25 kV proved to be optimum separation conditions. Further, Lämmerhofer et al. [83] studied the different types of silica particle (porous and nonporous) for the chiral resolution of N-derivatized amino acids using capillary electrochromatography. The authors used the same chiral selector (i.e., *tert*-butyl carbamoyl quinine) for these studies. They optimized the separation by varying pH, the types of buffer with different concentrations, the types of organic modifier with different concentrations, voltage, and temperature. The authors described the reproducibility of the results by presenting 2% as the relative standard deviation.

The chiral ligand-exchange methods are also applicable for the chiral resolution by thin-layer chromatography (TLC). In 1984, two independent groups, Günther et al. [84,85] and Weinstein [86], presented the chiral resolution by ligand-exchange TLC. The procedures of both the groups were similar except for the use of the chiral selectors. Günther et al. [84,85] used (2*S*,4*R*, 2′*RS*)-*N*-(2′-hydroxy-dodecyl)-4-hydroxyproline as the chiral selector for the enantiomeric resolution of amino acids. Weinstein [86] impregnated thin layers with the optically active copper complex of *N*,*N*-di-*n*-propyl-L-valine and the plates were used for the chiral resolution of amino acids using sodium acetate (0.3%)–acetonitrile (60 : 40, v/v) as the mobile phase. Further, Grinberg and Weinstein [87] presented a two-dimensional TLC method for the chiral resolution of dansyl amino acids. In another study, Grinberg [88] studied the effect of temperature on the chiral resolution of amino acids. Later, Marchelli et al. [89] studied the diaminodiamide copper(II) complex as the chiral selector for the enantiomeric resolution of dansyl amino acids. Under license from Degussa [90], Chiralplate was introduced as the chiral TLC plate, ready to use and based on ligand-exchange chromatography, was developed and commercialized in 1985 by Macherey-Nagel (Düren, Germany) [91]. Again in 1988, by license from

TABLE 9 Enantiomeric Resolution of Racemic Compounds Using Ligand-Exchange TLC

Racemic compound	CSP	Ref.
Amino acids	(2*S*,4*R*, 2′*RS*)-*N*-(2′-hydroxy-dodecyl)-4-hydroxyproline,	95
	N-(2-hydroxy-decyl)-L-hydroxyproline	96
	Chiralplate	13, 100
	Cu(II)–proline complex	99
	(2*S*,4*R*,2′*RS*)-4-hydroxy-1-(2′-hydroxydodecyl)-L-proline (Chiralplate)	101, 102
	Cu(II)–histidine complex	103
Derivatives of amino acids	(2*S*,4*R*,2′*RS*)-4-hydroxy-1-(2-hydroxydodecyl) proline	97
	Chiralplate	13
Dipetides	(2*S*,4*R*,2′*RS*)-*N*-(2′-hydroxy-dodecyl)-4-hydroxyproline	95
	Chiralplate	13
α-Hydroxy acids	(2*S*,4*R*,2′*RS*)-*N*-(2′-hydroxy-dodecyl)-4-hydroxyproline	95
	Chiralplate	13
Mandelic acid and derivatives	Chiralplate	98
Thyroxine	Chiralplate	104

Degussa, Merck Company (Darmstadt, Germany) developed and commercialized HPTLC plates. These plates are very successful for the chiral resolution of amino acids and related compounds. Luebben et al. [92] synthesized model chiral selectors and tested for the chiral resolution of different racemic compounds. Recently, the chiral resolution by TLC, including the use of ligand exchangers as chiral selectors, has been reviewed [93,94]. The chiral resolution of amino acids and other racemates using chiral ligand-exchange TLC is presented in Table 9.

7.7 CONCLUSION

With the development of the chiral ligand exchange chromatography by Davankov, this technique has been used frequently for the chiral resolution of racemic compounds containing electron-donating atoms. It is useful for providing the basic information on the chiral resolution and, hence, is still in use. In spite of this, there are some limitations with this chiral resolution technique. The most unpleasant limitation of this technique is its narrow range of applications. Davankov [8] concluded that the stability of the diastereoisomeric complexes formed in CLEC is higher than the stability of the diastereoisomeric complexes formed by many other chiral selectors in other equilibria. Therefore, the relatively high stabilities of the diastereoisomeric complexes may result in slow ligand exchange and reduce the plate count of CLEC columns, which result in poor chiral recognition. Also, the metal ions in the mobile phase absorb UV radiation, resulting into the increase of the limit of detection. In addition to these factors, sometimes the recognition capabilities of the columns containing Cu(II)-complexes as the chiral selectors is reduced, and, hence, regeneration of these columns is required. Therefore, the use of this technique, currently, in chiral resolution is very limited because of these drawbacks and the development of the new CSPs.

REFERENCES

1. Helfferich FG, Nature 189: 1001 (1961).
2. Rogozhin SV, Davankov VA, German Patent 1932190 (1969).
3. Davankov VA, Navratil JD, Walton HF. Ligand Exchange Chromatography, CRC Press, Boca Raton, FL (1988).
4. Davankov VA, Rogozhin SV, Piesliakas II, Vysokomolek Soed 14B: 276 (1972).
5. Davankov VA, Semechkin AV, J Chromatogr 141: 313 (1977).
6. Dvankov VA, Adv Chromatogr 18: 139 (1980).
7. Davankov VA, Kurganov A, Bochkov AS, Adv Chromatogr 22: 71 (1983).
8. Davankov VA, J Chromatogr 666: 55 (1994).
9. Rogozhin SV, Davankov VA, Chem Lett 490 (1971).
10. Davankov VA, Kurganov A, Bochkov AS, Resolution of racemates by high performance liquid chromatography, in Advances in Chromatography, Giddings JC, Grushka E, Cazes J, Brown PR (Eds.), Vol. 22, p. 71 (1983).
11. Davankov VA, Introduction to chromatographic resolution of enantiomers, in Chiral Separations by HPLC, Krstulovic A (Ed.), Ellis Horwood, Chichester, p. 175 (1989).
12. Davankov VA, Ligand exchange phases, in Chiral Separations by HPLC, Krstulovic A (Ed.), Ellis Horwood, Chichester, p. 447 (1989).
13. Günther K, Enantiomer separations, in Handbook of Thin Layer Chromatography, Sherma J, Fried B (Eds.), Marcel Dekker, New York, p. 541 (1991).
14. Kurganov A, J Chromatogr A 906: 51 (2001).
15. Feibush B, Cohen MJ, Karger BL, J Chromatogr 282: 3 (1983).
16. Zolotarev YA, Myasoedov NF, Penkina VI, Petrenik OR, Davankov VA, J Chromatogr 207: 63 (1981).
17. Davankov VA, Bochkov AS, Belov YP, J Chromatogr 218: 547 (1981).
18. Gübitz G, Jellenz W, J Chromatogr 203: 377 (1981).
19. Roumeliotis P, Unger KK, Kurganov AA, Davankov VA, J Chromatogr 255: 51 (1983).

20. Gübitz G, Jellenz W, Santi W, J Liq Chromatogr 4: 701 (1987).
21. Kurganov AA, Tevlin AB, Davankov VA, J Chromatogr 261: 223 (1983).
22. Petit MA, Jozefonvicz J, J Appl Polym Sci 21: 2589 (1977).
23. Jozefonvicz J, Petit MA, Szubarga A, J Chromatogr 147: 177 (1978).
24. Spassky N, Reix M, Quette J, Quette M, Sepulchre M, Blanchard J, Comp Rend 287: 589 (1978).
25. Tsuchida E, Nishikawa H, Terada E, Eur Polym J 12: 611 (1976).
26. Snyder RV, Angelichi RJ, Meck RB, J Am Chem Soc 94: 2660 (1972).
27. Lefebvre B, Audebert R, Quivoron C, Israel J Chem 15: 69 (1977).
28. Lefebvre B, Audebert R, Quivoron C, J Liq Chromatogr 1: 761 (1978).
29. Yamskov IA, Berezin BB, Davankov VA, Zolotarev YA, Dostovalon IN, Myasoedov NF, J Chromatogr 217: 539 (1981).
30. Watanabe N, Ohzeki H, Niki E, J Chromatogr 216: 406 (1981).
31. Davankov VA, Bochkov AS, Kurganov AA, Roumeliotis P, Unger KK, Chromatographia 13: 677 (1980).
32. Miyazawa T, Minowa H, Jamagawa K, Yamada T, Anal Lett 30: 867 (1997).
33. Oi N, Kitahara H, Akoi F, J Chromatogr A 707: 380 (1995).
34. Slivka M, Slebioda M, Kolodziejczyk A, Chem Anal (Warsaw) 42: 895 (1998).
35. Kurganov A, Davankov VA, Unger K, Eisenbeiss F, Kinkel J, J Chromatogr A: 99 (1994).
36. Gübitz G, Jellenz W, Lofler G, Santi W, J High Resolut Chromatogr 2: 145 (1979).
37. Bochkov AS, Zolotarev YA, Belov YP, Davankov VA, Second Danube Symposium on Progress in Chromatography, Paper B 3.23 (1979).
38. Sugdan K, Hunter C, Lloyd-Jones G, J Chromatogr 192: 228 (1980).
39. Gübitz G, Mihellyess S, Kobinger G, Wutt A, J Chromatogr A 666: 91 (1994).
40. Gübitz G, Vollman B, Cannazza G, Schmid MG, J Liq Chromatogr Relat Technol 19: 2933 (1996).
41. Yang M, Hsieh M, Lin J, Taiwan Kexue 50: 17 (1997).
42. Hyun MH, Han SC, Lee CW, Lee YK, J Chromatogr A 950: 55 (2002).
43. Lindner W, Naturwiss 67: 345 (1980).
44. Engelhardt H, Kromidas S, Naturwiss 67: 353 (1980).
45. Davankov VA, Kurganov AA, Chromatographia 17: 686 (1983).
46. Foucault A, Caude M, Oliveros L, J Chromatogr 185: 345 (1979).
47. Roeske R, J Org Chem 28: 1251 (1963).
48. Davankov VA, Rogozhin SV, Piesliakas II, Semechkin AV, Sachkova TP, Doklady Akad Nauk SSR 201: 854 (1971).
49. Roumeliotis P, Unger KK, Kurganov AA, Davankov VA, Angew Chem Int Ed Eng 21: 930 (1982).
50. Zolotarev YA, Myasoedov NF, Penkina VI, Dostovalov IN, Petrenik OV, Davankov VA, J Chromatogr 207: 231 (1981).
51. Busker E, Günther K, Martens J, J Chromatogr 350: 179 (1985).
52. Yuki Y, Saigo K, Kimoto H, Tachibana K, Hasegawa M, J Chromatogr 400: 65 (1987).
53. Takeuchi T, Asai H, Ishii D, J Chromatogr 407: 151 (1987).
54. Gübitz G, Juffmann F, Jellenz W, Chromatographia 16: 103 (1982).

55. Gübitz G, Juffmann F, J Chromatogr 404: 391 (1987).
56. Gübitz G, J Liq Chromatogr 9: 519 (1986).
57. Sinibaldi M, Carunchio V, Corradini C, Girelli AM, Chromatographia 18: 459 (1984).
58. Corradini C, Federici F, Sinibaldi M, Messina A, Chromatographia 23: 118 (1987).
59. Watanabe N, J Chromatogr 260: 75 (1983).
60. Ito SN, Takeuchi T, Miwa T, J Chromatogr A 864: 25 (1999).
61. Buglass AJ, Lee SH, J Chromatogr Sci 38: 207 (2000).
62. Schlauch M, Volk FJ, Fondekar KP, Wede J, Frahm AW, J Chromatogr A 897: 145 (2000).
63. Vidyasankar S, Ru M, Arnold FH, J Chromatogr A 775: 51 (1997).
64. Boue EJ, Audebert R, Quivoran C, J Chromatogr 204: 185 (1981).
65. Roumeliotis P, Kurganov AA, Davankov VA, J Chromatogr 266: 439 (1983).
66. Bruckner H, Bosch I, J Chromatogr 395: 569 (1987).
67. Galaverna G, Corradini R, Dossena A, Marchelli R, Dallavalle F, Chirality 8: 189 (1996).
68. Okubo S, Mashige F, Omori M, Hashimoto Y, Nakahara K, Kanazawa H, Matsushima Y, Biomed Chromatogr 14: 474 (2000).
69. Chilmonczyk Z, Ksycinska H, Cybulski J, Rydzewski M, Les A, Chirality 10: 821 (1998).
70. Aboul-Enein HY, Ali, Chromatographia 54: 200 (2001).
71. Shieh CH, Karger BL, Gelber LR, Feibush B, J Chromatogr 406: 343 (1987).
72. Chen Z, Uchiyama K, Hobo T, J Chromatogr A 942: 83 (2002).
73. Davankov VA, Zolotarev YA, Kurganov AA, J Liq Chromatogr 2: 1191 (1979).
74. Weinstein S, Tetrahedron Lett 25: 985 (1984).
75. Krasutskii PA, Rodionov VN, Tichonov VP, Yurchenko AG, Teoret Exper Khim 20: 58 (1984).
76. Gris GJ, Soerensen C, Su H, Porret J, Chromatographia 28: 337 (1989).
77. Perry JA, Rateike JD, Szczerba TJ, J Chromatogr 389: 57 (1987).
78. Remelli M, Fornasari P, Dandi F, Pulidori F, Chromatographia 37: 23 (1993).
79. Jin R, He B, J Liq Chromatogr 12: 501 (1989).
80. Cooke NHC, Viavattene RL, Eksteen R, Wong WS, Davies G, Karger BL, J Chromatogr 149: 391 (1978).
81. Lämmerhofer M, Lindner W, J Chromatogr 829: 115 (1998).
82. Tobler E, Lämmerhofer M, Lindner W, J Chromatogr 875: 341 (2000).
83. Lämmerhofer M, Tobler E, Lindner W, J Chromatogr 887: 421 (2000).
84. Günther K, Martens J, Schickedanz M, Angew Chem 96: 514 (1984).
85. Günther K, Martens J, Schickedanz M, Angew Chem Int Ed Engl 23: 506 (1984).
86. Weinstein S, Tetrahedron Lett 25: 985 (1984).
87. Grinberg N, Weinstein S, J Chromatogr 303: 251 (1984).
88. Grinberg N, J Chromatogr 333: 69 (1985).
89. Marchelli R, Virgili R, Armani E, Dossena A, J Chromatogr 355: 354 (1986).
90. Günther K, Martens J, Schickedanz M, EP-PS Patent 143147 (to Degussa AG, Hanau, Germany) (1985).

91. Günther K, Rausch R, Proceedings of Third International Symposium on Instrumental HPTLC (1985); Bad Dürkheim, Germany, p. 469 (1985).
92. Leubben S, Martens J, Hasse D, Phol S, Saak W, Tetrahedron Lett 31: 7127 (1990).
93. Aboul-Enein HY, El-Awady MI, Heard CM, Nicholls PJ, Biomed Chromatogr 13: 531 (1999).
94. Bhushan R, Martens J, Biomed Chromatogr 15: 155 (2001).
95. Günther K, J Chromatogr 448: 11 (1988).
96. Günther K, Schickedanz M, Martens J, Naturwiss 72: 149 (1985).
97. Gont LK, Neuendorf SK, J Chromatogr 391: 343 (1987).
98. Mack M, Hauck H, Herbert H, J Planar Chromatogr 1: 304 (1988).
99. Bhushan R, Reddy GP, Joshi S, J Planar Chromatogr 7: 126 (1994).
100. Brinkman UATh, Kamminga DK, J Chromatogr 330: 375 (1985).
101. Günther K, Schickedanz M, Drauz K, Martens J, Fresenius J Anal Chem 325: 297 (1986).
102. Günther K, Martens J, Schickedanz M, Fresenius J Anal Chem 322: 513 (1985).
103. Remelli M, Pizza R, Pulidori F, Chromatographia 32: 278 (1991).
104. Aboul-Enein HY, Serignese V, Biomed Chromatogr 8: 317 (1994).

8

Crown Ether–Based Chiral Stationary Phases

Crown ethers are synthetic macrocyclic polyethers with the appearance of a crown. The appearance of their molecular structures and ability to crown selectively with the cations give them their names "crown ethers." Pedersen [1,2] synthesized these polyethers in 1967. The ether oxygens, which are electron donors, remain in the inner wall of the crown cavity and are surrounded by methylene groups in a collar fashion. The IUPAC nomenclature of these ethers is complex and, hence, trivials names are common and in use [2]. For example, 2,3,11,12-dibenzo-1,4,7,10,13,16 hexaoxa-cyclo-octadeca-2,11-diene is called dibenzo-18-crown-6 ether, where dibenzo, 18, and 6 indicate the substituent groups, total number of atoms in the ring, and the number of the oxygen atoms, respectively. If oxygen atoms of the ether are replaced by nitrogen or sulfur atoms, the crown ether is called as aza or thia crown ether, respectively. The chirality in the crown ether is developed by introducing the chiral moieties and, hence, the developed crown ether is called as chiral crown ether (CCE). The most important chiral groups used for this purpose are binaphthyl [3–6], biphenanthryl [7–9], helicene derivatives [10], tartaric acid derivatives [11], carbohydrate moiety [12], chiral carbon atom with a bulky group directly incorporated in a crown ring [13,14], aromatic bicyclo derivatives [3.3.1] nonane derivatives [15,16] hexahydrochrysese or tetrahydroindenoinden [17], and 9,9′-spiro-bifluorene groups [18]. The capability of the crowning selectively of these CCEs and their stereospecific configurations make them suitable chiral selectors in liquid chromatography. Sousa et al. [19] reported the optical resolution of racemic amino compounds on chiral crown ethers using liquid chromatography. In 1974, Blasius et al. [20] reported the use of some chargeless crown ethers in liquid chromatography. This chapter discusses crown ethers as chiral stationary phases; however, their use as mobile phase additives will be covered in Chapter 10.

8.1 STRUCTURES AND PROPERTIES

The main skeleton of the chiral crown ether molecule is the cyclic structure containing oxygen and methylene groups alternatively. Other derivatives of crown ethers, containing alkyl or phenyl or other groups (as cited earlier) attached to methylene groups, have been reported. The structures of some of the important chiral crown ethers are given in Figure 1. Most of the crown ethers form complexes with metal ions (cations). However, some crown ethers were reported to form complexes with organic compounds, especially compounds having primary amino groups. The stability of the complex formed between guest (molecule containing primary amino group) and host (crown ether) depends on the molecular structure around the primary amino group of the guest molecule. A decrease in the stability of the guest–host complex was reported by the reduction

(18-Crown-6)-2,3,11,12-tetracarboxylic acid crown ether

R,R-Alkyl-pyridino-18-crown-6 crown ether
R = $-CH_3$ and $-C_6H_5$ separately

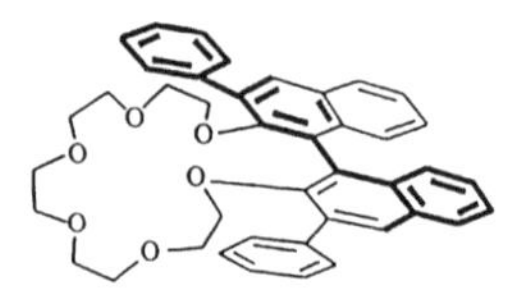

Trisubstituted-18-crown-6 crown ether

Disubstituted-18-crown-6 crown ether

Tetrasubstituted-18-crown-6 crown ether

FIGURE 1 Chemical structures of the chiral crown ethers.

in the number of hydrogen bonds available for guest–host interactions and an increase in the steric hindrance, because of the increase in bulky substituent groups attached to the amino group [21–23].

8.2 PREPARATION AND COMMERCIALIZATION

Various reports are available on the synthesis of CCEs [3–6,19,24–32]. Cram et al. [3–6,19,29–32] carried out a systematic and extensive study on the synthesis of CCEs. They synthesized a series of CCEs containing 1,1′-binaphthyl moiety such as CE-1 to CE-6 (Cram's crown ethers). There are well-known methods for the preparation of chiral crown ethers and the topic is beyond the scope of this chapter. However, recently, CCEs containing pyridine, acridine, and phenazine heterocyclic units were synthesized from tetraethylene glycols containing suitable substituents at the stereogenic centers, and diols with the same groups at the stereogenic centers. The diols containing side chains with terminal double bonds were also used for the synthesis of CCEs. Furthermore, in a new method, 1,9-dichlorophenazine was allowed to react with enantiomerically pure dimethyl-substituted tetraethylene glycol and diisobutyl-substituted diols containing side chains with a terminal double bond [25–28]. Although there are many CCEs available 18-crown-6 and its derivatives are still commonly used in liquid chromatography (Fig. 1).

The CSPs based on chiral crown ethers were prepared by immobilizing them on some suitable solid supports. Blasius et al. [33–35] synthesized a variety of achiral crown ethers based on ion exchangers by condensation, substitution, and polymerization reactions and were used in achiral liquid chromatography. Later, crown ethers were adsorbed on silica gel and were used to separate cations and anions [36–39]. Shinbo et al. [40] adsorbed hydrophobic CCE on silica gel and the developed CSP was used for the chiral resolution of amino acids. Kimura et al. [41–43] immobilized poly- and bis-CCEs on silica gel. Later, Iwachido et al. [44] allowed benzo-15-crown-5, benzo-18-crown-6 and benzo-21-crown-7 CCEs to react on silica gel. Of course, these types of CCE-based phases were used in liquid chromatography, but the column efficiency was very poor due to the limited choice of mobile phases. Therefore, an improvement in immobilization was realized and new methods of immobilization were developed. In this direction, CCEs were immobilized to silica gel by covalent bonds.

Machida et al. presented a method for the immobilization of CCE on silica gel in 1998 [45]. They reported the covalent binding of (+)-(18-crown-6)-tetracarboxylic acid to 3-aminopropylsilanized silica gel and the prepared CSP was tested to separate enantiomers of amino acids, amino alcohols and other drugs (containing primary amino group). In 1998, the same CSP was prepared by Hyun et al. [46]. The developed CSP was used extensively for the chiral resolution of a variety of racemic compounds having a primary amino group

[46–51]. The (+)-(18-crown-6)-2,3,11,12-tetracarboxylic acid and (+)-(18-crown-6)-dinaphthyl CCEs were immobilized to 3-aminopropyl and (3-aminopropyl)-ethoxy silica gels separately and respectively. The immobilization was carried out in two steps. The chiral (18-crown-6)-2,3,11,12-tetracarboxylic acid was converted into dianhydride by treating with acetyl chloride via a known

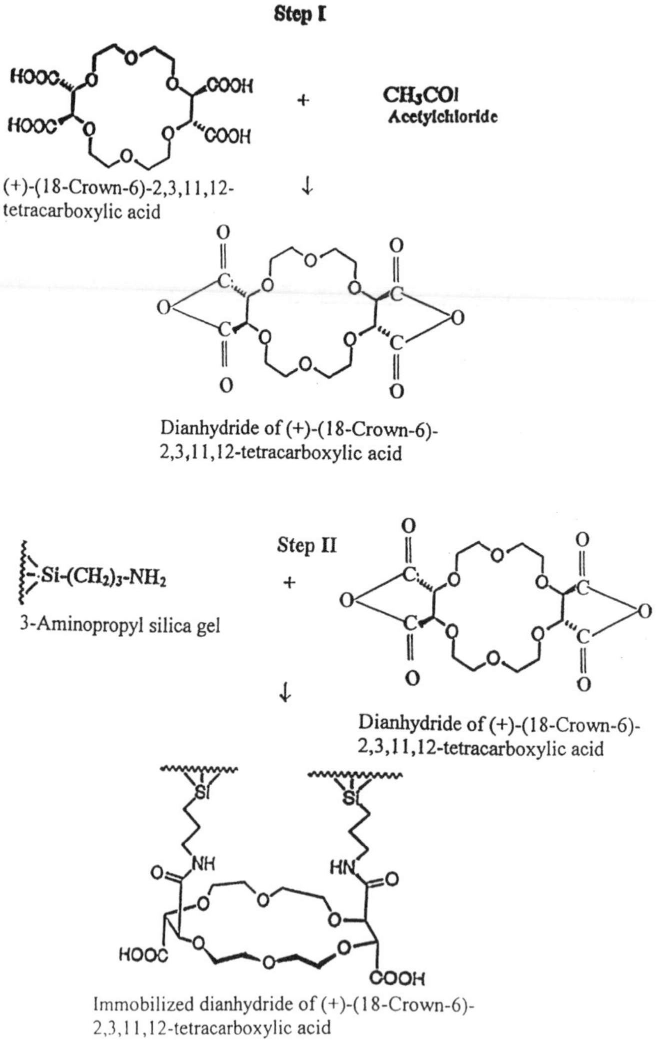

Step I

(+)-(18-Crown-6)-2,3,11,12-tetracarboxylic acid + CH_3COCl Acetylchloride

Dianhydride of (+)-(18-Crown-6)-2,3,11,12-tetracarboxylic acid

Step II

Si-$(CH_2)_3$-NH_2

3-Aminopropyl silica gel + Dianhydride of (+)-(18-Crown-6)-2,3,11,12-tetracarboxylic acid

Immobilized dianhydride of (+)-(18-Crown-6)-2,3,11,12-tetracarboxylic acid

FIGURE 2 Chemical pathway for covalent bonding of the chiral crown ether with silica gel.

TABLE 1 Various CCE-Based Commercial CSPs

Commercialized CSP	Company
Crownpak CR	Chiral Technologies Inc., Exton, PA, USA
Crownpak CR	Separations Kasunigaseki-Chrome, Tokyo, Japan
Crownpak CR	Daicel Chemical Industries, Tokyo, Japan
Opticrown RCA	USmac Corporation, Winnetka, Glenview, USA
Chiralhyun-CR-1	K-MAC (Korea Materials & Analysis Corp.), South Korea
Chirosil CH RCA	Rstech, Corporation, Daedeok, Daejon, South Korea

procedure (step I) [11]. In step II, dianhydride of CCE was treated in dry methylene chloride at 0°C under an argon atmosphere for 2 days with 3-aminopropyl silica gel containing triethylamine, which was dried in advance by azeotropic removal of water in refluxing benzene. The product was washed with methanol, water, 1 N HCl, water, methanol, dichloromethane, and hexane, in that order, and dried under high vacuum. The slurry of this product was prepared in methanol and then packed into stainless-steel columns of the desired dimensions. The protocol of this procedure is shown in Figure 2. The CCEs were commercialized and are available in stainless-steel columns with specific dimensions. The various commercial CSPs supplied by different companies are summarized in Table 1.

8.3 APPLICATIONS

The chiral crown ethers have a good tendency to complex with compounds having primary amino groups. Therefore, all of the chiral separations on these CSPs belong to racemic compounds containing primary amino groups. Mostly racemic compounds containing a primary amino group attached to a chiral center have been resolved. However, some reports indicate chiral resolution for compounds of which the primary amino group is not located at the chiral center. Recently, Steffeck et al. [52] reported the chiral resolution of secondary amines. The structures of some of the resolved racemic compounds on CCE-based CSPs are shown in Figure 3. In spite of the availability of different types of CCE, 18-crown-6 and its derivatives were used for the successful chiral resolution in liquid chromatography. Cram et al. [4–6,19] achieved the chiral resolution of certain amino acids and amino esters on CCE-based CSPs and the technique is called guest–host complexation chromatography. The CCEs were adsorbed on silica gel and the developed CSPs were used in normal and reversed-phase modes. In 1987, Shinbo et al. [40] coated 18-crown-6 crown ether on ODS silica gel and the CSP was tested for the chiral resolution of amino acids and amines

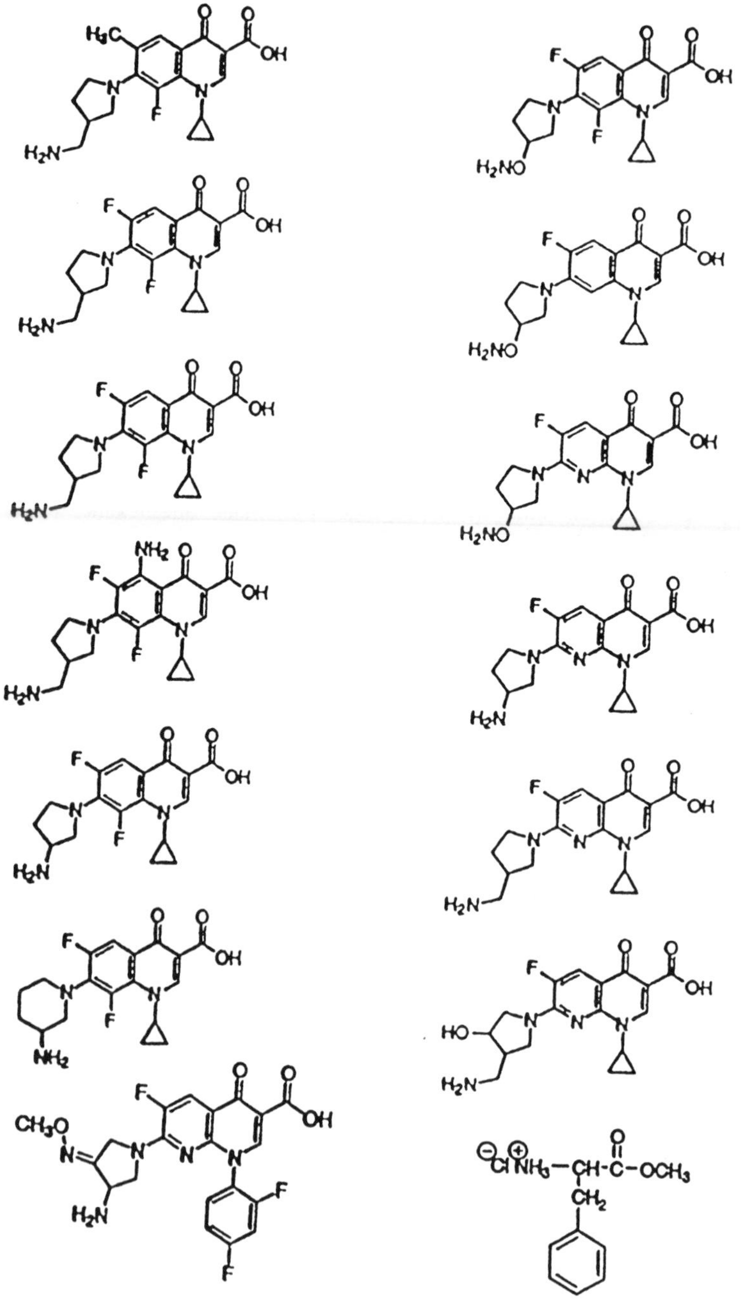

FIGURE 3 Structures of some racemic compounds having primary amino groups.

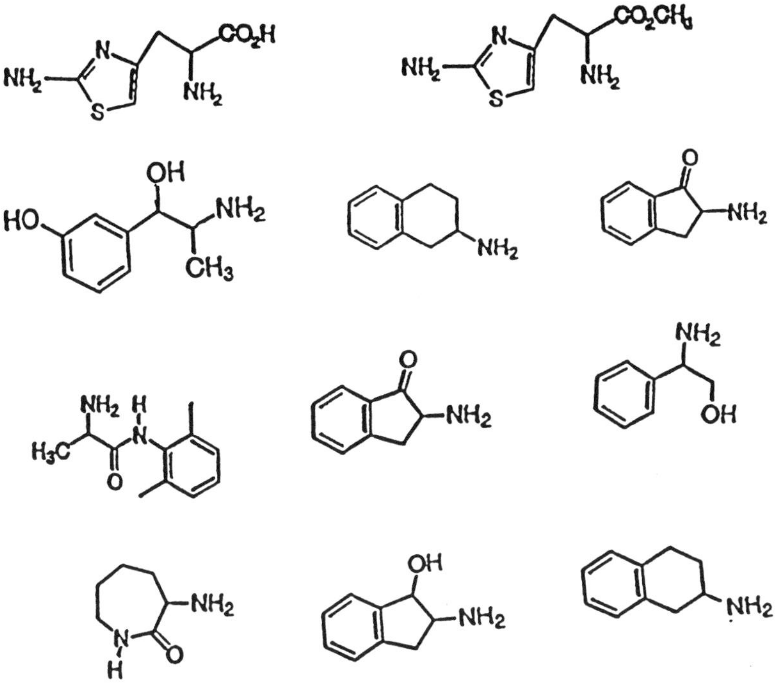

FIGURE 3 Continued.

under the reversed-phase mode using acidic mobile phases. Accordingly, in the late 1980s and early 1990s, adsorbed CSPs were developed by different groups and used for chiral resolution in liquid chromatography [53–63]. The typical chromatograms of this work are shown in Figure 4.

In 1998, Machida et al. [45] and Hyun et al. [46] developed a new CCE-based CSP (covalently bonded to silica gel; see Sect. 8.2). This CSP was used successfully for the chiral resolution of certain racemic compounds using a variety of mobile phases. The most important applications of this CSP are for the resolution of amino acids, amino esters, amino alcohols, amines, amides, quinolone antibacterials, and other drugs having primary amino groups [46–51,64,65]. The typical chromatograms of the chiral resolution of amino acids on (+)-(18-crown-6)-2,3,11,12-tetracarboxylic acid CSP are shown in Figure 4. The enantiomeric resolution of the racemic compound on CCE-based CSPs are listed in Table 2. There is no report available on the chiral separations at the preparative scale using these CSPs.

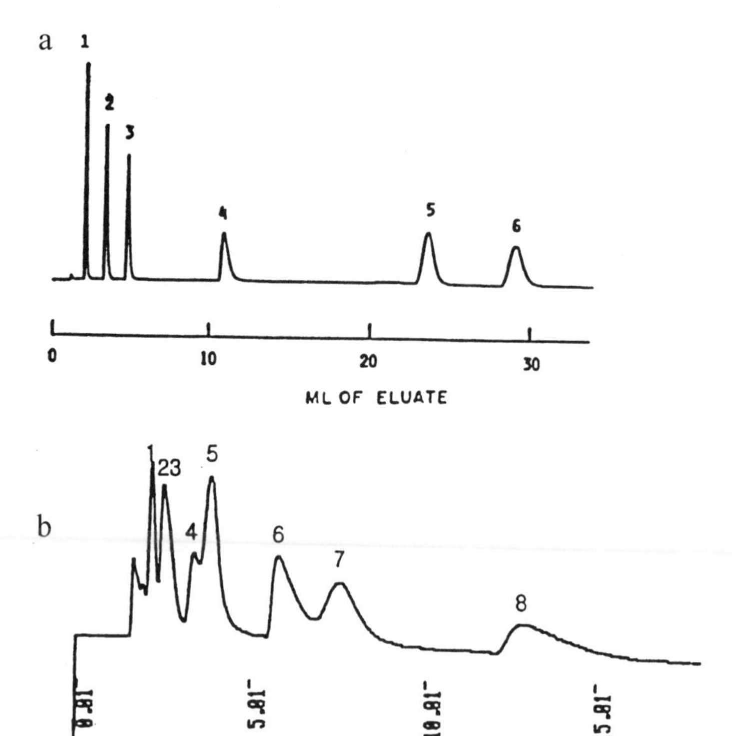

FIGURE 4 Chromatograms of the chiral resolution of (a) D-arginine (1), L-arginine (2), D-methionine (3), L-methionine (4), D-tryptophan (5), and L-tryptophan (6) on a LiChrosorb RP-18 column coated with 49 mg chiral 18-crown-6 ether using 10 mM perchloric acid as the mobile phase and (b) L-alanine (1), D-alanine (2), L-valine (3), D-valine (4), L-methionine (5), D-methionine (6), L-leucine (7), and D-leucine (8) on (+)-(18-crown-6)-2,3,11,12 tetracarboxylic acid crown ether using 10 mM perchloric acid as the mobile phase. (From Refs. 40 and 50.)

8.4 OPTIMIZATION OF HIGH-PERFORMANCE LIQUID CHROMATOGRAPHIC CONDITIONS

The applications of the CCE-based CSPs are very limited, as these are used only for the chiral resolution of the compounds having primary amino groups. Aqueous mobile phases containing organic modifiers and acids have been used on these CSPs and, therefore, composition of the mobile phase is the key parameter for the optimization of chiral resolution. However, in some reports, the effects of temperature, pH, and flow rate were also studied. The optimization of these parameters are discussed herein in detail.

TABLE 2 Enantiomeric Resolution of Racemic Compounds on CCE-Based CSPs

Racemic compound	CSP	Ref.
Amino acids	(+)-(18-Crown-6)-2,3,11,12-tetracarboxylic acid	45, 47, 51, 52
	Binaphthyl-(+)-(18-crown-6)-2,3,11,12-tetracarboxylic acid	4–6
	18-Crown-6	40
	Crownpak CR	59
	Diphenyl-substituted 1,1′-binaphthol	67
	Crownpak CR	48, 69, 70
Amino alcohols	(+)-(18-Crown-6-)-2,3,11,12-tetracarboxylic acid	45, 48
Amino esters	Binaphthyl-18-crown-6	4–6
	Crownpak CR	59
Amines	Binaphthyl-18-crown-6	4–6
	Binaphthyl-18-crown-6	40
	(+)-(18-Crown-6)-2,3,11,12-tetracarboxylic acid	48
	Crownpak CR	55
	Crownpak CR	72, 74
	18-Crown-6	40
Amides	Crownpak CR	68
Amino carbonyls	(+)-(18-Crown-6)-2,3,11,12-tetracarboxylic acid	48
Quinolones	(+)-(18-Crown-6)-2,3,11,12-tetracarboxylic acid	46
Fluoroquinolones	(+)-(18-Crown-6)-2,3,11,12-tetracarboxylic acid	49
Dipeptides	Crownpak CR	57
Miscellaneous racemic compounds with primary amino groups	(+)-(18-Crown-6)-2,3,11,12-tetracarboxylic acid	50, 66, 70
	Crownpak CA	56, 70, 71, 73

8.4.1 Composition of the Mobile Phase

In all of the applications, aqueous and acidic mobile phases are used. The most commonly used mobile phases are aqueous perchloric acid, aqueous methanol containing sulfuric or perchloric acids separately. The compounds with higher hydrophobicity in general have the longer retention times on CCE-based CSPs and, therefore, organic modifiers are used to optimize the resolution [58]. Optimization of the resolution is carried out by adjusting the amounts of methanol, sulfuric acid, and perchloric acid separately. Generally, the separation increased with an increase in methanol and decrease in acid concentrations. The other organic modifiers used were ethanol, acetonitrile, tetrahydrofuran (THF), and trifluoroacetic acid, but methanol was found to be the best one [47–49]. Hyun et al. [47–49] studied the effect of methanol concentration on the chiral resolution of amino acids, amino alcohols, amines, and fluoroquinolones. It has been observed that the retention increased with an increase of methanol concentration but there was little effect on the separation of amino acids. However, a marked effect on the resolution of fluoroquinolones has been observed as the resolution increased by an increase in methanol contents [49]. It is very interesting to note that the retention of some compounds decreased whereas separation increased with an increase of methanol concentration [48]. Therefore, the effect of methanol concentration varies from one analyte to another. Recently, Hyun et al. [51] studied the effect of methanol and triethylamine concentrations on the chiral resolution of certain racemates containing a primary amino group. The authors reported an increase in the retention times with increasing amounts of these two organic modifiers. The effect of methanol, ethanol, and acetonitrile concentrations is presented in Table 3 [47].

The effect of the concentration of sulfuric acid was also studied in detail and the best concentrations were found in the range of 10–20 mM [47–49]. As an example, the effect of acids (sulfuric acid, perchloric acid, and trifluoroacetic acid) on the chiral resolution of leucine and phenylglycine are given in Table 3 [47]. This table shows that the best mobile phases were 80% methanol containing 10 mM sulfuric acid and 100% methanol having 10 mM sulfuric acid for the chiral resolution of leucine and phenylglycine separately and respectively. Recently, Aboul-Enein et al. [75] also studied the effect of the concentrations of sulfuric acid on the chiral resolution of thyroxine and tocainide racemates. It has been observed that the separation decreased by increasing the amount of sulfuric acid. However, there was no chiral resolution for tocainide at higher concentrations of sulfuric acids, which may be the result of the higher polarity of the mobile phase (Table 4). Furthermore, the authors have achieved the chiral resolution of these drugs without using sulfuric acid. Therefore, 80% methanol containing 10 mM sulfuric may be a suitable starting mobile phase. Optimization can be carried out on the basis of the results obtained. A simplified protocol for

TABLE 3 Effect of Organic Modifiers and Sulfuric Acid on the Chiral Resolution of Leucine and Phenylglycine

	Leucine			Phenylglycine		
Mobile phase	k_1	α	R_s	k_1	α	R_s
20% $CH_3OH + H_2SO_4$ (10 mM)	0.32	1.34	1.46	1.02	1.96	3.16
40% $CH_3OH + H_2SO_4$ (10 mM)	0.44	1.39	1.32	1.43	2.09	4.89
80% $CH_3OH + H_2SO_4$ (10 mM)	0.73	1.32	1.42	2.07	2.25	6.46
100% $CH_3OH + H_2SO_4$ (10 mM)	0.99	1.21	0.72	3.48	2.49	9.60
80% $CH_3OH + H_2SO_4$ (1 mM)	0.66	1.45	1.71	3.23	2.23	4.54
80% $CH_3OH + H_2SO_4$ (5 mM)	0.68	1.40	1.63	2.16	2.26	5.74
80% $CH_3OH + H_2SO_4$ (20 mM)	0.80	1.28	1.36	2.09	2.25	7.02
20% EtOH+H_2SO_4 (10 mM)	0.32	1.00	0.00	0.85	1.94	3.42
40% EtOH+H_2SO_4 (10 mM)	0.49	1.00	0.00	1.33	2.07	4.65
80% EtOH+H_2SO_4 (10 mM)	1.94	1.09	0.32	4.84	2.04	6.58
20% $CH_3CN + H_2SO_4$ (10 mM)	0.19	1.00	0.00	0.49	1.92	2.48
40% $CH_3CN + H_2SO_4$ (10 mM)	0.21	1.00	0.00	0.45	2.14	3.04
80% $CH_3CN + H_2SO_4$ (10 mM)	0.46	1.06	0.40	1.63	2.09	8.93
80% $CH_3OH + HClO_4$ (10 mM)	0.52	1.33	1.25	1.49	2.19	5.10
80% $CH_3OH + CF_3Ac$ (10 mM)	0.35	1.00	0.00	1.02	2.24	3.96

Source: Ref. 47.

the development and optimization of the mobile phases on CCE-based CSPs is presented in Scheme 1.

8.4.2 Other Parameters

In addition to the mobile phase composition, the effect of other parameters such as temperature, flow rate, pH, and structure of the analytes were also studied, but only a few reports were available in the literature. In 1995, Lin and Maddox [66] studied the effect of temperature on the chiral resolution of amino acids and esters. The temperature was varied from 5°C to 25°C and it was reported that the resolution improved at low temperature. Hyun et al. [48–50,67] carried out the effect of temperature on the chiral resolution of amino alcohols, amines, fluoroquinolones, and other drugs. Again, lowering of temperature resulted in better resolution. The effect of temperature on the chiral resolution of phenylalanine, phenylglycine, and 2-hydroxy-2-(4-hydroxy-phenyl)-ethyl amine is shown in Table 5 [50], which indicates an increase in retention factors at lower temperature, but the best separation occurred at 20°C. These experiments indicated the exothermic nature of chiral resolution on CCE-based CSPs. Lin and Maddox [66] also studied the effect of flow rate on the chiral resolution of

TABLE 4 Effect of the Concentrations of Sulfuric Acids on the Chiral Resolution of Thyroxine and Tocainide

H_2SO_4 (mM)	k_1 (−)	k_2 (+)	α	R_s
Thyroxine				
0	4.50	14.00	3.11	2.60
5	2.50	6.30	2.51	2.33
10	1.56	3.77	2.41	2.06
20	1.32	2.88	2.18	2.00
30	1.10	2.23	2.10	1.56
40	0.94	1.97	2.08	1.00
50	0.94	1.97	2.08	1.00
Tocainide				
0	2.18	2.74	1.26	1.30
5	0.44	0.53	1.20	0.20
10	0.32	0.41	1.18	0.15
20	0.29	0.35	1.16	0.12
30	0.24	0.30	1.13	0.10
40	nr			
50	nr			

Note: nr = not resolved.
Source: Ref. 75.

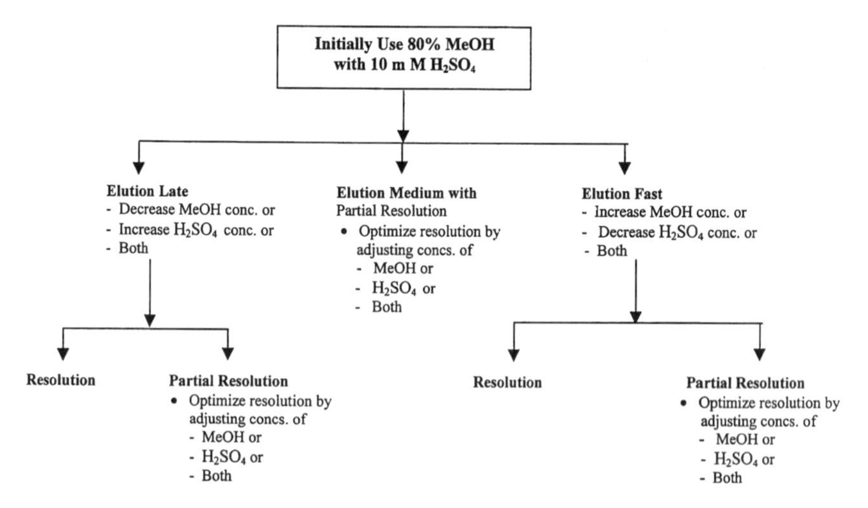

SCHEME 1 Protocol for the development and optimization of mobile phases on CCE-based CSPs.

TABLE 5 Effect of Temperature on the Chiral Resolution of Phenylalanine, Phenylglycine, and 2-Hydroxy-2-(4-hydroxy-phenyl)-ethyl Amine

Temp.	Phenylalanine			Phenylglycine			2-Hydroxy-2-(4-hydroxy-phenyl)-ethyl amine		
(oC)	k_1	α	R_s	k_1	α	R_s	k_1	α	R_s
10	3.54	2.63	2.40	1.66	4.79	3.03	3.89	1.86	1.23
20	2.71	2.55	2.50	1.40	4.32	3.98	2.67	1.67	1.08
25	2.24	2.51	2.37	1.05	4.04	3.24	2.06	1.62	0.97
30	1.78	2.48	2.32	0.91	3.84	3.07	1.66	1.57	0.87

amino acids and esters. They varied the flow rate from 0.3 to 0.7 mL/min and observed a small effect on the chiral resolution, with the best resolution at 0.3 mL/min (Fig. 5). Because all of the mobile phases are acidic in nature, the effect of pH on the chiral resolution is not significant. However, Kersten [59] studied the effect of pH on the chiral resolution of amino acids and their esters and found the pH 1.0 is the optimum. Lin and Maddox [66] carried out the chiral resolution of amino acids and their esters at 1.0, 1.3, and 1.6 and reported the best resolution at pH 1.0 (Fig. 5).

Because the steric effect contributes to the complex formation between guest and host, the chiral resolution on these CSPs is affected by the structures of the analytes. Amino acids, amino alcohols, and derivatives of amines are the best classes for studying the effect of analyte structures on the chiral resolution. The effect of analyte structures on the chiral resolution may be obtained from the work of Hyun et al. [47,48]. The authors studied the chiral resolution of amino alcohols, amides, amino esters, and amino carbonyls. The effects of the substituents on the chiral resolution of some racemic compounds are shown in Table 6. A perusal of this table indicates the dominant effect of steric interactions on chiral resolution. Furthermore, an improved resolution of the racemic compounds, having phenyl moieties as the substituents, may be observed from this Table 6. It may be the result of the presence of π–π interactions between the CCE and racemates. Generally, the resolution decreases with the addition of bulky groups, which may be caused by the steric effects. In addition, some anions have been used as the mobile phase additives for the improvement of the chiral resolution of amino acids [76]. Recently, Machida et al. [69] reported the use of some mobile phase additives for the improvement of chiral resolution. They observed an improvement in the chiral resolution of some hydrophobic amino compound using cyclodextrins and cations as mobile phase additives.

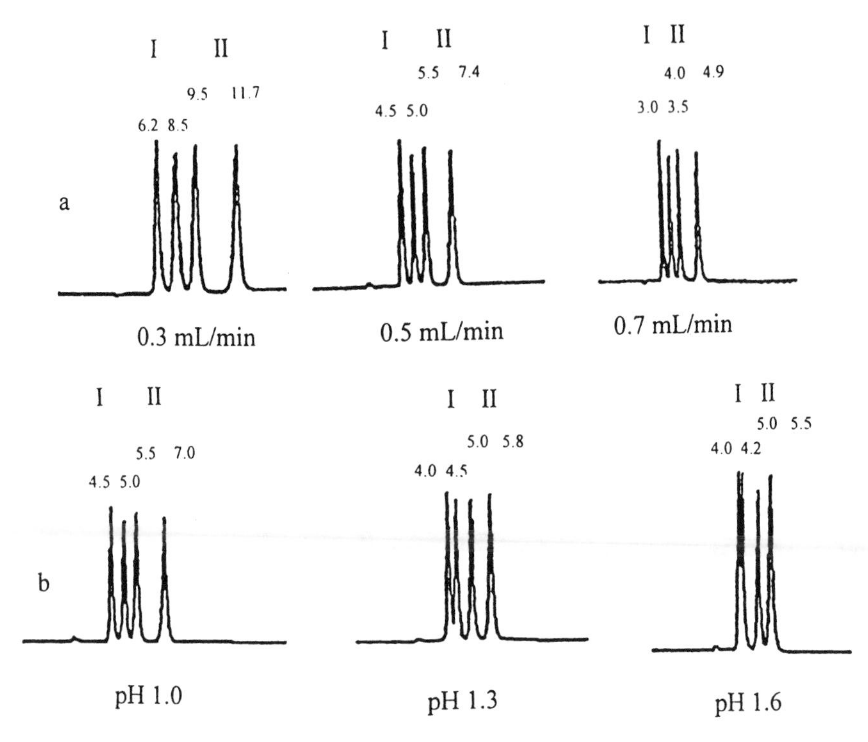

FIGURE 5 Effect of (a) flow rate and (b) pH on the chiral resolution of 2-(2-amino-1,3-thiazol-4-yl)-methylglycine [I] and 2-(2-amino-1,3-thiazol-4-yl)-methylglycine methyl ester [II] on a Crownpak CR column using aqueous perchloric acid as the mobile phase. (From Ref. 66.)

8.5 CHIRAL RECOGNITION MECHANISM

Only racemic compounds having primary amino groups can be resolved on these CSPs. Moreover, Hyun et al. reported no resolution for proline in which primary amino group is not free [47]. This means that primary amino group is essential for the chiral resolution and it plays an essential role in the chiral resolution and in the formation of the guest–host complex. Pedersen [1] calculated the values of distribution constants and free energies of the enantiomers of some amino acids on a variety of CCEs. The author reported the different values of these parameters for the two enantiomers. Furthermore, Cram et al. [3–6] determined the stability of guest–host complex formation and postulated the importance of hydrogen, π–π interaction, and steric forces in the guest–host complex formation. Pocsfalvi et al. [24] described the guest–host complex formation of the enantiomers of two ammonium cations on CCE and observed different stabilities of the complexes

TABLE 6 Effect of Substituents on the Chiral Resolution of Some Racemic Compounds

$RCH(NH_2)COY$					
R	Y	k_1	k_2	α	R_s
$-CH_3$	$-NYH(CH_2)_3CH_3$	1.60	2.25	1.41	2.34
	$-NHC(CH_3)_3$	1.39	1.98	1.42	2.32
	$-N(CH_2CH_3)_2$	2.13	2.13	1.00	0.00
$-CH(CH_3)_2$	$-NH(CH_2)_3CH_3$	0.28	0.45	1.64	1.32
	$-NHC(CH_3)_3$	0.25	0.39	1.59	1.11
	$-N(CH_2CH_3)_2$	0.21	0.21	1.00	0.00
$-C_6H_5$	$-NH(CH_2)_3CH_3$	1.55	3.82	2.46	7.27
	$-NHC(CH_3)_3$	1.28	3.42	2.67	6.32
	$-N(CH_2CH_3)_2$	1.40	4.40	3.15	9.77
$-CH_2C_6H_5$	$-NH(CH_2)_3CH_3$	1.94	4.74	2.45	6.99
	$-NHC(CH_3)_3$	2.06	4.68	2.28	7.36
	$-N(CH_2CH_3)_2$	1.88	1.88	1.00	0.00

Source: Ref. 47.

formed between the two enantiomers and CCE. The authors also described hydrogen, π–π, and steric interactions as the forces responsible for the complex formation. Recently, Machida et al. [77] studied the guest–host complex formation by x-ray analysis. They reported the presence of the above-cited forces between the chiral selectors and the analytes. It has also been indicated that the presence of carboxylic groups on analytes enhanced hydrogen-bondings, which resulted in an improved chiral resolution [76]. The best chiral resolution has been observed when the functional and alkyl groups were located at the chiral center of the analyte. Therefore, CCEs involve a simple chiral recognition mechanism. The two enantiomers fit stereogenically at different extents into the chiral cavity of CCE which are stabilized by the various interactions (as cited earlier) at different magnitudes; hence, the chiral resolution occurs.

The primary amino group of the analyte ionizes in the presence of acid and forms ammonium ion ($-NH_3^+$). The ammonium ions form strong hydrogen-bondings with the oxygen atoms of CCE. Therefore, the presence of an acid in the mobile phase is essential to achieve the chiral resolution on these CSPs. However, recently, Aboul-Enein et al. [75] observed very interesting results for the chiral resolution of thyroxine and tocainide racemates on the (+)-(18-crown-6)-2,3,11,12-tetracarboxylic acid CCE. The authors reported the chiral resolution of these molecules without using an acid in the mobile phase. Moreover, they

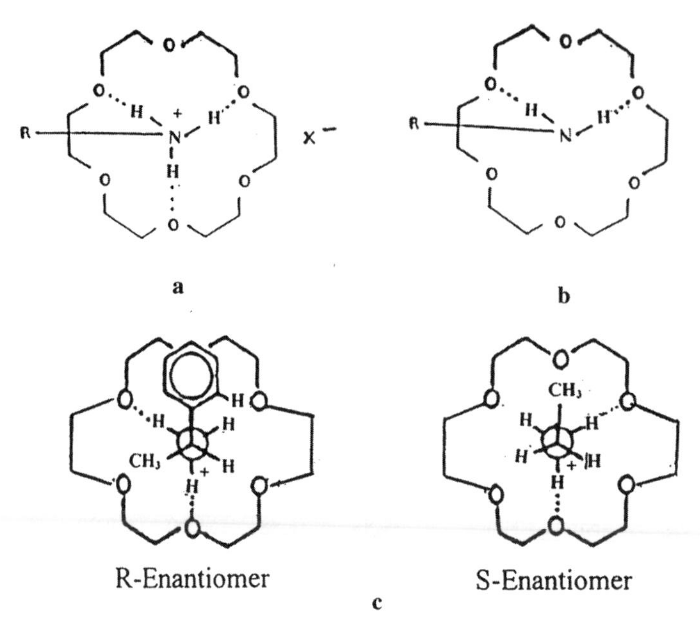

FIGURE 6 Graphical representation of the guest–host diastereoisomeric complex formation (a) in the presence of acid in the mobile phase and (b) in the absence of acid in the mobile phase. (c) Three-dimensional structures of the guest–host complexes formed between *R*- and *S*-enantiomers of α-phenylethylamine and chiral 18-crown-6 ether CSP.

have also achieved the chiral resolution of thyroxine using triethylamine (a base) as an organic modifier. Therefore, the acid is not essential for the chiral resolution of all of the racemates containing primary amino groups. It may be concluded that the acid is not required for the chiral resolution of the molecules that have a high hydrogen-bonding capacity through the primary amino group, whereas it is required for those molecules having a low ability of hydrogen-bonding. Therefore, the acid is used to convert primary amino group into ammonium ion because the latter ($-NH_3^+$) has a stronger hydrogen ability than the former ($-NH_2$). Therefore, the concept of essentiality of the addition of an acid in the mobile phase is not always required. However, the presence of primary amino group is an essential feature for the chiral resolution on CCE-based CSPs. To make this concept clear, the structure of the guest–host complex is shown in Figure 6.

8.6 MISCELLANEOUS APPLICATIONS

CCE-based CSPs have limited application in the chiral resolution of racemic compounds using HPLC. There is only one report on the use of these CSPs in capillary electrochromatography [78]. Koide and Ueno [78] synthesized 2-allyl esters of (+)-tetraallyl-18-crown-6 carboxylate and (+)-18-crown-6 tetracarboxylic acid CCEs. These esters were allowed to bind to a negatively charged polyacrylamide gel and this treated gel was packed into the capillaries. The developed CSPs were called monolithic chiral stationary phases. These CSPs were used for the chiral resolution of some amino compounds using a 200 mM triethanolamine–300 mM boric acid buffer as the mobile phase.

8.7 CONCLUSION

CCE-based CSPs are used only for the chiral resolution of racemic compounds containing primary amino groups. Therefore, these CSPs are suitable for the chiral resolution of amino acids, amino alcohols, amino esters, amino carbonyls, amines, amides, and other compounds having primary amino groups. The experimental cost of these CSPs is slightly low due to the inexpensive nature of a mobile phase having water as the major component. However, there are some serious limitations of these CSPs because their applications are very limited to a particular class of compounds (i.e., racemates) having primary amino groups. In addition, no report has been published on the chiral resolution at a preparative scale. Additionally, the recovery of the resolved enantiomers requires special attention to the complex procedure of extraction because of the presence of an acid in the mobile phase. In view of these points, it may be concluded that these CSPs cannot be used for other classes of racemic compounds (compounds not having primary groups). For these reasons, these CSPs are not in their full swing and hence require more development. More substituted CCEs should be prepared and tested for the chiral resolution of a wider range of racemic compounds. We think that the development of these CSPs is underway and, in the future, CCE-based CSPs could be used for the chiral resolution of a wide variety of racemates, including preparative separation.

REFERENCES

1. Pedersen CJ, J Am Chem Soc 89: 2495 (1967).
2. Pedersen CJ, J Am Chem Soc 89: 7017 (1967).
3. Kyba EP, Timko JM, Kaplan JL, de Jong F, Gokel GW, Cram DJ, J Am Chem Soc 100: 4555 (1978).
4. Peacock SC, Domeier LA, Gaeta FCA, Helgeson RC, Timko JM, Cram DJ, J Am Chem Soc 100: 8190 (1978).
5. Peacock SC, Walba DM, Gaeta FCA, Helgeson RC, Timko JM, Cram DJ, J Am Chem Soc 102: 2043 (1980).
6. Linggenfelter DS, Helgeson RC, Cram DJ, J Org Chem 46: 393 (1981).
7. Yamamoto K, Yumioka H, Okamoto Y, Chikamatsu H, Chem Commun 168 (1987).

8. Yamamoto K, Kitsuki T, Okamoto Y, Bull Chem Soc Jpn 59: 1269 (1986).
9. Yamamoto K, Noda K, Okamoto Y, Chem Commun 1065 (1985).
10. Nakazaki M, Yamamoto K, Ikeda T, Kitsuki T, Okamoto Y, Chem Commun 787 (1983).
11. Behr JP, Girodeau JM, Hayward RC, Lehn IM, Sauvage JP, Helv Chim Acta 63: 2096 (1980).
12. Gehin D, Cesare PD, Gross B, J Org Chem 51: 1906 (1986).
13. Davidson RB, Bradshaw JS, Jones BA, Dalley NK, Christensen JJ, Izatt RM, Morin FG, Grant DM, J Org Chem 49: 353 (1984).
14. Chadwick JD, Cliffe IA, Sutherland IO, Newton RF, J Chem Soc Perkin Trans I 1707 (1984).
15. Naemura K, Fukunaga R, Chem Lett 1651 (1985).
16. Naemura K, Fukunaga R, Yamanaka M, Chem Commun 1560 (1985).
17. Naemura K, Komatsu M, Adachi K, Chikamatsu H, Chem Commun 1675 (1986).
18. Thoma AP, Viviani-Nauer A, Schellenberg KH, Bedekovic D, Pretsch E, Prelog V, Simon W, Helv Chim Acta 62: 2303 (1979).
19. Sousa LR, Hoffman DH, Kaplan L, Cram DJ, J Am Chem Soc 96: 7100 (1974).
20. Blasius E, Adrian W, Jansen KP, Klaute G, J Chromatogr 96: 89 (1974).
21. Izatt RM, Lamb JD, Izatt NE, Rossiter BE, Christensen JJ, Haymore BL, J Am Chem Soc 101: 6273 (1979).
22. Dietrich B, Lehn JM, Sanvage JP, Tetrahedron Lett 2885 (1969).
23. Wada F, Wada Y, Kikukawa K, Matsuda T, Bull Chem Soc Jpn 54: 458 (1981).
24. Pocsfalvi G, Liptak M, Huszthy P, Bradshaw JS, Izatt RM, Vekey K, Anal Chem 68: 792 (1996).
25. Gérczei T, Böcskei Z, Miklós Keserû G, Samu E, Huszthy P, Tetrahedron: Asymm 10: 1995 (1999).
26. Samu E, Huszthy P, Somogyi L, Hollósi M, Tetrahedron: Asymm 10: 2775 (1999).
27. Samu E, Huszthy P, Horváth G, Szöllõy A, Neszmélyi A, Tetrahedron: Asymm 10: 3615 (1999).
28. Huszthy P, Samu E, Vermes B, Mezey-Vándor G, Nógrádi M, Bradshaw JS, Izatt RM, Tetrahedron 55: 1491 (1999).
29. Sogah GDY, Cram DJ, J Am Chem Soc 97: 1259 (1975).
30. Sogah GDY, Cram DJ, J Am Chem Soc 98: 3038 (1976).
31. Sousa LR, Sogah GDY, Hoffman DH, Cram DJ, J Am Chem Soc 100: 4569 (1975).
32. Sogah GDY, Cram DJ, J Am Chem Soc 101: 3035 (1979).
33. Blasius E, Heo GS, Smid J, Yagi K, Talanta 27: 107 (1980).
34. Blasius E, Papadoya IN, Smid J, Yagi K, Talanta 27: 127 (1980).
35. Blasius E, Akelah M, Nakajima M, J Chromatogr 201: 147 (1980).
36. Igawa M, Saito K, Tsukamoto J, Tanaka M, Anal Chem 53: 1942 (1981).
37. Igawa M, Saito K, Tanaka M, Yamabe T, Bunseki Kagaku 32: E137 (1983).
38. Kimura K, Hayata E, Shono T, Chem Commun 271 (1984).
39. Kimura K, Harino H, Hayata E, Shono T, Anal Chem 58: 2233 (1986).

40. Shinbo T, Yamaguchi T, Nishimura K, Sugiura M, J Chromatogr 405: 145 (1987).
41. Kimura K, Nakajima M, Shono T, Anal Lett 13: 741 (1980).
42. Nakajima M, Kimura K, Shono T, Anal Chem 55: 463 (1983).
43. Nakajima M, Kimura K, Shono T, Bull Chem Soc Jpn 56: 3052 (1983).
44. Iwachido T, Naito H, Samukawa F, Ishimaru K, Toei K, Bull Chem Soc Jpn 59: 1475 (1986).
45. Machida Y, Nishi H, Nakamura K, Nakai H, Sato T, J Chromatogr A 805: 85 (1998).
46. Hyun MH, Jin JS, Lee W, Bull Korean Chem Soc 9: 819 (1998).
47. Hyun MH, Jin JS, Lee W, J Chromatogr A 822: 155 (1998).
48. Hyun MH, Jin JS, Koo HJ, Lee W, J Chromatogr A 837: 75 (1999).
49. Hyun MH, Han SC, Jin JS, Lee W, Chromatographia 52: 473 (2000).
50. Hyun MH, Koo HJ, Jin JS, Lee W, J Liq Chromatogr Relat Technol 23: 2669 (2000).
51. Jin JS, Stalcup AM, Hyun MH, J Chromatogr A 933: 83 (2001).
52. Steffeck RJ, Zelechonok Y, Gahm KH, J Chromatogr A 947: 301 (2002).
53. Udvarhelyi PM, Watkins JC, Chirality 2: 200 (1990).
54. Motellier S, Wainer IW, J Chromatogr 516: 365 (1990).
55. Hilton M, Armstrong DW, J Liq Chromatogr 14: 9 (1991).
56. Aboul-Enein HY, Bakr SA, Islam R, Rothchild R, J Liq Chromatogr 14: 3475 (1991).
57. Hilton M, Armstrong DW, J Liq Chromatogr 14: 3673 (1991).
58. Shinbo T, Yamaguchi T, Yanagishita H, Kitamoto D, Sakaik K, Sugiura M, J Chromatogr 625: 101 (1992).
59. Kersten BS, J Liq Chromatogr 17: 33 (1994).
60. Okamoto M, Takahashi KI, Doi T, J Chromatogr 675: 244 (1994).
61. Yamaguchi T, Kagaku 42: 255 (1989).
62. Walbroehl Y, Wanger J, J Chromatogr A 680: 253 (1994).
63. Walbroehl Y, Wanger J, J Chromatogr A 685: 321 (1994).
64. Török G, Peter A, Fulop F, Chromatographia 48: 20 (1998).
65. Machida Y, Nishi H, Nakamura K, Chromatographia 49: 621 (1999).
66. Lin S, Maddox NJ, J Liq Chromatogr 18: 1947 (1995).
67. Hyun MH, Han SC, Lipshutz BH, Shin JS, Welch CJ, J Chromatogr A 910: 359 (2001).
68. Peter A, Olajos E, Casimir R, Tourwe D, Broxterman QB, Kaptein B, Armstrong DW, J Chromatogr A 871: 105 (2000).
69. Machida Y, Nishi H, Nakamura K, J Chromatogr A 830: 311 (1999).
70. Peter A, Torok G, Toth G, Den Nest WV, Laus G, Tourwe D, J Chromatogr A 797: 165 (1998).
71. Aboul-Enein HY, Serignese V, Biomed Chromatogr 9: 101 (1995).
72. Aboul-Enein HY, Serignese V, Biomed Chromatogr 11: 101 (1997).
73. Breda M, Sarati S, Basileo G, Dostert P, Chirality 9: 133 (1997).
74. Van Dort ME, Chirality 11: 684: (1999).
75. Aboul-Enein HY, Ali I, Hyun MH, Cho YJ, Jin JS, J Biochem Biophys Methods 54: 407 (2002).

76. Shibukawa A, Nagakawa T, The use of chiral crown ethers in liquid chromatography, in Chiral Separations by HPLC, Krstulovic AM (Ed.), Ellis Horwood, New York, p. 477 (1989).
77. Machida Y, Nishi H, Nakamura K, Chirality 11: 173 (1999).
78. Koide T, Ueno K, J Chromatogr A 909: 305 (2001).

9

Miscellaneous Types of Chiral Stationary Phase

The most popular and commonly used chiral stationary phases (CSPs) are polysaccharides, cyclodextrins, macrocyclic glycopeptide antibiotics, Pirkle types, proteins, ligand exchangers, and crown ether based. The art of the chiral resolution on these CSPs has been discussed in detail in Chapters 2–8, respectively. Apart from these CSPs, the chiral resolutions of some racemic compounds have also been reported on other CSPs containing different chiral molecules and polymers. These other types of CSP are based on the use of chiral molecules such as alkaloids, amides, amines, acids, and synthetic polymers. These CSPs have proved to be very useful for the chiral resolutions due to some specific requirements. Moreover, the chiral resolution can be predicted on the CSPs obtained by the molecular imprinted techniques. The chiral resolution on these miscellaneous CSPs using liquid chromatography is discussed in this chapter.

9.1 ALKALOID-BASED CSPs

Most of the alkaloids are chiral molecules having many chiral centers. Therefore, these molecules have been used to prepare effective and successful CSPs. Cinchona alkaloids (i.e., quinine, quinidine, epiquinine, epiquinindine, cinchonine, and cinchonidine), which differ in their absolute configuration and alkyl side chains, have been used as the starting materials for the preparation of CSPs

[1–4]. Among these, the CSPs based on quinine, quinidine, and their derivatives have proven to be most effective for chiral resolution. In addition, CSPs based on alkaloids having bulky carbamate residues like *tert*-butyl or 1-adamantyl groups have shown higher enantioselectivities in almost all cases. Contrarily, the CSPs with aromatic carbamate residues such as 2,6-diisopropylphenyl, trityl, and 3,5-dinitrophenyl groups are complementary in their enantioselective profile, exhibiting higher enantioselectivities. *tert*-Butylcarbamoylated quinine linked to mercaptopropylsilica was used for the chiral resolution of *N*-3,5-dinitrobenzoylated leucine and phenylalanine with α values of 15.87 and 10.78, respectively, using a buffer as the mobile phase [5]. The authors also observed the decrease in the chiral resolution of leucine and phenylalanine by replacing the *N*-3,5-dinitrobenzoyl group by the 2,4-dinitrophenyl ring. The reversible enantiomerization of axially chiral 2-dodecycloxy-6-nitrophenyl-2-carboxylic acid was studied on *O*-(*tert*-butylcarbamoyl) quinine as a chiral selector by stopped-flow high-performance liquid chromatography (sfHPLC) [6].

Franco et al. [7] synthesized nine new quinine carbamate dimers and immobilized them on silica gel. The role of the presence of a second quinine subunit on the CSPs, as well as the influence of the structure and length of the spacer, on the overall chiral recognition of a set of N-derivatized amino acids and other acidic drugs was also investigated. The bulkiness of the intermediate spacer tuned the chiral recognition abilities of these CSPs, with the 1,3-adamantyl-derived CSP, which led to the best separations. Shorter spacers reduced the chiral discrimination abilities of the dimeric selectors, with the *n*-hexylene bridge being the most favorable distance to allow a nearly independent interaction of the two quinine subunits with the racemic analytes. The comparison of five monomeric CSPs showed that the dimeric ones usually retain the chiral analytes more strongly, although the enantioseparation is not improved. Nevertheless, the exceptional resolution abilities of dimeric CSPs with a *trans*-1,2-diaminocyclohexylene-bridge for the separation of 2,4-dinitrophenyl (DNP) derivatives of amino acids and certain acidic drugs of therapeutical interest (e.g., profens) seemed to be superior in comparison to other CSPs [7]. The structures of some of the cinchona alkaloid-based CSPs are shown in Figure 1 and the chiral resolution of some amino acids derivatives on these CSPs is given in Table 1 [8].

The effect of mobile phase composition, including pH and organic modifiers, was carried out on the chiral resolution of leucine derivatives on the *tert*-butyl carbamoylated quinine-based CSP [2]. The results of these findings are given in Table 2. This table shows that the best resolution was obtained at pH 5, 2 mM concentration of buffer, 60% methanol, and 80% acetonitrile concentrations, separately. In another study, the same authors [4] studied the influence of the mobile phase, pH, and temperature on the chiral resolution of leucine derivatives. The effect of temperature on the chiral resolution of leucine derivatives is shown in Figure 2. It is clear from this figure that the chiral

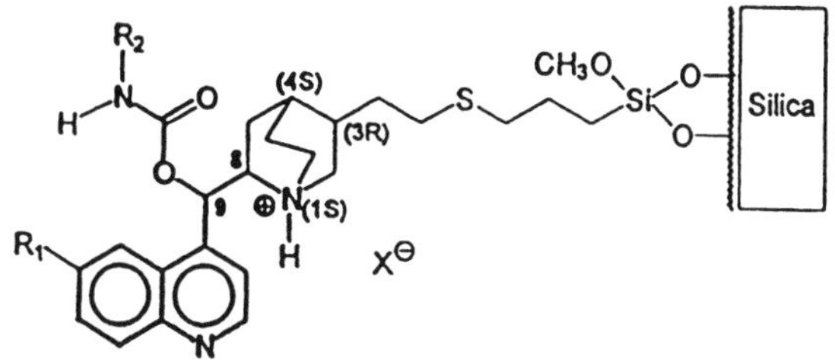

CSP	parent alkaloid	configuration of C8	configuration of C9	R_1	R_2
I	quinine (QN)	8S	9R	CH_3O-	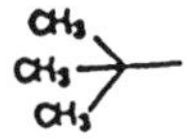
II	quinidine (QD)	8R	9S	CH_3O-	
III	epiquinine (EQN)	8S	9S	CH_3O-	
IV	epiquinidine (EQD)	8R	9R	CH_3O-	
V	quinine (QN)	8S	9R	CH_3O-	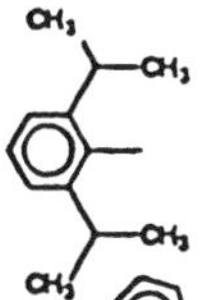
VI	cinchonine (CN)	8R	9S	H	
VII	quinine (QN)	8S	9R	CH_3O-	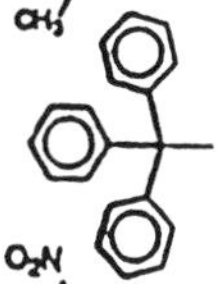
VIII	quinine (QN)	8S	9R	CH_3O-	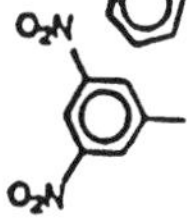
IX	quinidine (QD)	8R	9S	CH_3O-	

FIGURE 1 Structures of some CSPs based on cinchona alkaloids. (From Ref. 8.)

resolution decreases with increasing temperature. As a typical example, the chiral resolutions of 3,5-dinitrobenzyloxycarbonyl (DNZ)–leucine, DNZ–glutamic acid, and DNZ–phenylalanine are shown in Figure 3.

The ergot alkaloids were used to prepare different types of CSP and good resolutions were obtained for acidic compounds in buffered aqueous media [9–11]. The enantiomers of 2-aryloxypropionic acid, chrysanthemic acid, and analogs and profens were also resolved on CSPs based on ergot alkaloids. The most commonly used CSP, based on ergot alkaloid is 1-allyl terguride [12–14]. The tertiary amine of the methylergoline moiety represents the fixed charge of this CSP. The urea group adjacent in the β-position to the primary ionic interaction site is able to form hydrogen bonds. An interesting feature of the terguride CSP is its self-recognition ability; that is, the terguride CSP can be used

TABLE 1 Chiral Resolution of Amino Acid Derivatives on Cinchona Alkaloid–Based CSPs

Amino acid	Derivative	CSP	*k*	α
Leucine	DNB	I	11.74	15.88
Leucine	DNB	II	8.0	12.46
Leucine	DNB	III	11.38	1.16
Leucine	DNB	IV	9.41	1.15
Leucine	DNB	V	7.43	3.49
Leucine	DNB	VI	2.76	1.41
Leucine	DNB	V	13.75	1.74
Leucine	DNB	I	15.41	1.39
3-(4-Pyridyl)alanine	Bz	I	5.11	1.94
3-(2-Thienyl)alanine	Ac	I	4.65	1.41
tert-Leucine	For	I	2.67	1.25
Arginine	FMOC	I	1.14	1.68
Proline	DNZ	I	7.52	1.21
3-Amino-3-phenyl-propionic acid	DNZ	I	8.73	1.83
2-Methyltaurine	DNZ	I	5.00	1.92
Glutamic acid	DNZ	I	34.73	1.36
Citrulline	DNZ	I	3.89	1.49
Cysteic acid	DNZ	I	34.45	1.14
α-Aminopropyl phosphonic acid	DNZ	I	35.43	1.14
3-Aminobutyric acid	PNZ	I	4.06	1.32
Valine	NVOC	I	6.95	3.25
Serine	Z	I	3.61	1.21
3-Amino-3-phenyl-propionic acid	Z	VII	5.44	1.23
3-Aminobutyric acid	Z	VII	2.37	1.21
Tyrosine	BOC	VII	3.09	1.44
Tryptophan	DNS	V	34.88	2.97

Abbreviations: DNP: 2,4-dinitrophenyl; Bz: benzoyl; Ac: acetyl; For: formyl; FMOC: 9-fluorenyl-methoxycarbonyl; DNZ: 3,5-dinitrobenzyloxycarbonyl; PNZ: 4-nitrobenzyloxycarbonyl; NVOC: 6-nitroveratryloxycarbonyl; Z: benzyloxycarbonyl; BOC: tert-butoxycarbonyl; DNS: dansyl. CSPs: I–VII, see Fig. 1.
Source: Ref. 8.

to separate the enantiomers of terguride [12]. The semipreparative separation of fenoprofen on the terguride-based CSP has been described [11]. Blaschke et al. [15] reported the analytical and preparative chiral resolutions of zopiclone enantiomers on polymethacrylol-1*R*,2*S*-norephedrine CSP using 0.01 M phosphate buffer (pH 5.5) with 1.2% 2-propanol. The effect of pH and concentrations of buffer and acetonitrile was studied on the chiral resolution of some profens on the terguride-based CSP [11]. The results of these findings are given in Figures 4a

TABLE 2 Effect of Mobile Phase Composition (pH and Organic Modifiers) on the Chiral Resolution of Leucine Derivatives

Leucine derivative	k	α	R_s	k	α	R_s	k	α	R_s
Effect of pH		pH 6			pH 5			pH 4	
DNB–leucine	0.39	7.31	5.38	0.39	9.71	8.86	0.25	5.5	4.23
DNP–leucine	0.59	1.22	0.50	0.95	1.35	1.27	0.71	1.23	0.50
DNZ–leucine	0.36	2.13	1.24	0.38	2.07	1.39	0.16	1.54	0.50
Effect of buffer conc.		20 mM			10 mM			2 mM	
DNB–leucine	0.43	4.12	9.26	0.44	9.75	9.23	0.81	10.2	12.30
DNP–leucine	0.83	1.12	1.04	0.84	1.32	1.03	0.57	1.27	1.37
DNZ–leucine	0.43	0.90	1.57	0.46	2.12	1.83	0.86	2.20	3.60
Effect of methanol conc.		80%			60%			40%	
DNB–leucine	0.78	10.21	12.23	1.59	8.31	12.43	4.05	6.24	11.8
DNP–leucine	1.46	1.28	1.48	3.58	1.25	2.05	13.38	1.21	2.11
DNZ–leucine	0.81	2.22	3.59	1.95	2.00	4.55	6.31	1.76	5.19
Effect of acetonitrile conc.		80%			60%			40%	
DNB–leucine	0.95	6.55	12.66	0.94	5.60	9.97	1.48	5.48	11.42
DNP–leucine	1.98	1.20	1.36	1.57	1.25	0.50	3.05	1.16	1.21
DNZ–leucine	0.77	1.84	2.51	0.97	1.78	2.62	1.81	1.77	3.63

Abbreviations: DNB: dinitrobenzoyl; DNP: 2,4-dinitrophenyl; DNZ: 3,5-dinitrobenzyloxycarbonyl. CSP: *tert*-butyl carbamoylated quinine based; buffer: ammonium acetate.
Source: Ref. 2.

(pH), 4b (buffer conc.), and 4c (acetonitrile conc.), respectively. It is clear from these figures that the enantioselectivity increased by decreasing pH and concentrations of buffer and acetonitrile, respectively. Flieger et al. [12] studied the effect of pH and the concentration of methanol on the chiral resolution of terguride on the 1-(3-aminopropyl) derivative of (+)-(5*R*,8*S*,10*R*)-terguride-based CSP. The effect of pH and the concentration of methanol on the chiral resolution of terguride on 1-(3-aminopropyl) derivative of (+)-(5*R*,8*S*,10*R*)-terguride-based CSP is shown in Figure 5, which shows that the lower concentration of methanol favored the enantioselectivity. The chromatograms of the chiral resolution of fenoprofen, naproxen, ketoprofen, and flobuprofen on terguride-based CSP are shown in Figure 6. The structures of some of the ergot alkaloids based CSPs are shown in Figure 7.

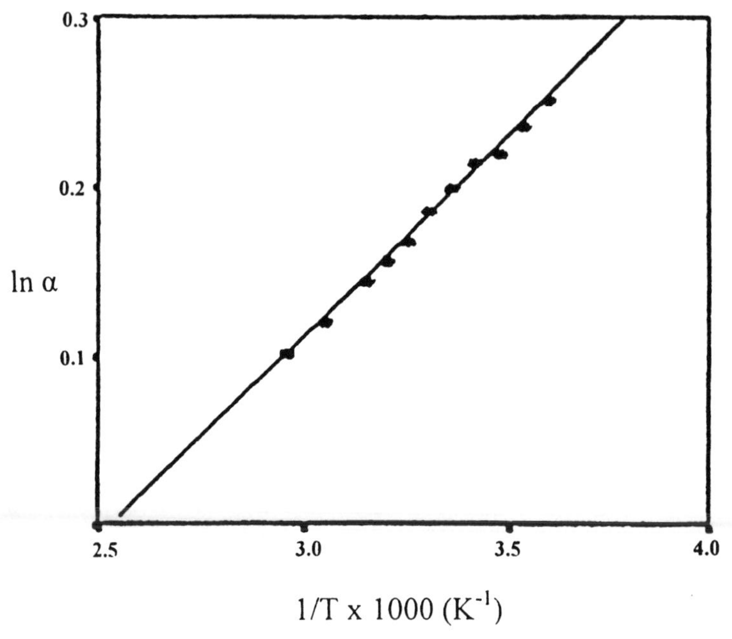

FIGURE 2 Effect of temperature on the chiral resolution of dansyl valine amino acids on CSP I (see Fig. 1) using methanol–0.1 M ammonium acetate (80 : 20, v/v). (From Ref. 8.)

9.2 AMIDE- AND AMINE-BASED CSPs

Many of the chiral molecules containing amide groups were bonded to a solid support for the preparation of CSPs [16–19]. The racemic compounds resolved on these CSPs include α-hydroxycarbonyls, β-hydroxycarbonyls, amino acids, amino alcohols, amine, and derivatized and underivatized diols. The preliminary chiral diamide phase [(*N*-formyl-L-valyl)aminopropyl)silica gel] has sufficient separability for racemic *N*-acylated α-amino acid esters but not in other types of enantiomer [16]. Most of the eluents used with these CSPs are of normal phase mode, including *n*-hexane, 2-propanol, chlorinated organic solvents, and acetonitrile.

Allenmark et al. [20] prepared a CSP consisting of *N,N*-diallyl-(*R,R*)-tartaric acid diamide as chiral monomer. These monomers were cross-linked with multifunctional hydrolisanes, yielding a network polymer which was used for the chiral resolution of amino alcohols, profens, β-blockers, benzodiazepinones, and benzothiadiazines. A highly sensitive high-performance liquid-chromatographic (HPLC) method was developed for the separation of enantiomeric 1,2-diacyl-*sn*-glycerols (*S* configuration), 2,3-diacyl-*sn*-glycerols (*R* configuration), and regio-

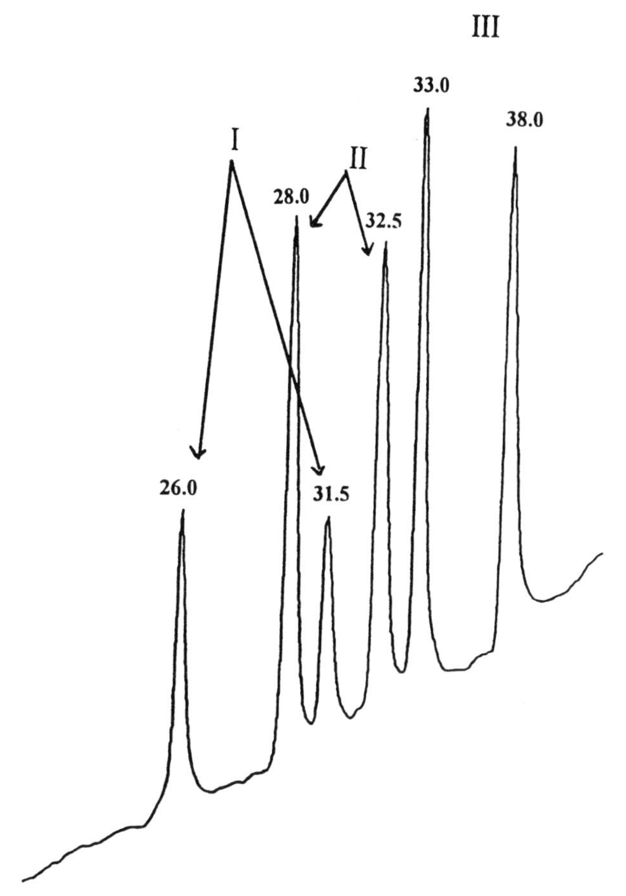

FIGURE 3 Chromatograms of the chiral resolution of (I) DNZ–leucine, (II) DNZ–glutamine, and (III) DNZ-phenylalanine on *tert*-butyl carbamoylated quinine CSP using (a) ammonium acetate (1 mM, pH 5.45) and (b) acetonitrile containing 1% ammonium acetate with gradient elution (DNZ: 3,5-dinitrobenzyloxycarbonyl). (From Ref. 2.)

isomeric 1,3-diacyl-*sn*-glycerols. For this purpose, the diacylglycerols were converted to 2-anthrylurethanes and subjected to chiral phase HPLC with fluorescence detection. Satisfactory resolution of the enantiomers and regio-isomers was achieved on a (*R*)-1-(1-naphthyl)ethylamine polymeric phase, using a mixture of *n*-hexane–dichloromethane–ethanol (150 : 10 : 1, v/v/v) as the mobile phase [21].

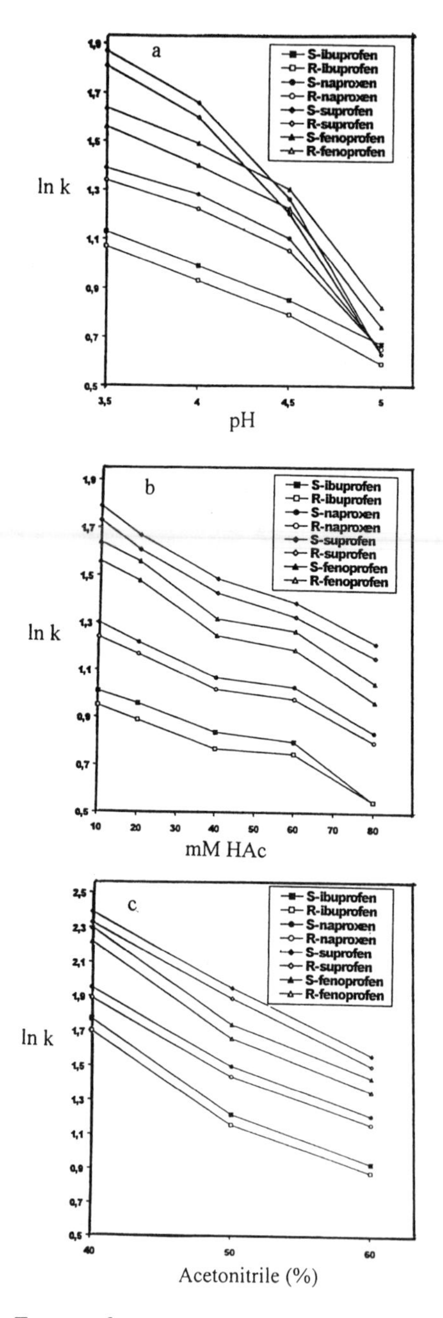

FIGURE 4 Effect of (a) pH, (b) concentration of buffer, and (c) concentration of acetonitrile on the chiral resolution of some profens on terguride-based CSP. (From Ref. 11.)

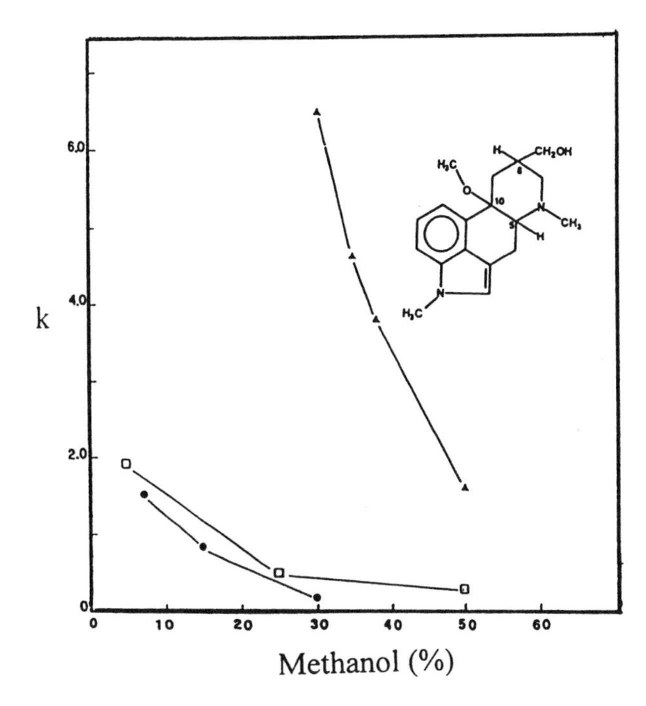

FIGURE 5 Effect of methanol concentration on the chiral resolution of ergot alkaloid on terguride-based CSP using 0.05 M ammonium acetate as the mobile phase at pH 7.2 (▲), 6.1 (□), and 4.6 (●). (From Ref. 12.)

Abou-Basha and Aboul-Enein [22] presented an isocratic and simple HPLC method for the direct resolution of the clenbuterol enantiomers. The method involved the use of a urea-type CSP made of (*S*)-indoline-2-carboxylic acid and (*R*)-1-(naphthyl) ethylamine known as the Chirex 3022 column. The separation factor (α) obtained was 1.27 and the resolution factor (R_s) was 4.2 when using a mobile phase composed of hexane–1,2-dichloroethane–ethanol (80 : 10 : 10, v/v/v). The (+)-enantiomer eluted first with a capacity factor (k) of 2.67 followed by a (−)-enantiomer with a k of 3.38. Biesel et al. [23] resolved 1-benzylcyclohexane-1,2-diamine hydrochloride on a Chirex D-penicillamine column. Gasparrini et al. [24] synthesized a series of the chiral selectors based on *trans*-1,2-diaminocyclohexane. The developed CSPs were used for the chiral resolution of arylacetic acids, alcohols, sulfoxides, selenoxides, phosphinates, tertiary phosphine oxides, and benzodiazepines. In another study, the same authors [25] described the chiral resolution of β-aminoesters enantiomers on synthetic CSPs based on a π-acidic derivatives of *trans*-1,2-diaminocyclohexane

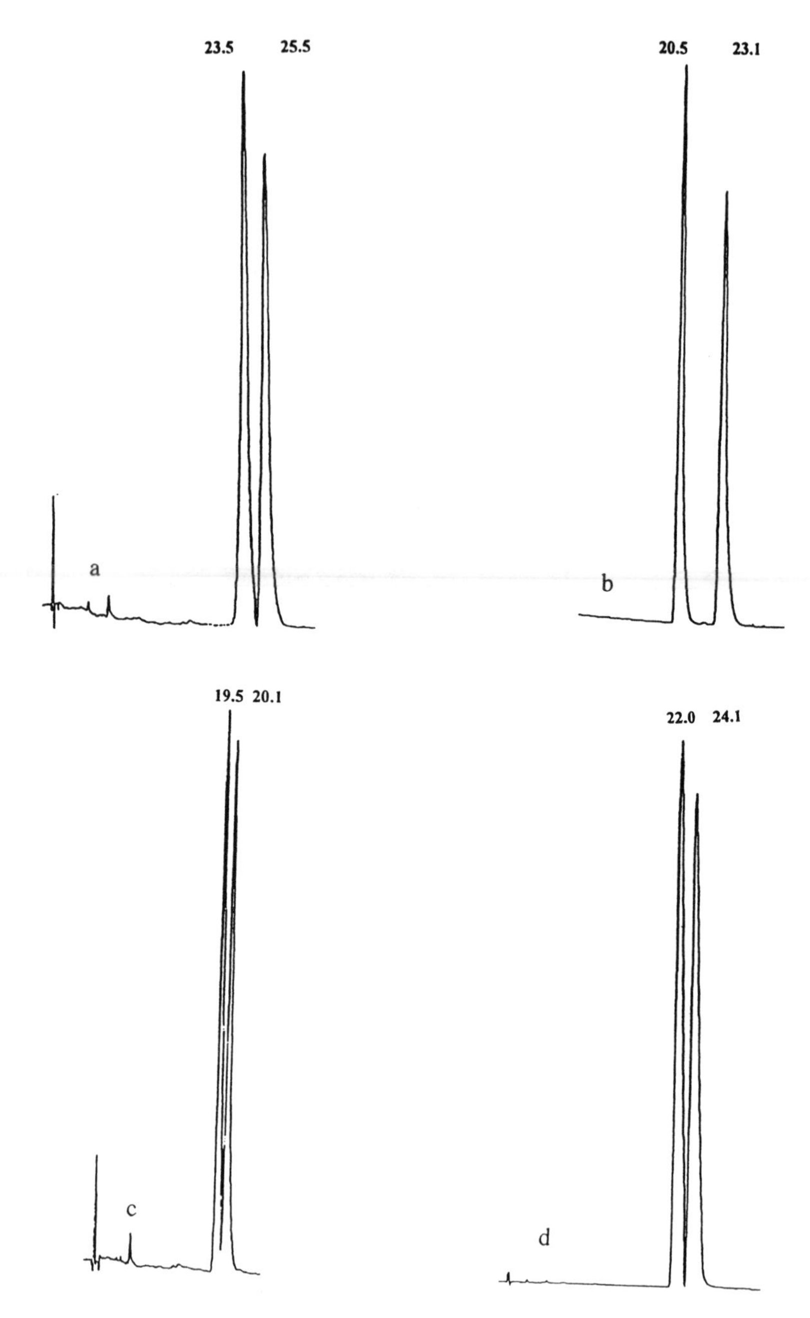

FIGURE 6 Chromatograms of the chiral resolution of (a) fenoprofen, (b) flobufen, (c) naproxen, and (d) ketoprofen on 1-allyl terguride-based CSP using 20 mM potassium acetate (pH 3.6)–acetonitrile (50 : 50, v/v). (From Ref. 11.)

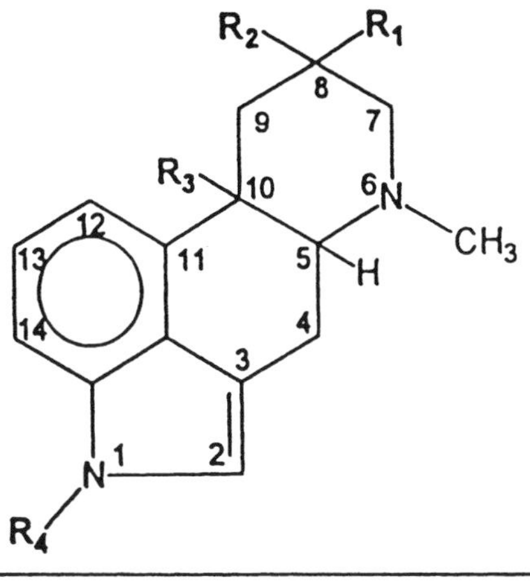

COMPOUNDS	R_1	R_2	R_3	R_4
I-II) (+) (5R, 8S, 10R)- and (-) (5S, 8R, 10S)- terguride	-H	$-NH-CO-N(C_2H_5)_2$	-H	-H
III-IV) (+) (5R, 8S)- and (-) (5S, 8R)- lisuride	-H	$-NH-CO-N(C_2H_5)_2$	C(9)=C(10)	-H
V-VI) (+) (5R, 10S)- and (-) (5S, 10R)- meluol	$-CH_2OH$	-H	CH_3O-	$-CH_3$
VII) (+) (5R, 8S, 10R) 1-(3-aminopropyl)- terguride	-H	$-NH-CO-N(C_2H_5)_2$	-H	$-(CH_2)_3-NH_2$

FIGURE 7 Structures of some ergot alkaloids used for the preparation of CSPs. (From Ref. 12.)

as the chiral selector. The application of different columns containing the stationary phase with opposite configuration and in the racemic form for the determination of enantiomeric excess in chemically impure samples is described. The structures of some amide-based CSPs are given in Figure 8.

9.3 ACID-BASED CSPs

Cholic acid and 3-phenylcarbamoyl cholic acid allyl esters were grafted to hydride-activated silica gel and the developed CSPs were used for the chiral resolution of derivatized amino acids, amines, alcohols, hydantoins, and 2,2′-

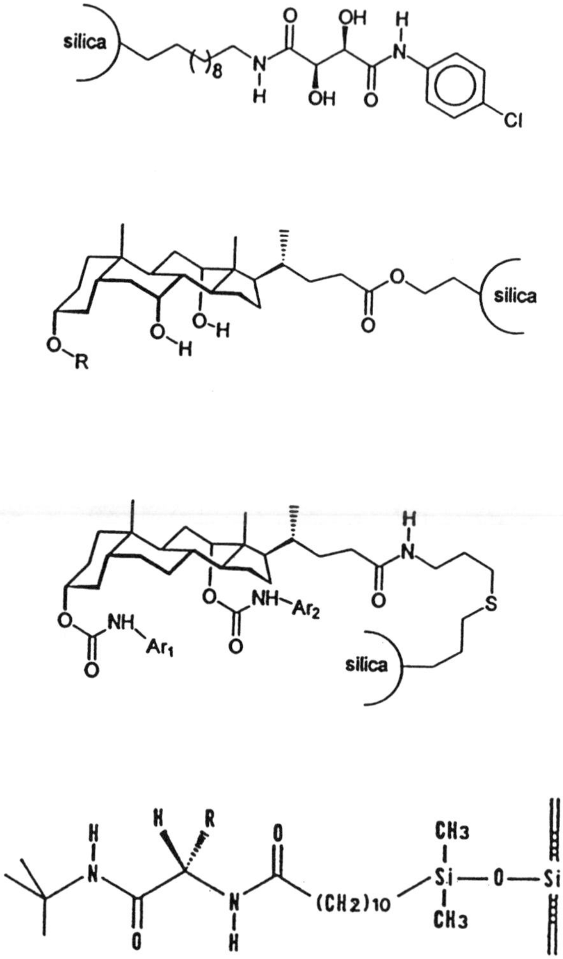

CSP 1, R=i-Pr; CSP 2, R=i-Bu; CSP 3, R=t-Bu

FIGURE 8 Structures of some amide-based CSPs.

dihydroxy-binaphthyl [26,27]. Deoxycholic acid derivatives were linked to silica gel by the conversion of acid to the allylamide, followed by the introduction of two identical or two different arylcarbamate groups [28]. These CSPs were utilized for the chiral resolution of amines, amino acids derivatives, acids, and 3-hydroxy benzodiazepin-2-ones. The enantiomer separation of C_{60} fullerene derivatives on the CSP of (*R*)-(−)-2-(2,4,5,7-tetranitro-9-fluorenylideneaminooxy) propionic acid (TAPA) was achieved. A C_{60} fullerene *tris*- and hexakis-adducts with achiral addends and an inherent chiral addition pattern were separated on a microbore column containing *R*-(−)-TAPA bonded to silica gel as the CSP. The stationary phase exhibited an intermediate polarity, and separations were performed using the normal and reversed-phase modes. With hexakis-adduct, near-baseline separation was achieved. The *tris*-adduct showed a strong interaction with the CSP expressed by a high retention factor but inferior enantiomer separation [29]. The synthesis of (+)- and (−)-hexahelicen-7-yl acetic acid-bonded phases, using two bonding reagents, was conducted. The relationship between the structure of a series of dinitrobenzoyl (DNB) derivatives of chiral aryl alcohols and their enantioselectivity on the (+)-hexahelicen-7-yl acetic acid bonded phase was described. The DNB derivatives were selected to reflect a range of alkyl side-chain lengths, spacer chain lengths between the chiral center and the aromatic group, and the aromatic group size. The chiral resolution increased with increasing length of the alkyl chain at the chiral center and decreased on increasing the distance between the aromatic groups at the chiral center. It was also reported that the enantiomer elution order reversed when changing from the (+)-hexahelicene phase to the (−)-hexahelicene phase [30].

9.4 SYNTHETIC POLYMER–BASED CSPs

The synthesis of optically active polymers is an important area in macromolecular science, as they have a wide variety of potential applications, including the preparation of CSPs [31–37]. Many of the optically active polymers with or without binding to silica gel were used as CSPs and commercialized [38]. These synthetic polymers are classified into three groups according to the methods of polymerization: (1) addition polymers, including vinyl, aldehyde, isocyanide, and acetylene polymers, (2) condensation polymers consisting of polyamides and polyurethanes, and (3) cross-linked gels (template polymerization). The art of the chiral resolution on these polymer-based CSPs is described herein.

9.4.1 Addition Polymers

Various addition polymers were synthesized and used for the chiral resolution purposes, but helical polymethacrylates and polymethacrylamides are more effective and useful for chiral resolution. However, the chiral resolution on other polymers is also discussed.

9.4.1.1 Helical Polymethacrylates

The helical polymethacrylates show a chiral recognition capability for different racemic compounds, including macromolecules [39–41]. Among the various

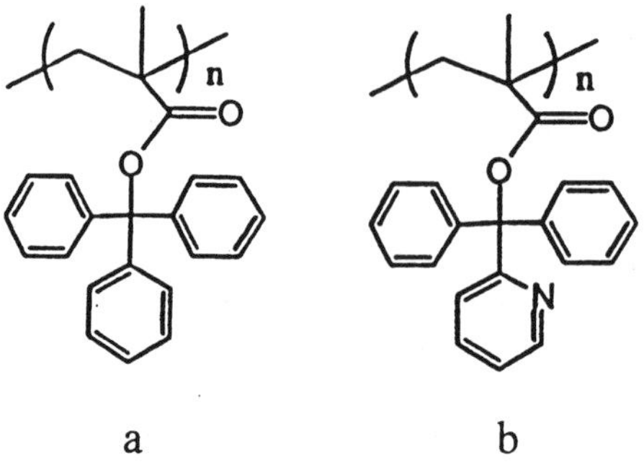

FIGURE 9 Structures of (a) poly(triphenylmethyl methacrylate) [poly(TrMA)] and (b) diphenyl-2-pyridyl methacrylate [poly(D2pyMA)] chiral selectors used for the preparation of CSPs. (From Ref. 37.)

polymers, poly(triphenylmethyl methacrylate) [(poly(TrMA)] and diphenyl-2-pyridyl methacrylate [(D2PyMA)] (Fig. 9) exhibit a high resolving power toward various racemates.

Poly(triphenylmethyl methacrylate). The resolution of Tröger's base, hexahelicene, and cyclophanes were successfully resolved on the CSP based on poly(TrMA) [37]. A significant improvement in the chiral resolution of poly(TrMA) was achieved by coating onto macroporous silica gel [40]. This coated CSP was very successful for the chiral resolution of a wide variety of racemic compounds [32–46]. The structures of some of the racemic compounds resolved on this CSP are given in Figures 10 and 11. The main characteristic feature of this CSP is that it can resolve the racemic compounds, even those lacking any functional groups. Moreover, the racemic compounds, difficult to resolve on other CSPs, can be resolved on this CSP easily. The best mobile phase for this CSP is the use of polar eluents having methanol and water mixtures, which suggests that resolution takes place through hydrophobic interactions between nonpolar groups of the solute and side groups of poly(TrMA) CSP.

It has also been observed that the concentration ratio of poly(TrMA) to silica plays an important role for the chiral resolution. Generally, a high ratio of poly(TrMA) to silica gel was found suitable for the chiral resolution of different racemic compounds [39]. This observation may be explained on the basis of the aggregation state of poly(TrMA) chains. At a higher concentration of the polymer, the helical chains aggregate in an ordered form, which may create new chiral spaces between closely packed polymer chains whose shapes may be different from those of the space around isolated helical chains, and, hence, greater chiral recognition is observed at a high concentration of poly(TrMA) onto the silica gel. This coated CSP suffers from the drawback that it could not be used

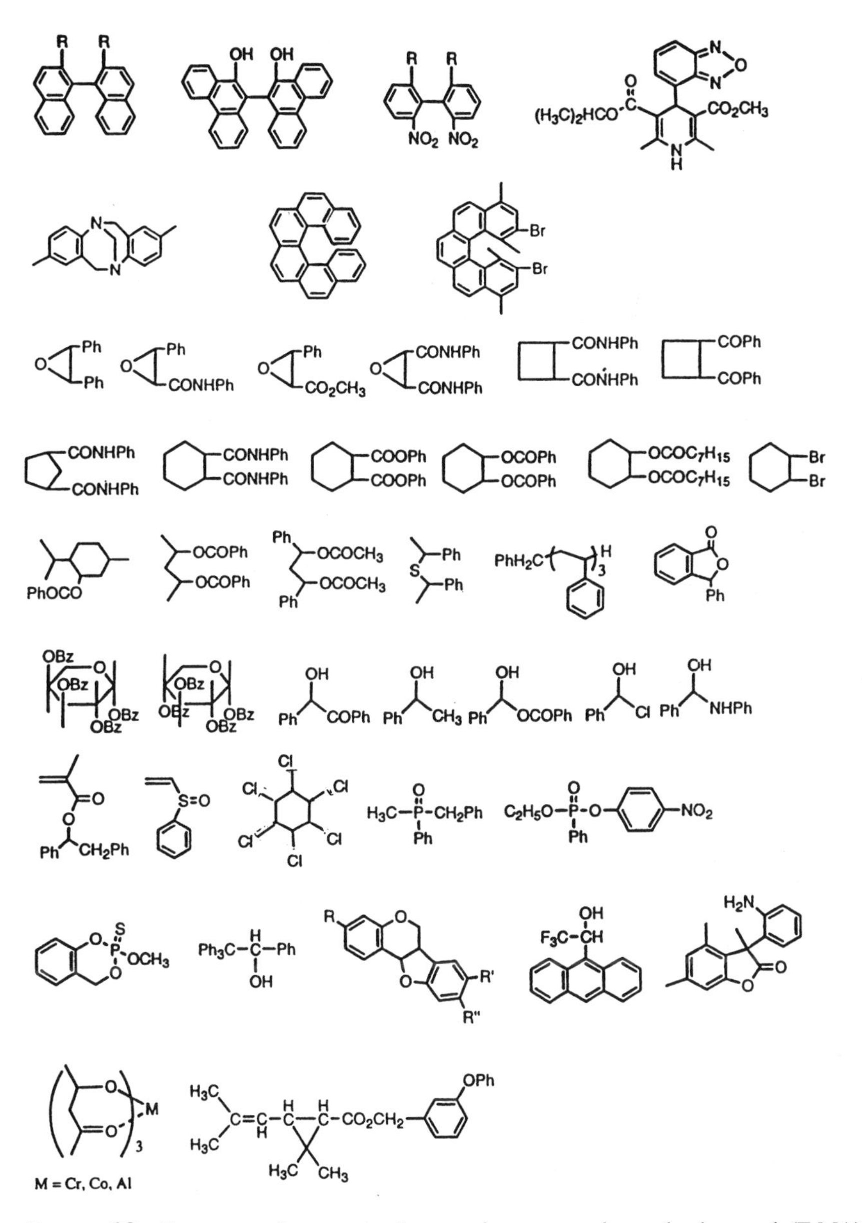

FIGURE 10 Structures of some simple racemic compounds resolved on poly(TrMA) CSP. (From Ref. 37.)

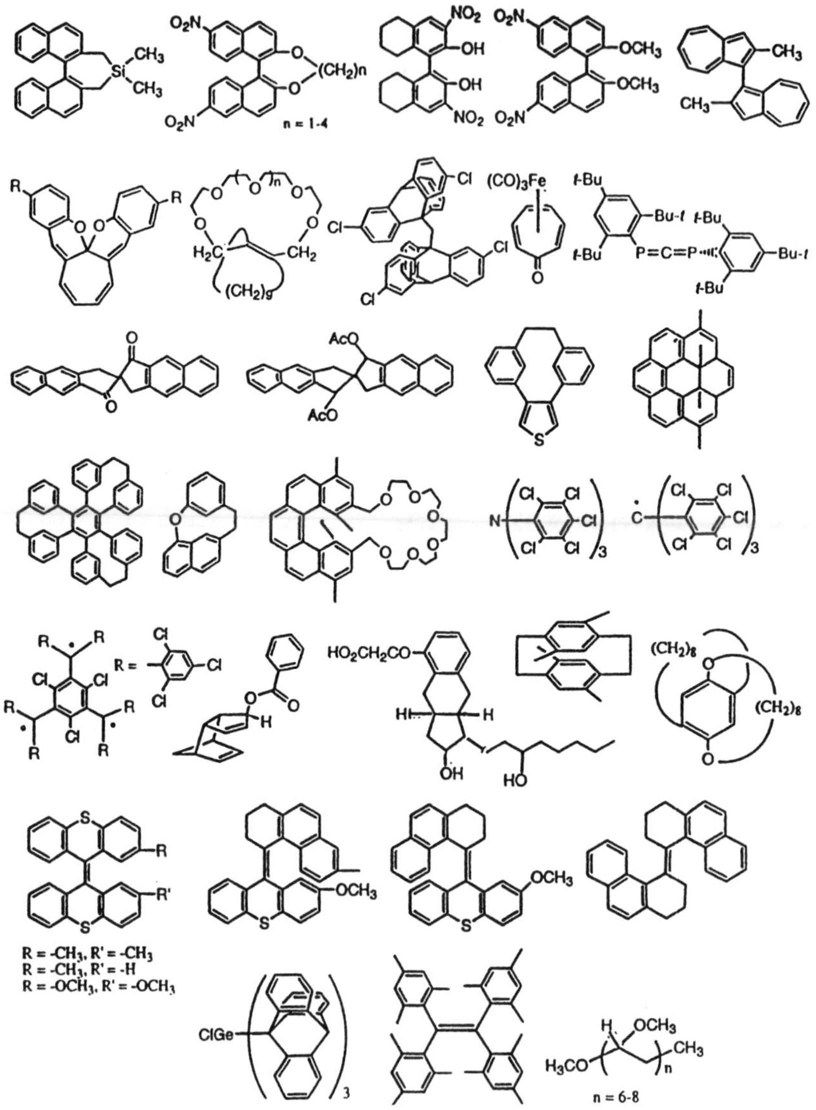

FIGURE 11 Structures of some complex racemic compounds resolved on poly(TrMA) CSP. (From Ref. 37.)

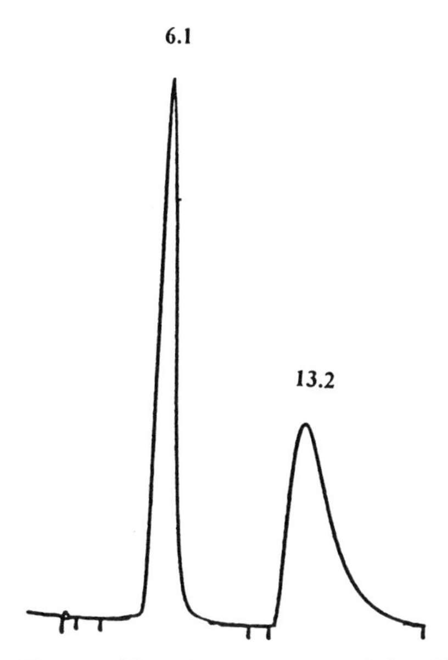

FIGURE 12 Chromatograms of the chiral resolution of the perchlorotriphenylmethyl radical on a Chiralpak OT (+) column using methanol as the eluent. (From Ref. 46.)

with the eluent containing some organic solvents like chloroform and tetrahydrofuran. Therefore, Okamoto et al. [47] prepared a chemically bonded poly(TrMA) CSP. The chemical binding of poly(TrMA) with silica gel was achieved by (1) the reaction of poly(TrMA) and 3-trimethoxysilylpropyl methacrylate with silica gel or (2) the reaction of poly(TrMA) having a PhNH–CH_2CH_2–N–(Ph)– terminal group with silica gel pretreated with (3-aminopropyl)-triethoxysilane and toulene-2,4-diyldiisocyanate. The chiral recognition capacities of both CSPs were comparable. The CSP obtained by this polymer was commercialized under the name Chiralpak OT (+) by Daicel Chemical Industries (Tokyo, Japan). The chiral resolution of the perchlorotriphenylmethyl radical on Chiralpak OT (+) CSP is shown in Figure 12 [46].

Diphenyl-2-pyridyl methacrylate. Diphenyl-2-pyridyl methacrylate (D2PyMA) was more stable than the poly(triphenylmethyl methacrylate) polymer [poly(TrMA)] and the chiral recognition of D2PyMA was comparable to the chiral resolution of the poly(TrMA) polymer [47]. The chiral resolution of several racemic compounds, including calixarene derivatives and Tröger's base and analogs has been achieved on the D2PyMA CSP [48]. In most of the cases, chiral resolution was carried out using methanol, which indicates that hydrophobic interactions are controlling the chiral resolution on this CSP.

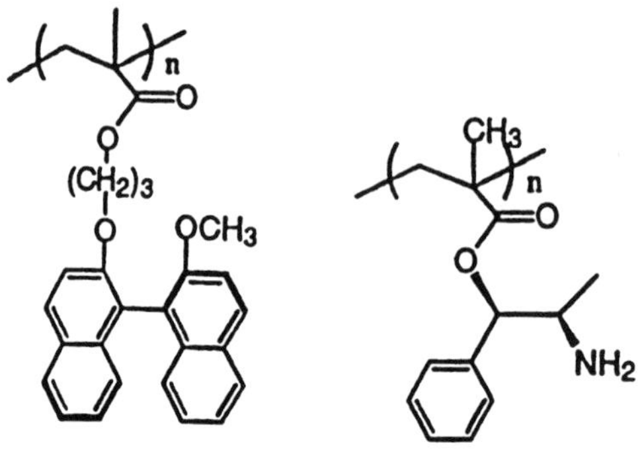

a b

FIGURE 13 Structure of (a) polymethacrylate having binaphthol moiety and (b) simple polymethacrylate CSPs. (From Ref. 37.)

In addition to these two CSPs, their derivatives were also synthesized and tested for chiral resolution. The most important derivatives are poly(*m*-Cl-TrMA), poly(*m*-F-TrMA), poly(*m*-Cl_3-TrMA), poly(*m*-Me_3-TrMA), poly(PB2PyPMA), poly(D3PyMA), and poly(MPyDMA). All of these derivatives show slightly lower chiral recognition capability than poly(TrMA) and D2PyMA CSPs, whereas no chiral recognition was shown by poly(*m*-Me_3-TrMA) CSP [49–52]. An optically active nonhelical polymethacrylate having binaphthol moiety in the side chain (Fig. 13a) was synthesized and showed chiral recognition capacities for 3,5-dinitrophenylcarbamates derivatives of 1,2-diols, 1,3-diols, and 1-phenylalkanols [53]. Similarly, an optically active polymethacrylate (Fig. 13b) showed chiral resolution capacities for some drugs [15]. The CSP obtained from D2PyMA was commercialized by Daicel Chemical Industries (Tokyo, Japan) under the name of Chiralpak OP (+).

9.4.1.2 Polyacrylamide and Polymethacrylamide

Blaschke et al. [54–56] synthesized polyacrylamide and polymethacrylamide containing chiral side chains. In order to make CSPs, these polymers were bonded to silica gel chemically [54–56]. The CSP obtained by *N*-acrylol-(*S*)-phenylalanine ethyl ester was commercialized by Merck Chemical Company by the trade name ChiraSpher. The racemic compounds resolved are those capable of forming hydrogen-bondings (i.e., amides, imides, carboxylic acids, and alcohols). It has been reported that nonpolar solvents like benzene and toluene individually or their mixtures were the best mobile phases. In addition to these CSPs, other amide CSPs were prepared and tested for the chiral resolution [57,58].

9.4.1.3 Other Addition Polymers

In addition to the above-mentioned polymers, other addition polymers such as polyolefin, polystyrene, polyvinylethers, polychloral, polyisocyanides, polyacetylene, and polyethers were synthesized and evaluated as the precursors for the preparation of CSPs. Some of them were coated or chemically bonded to silica gel and tested for the chiral resolution of different racemic compounds.

Pino et al. [59] resolved the enantiomers of poly[(*R*)-4-methyl-1-hexene] and poly[*S*-4-methyl-1-hexene] on crystalline and insoluble (+)-poly[(*S*)-3-methyl-1-pentene]. An optically active styrene derivative was prepared and ground to powder. The powder was packed into a column and used as the CSP for the chiral resolution of several amines and alcohols having aryl groups. However, this CSP could not resolve aliphatic alcohols and amines [60]. Kakuchi et al. [61] resolved the enantiomers of phenylglycine, valine, and methionine on CSP containing cyclopolymerized chiral divinyl ether. Ute et al. [62] resolved the enantiomers of *trans*-stilbene oxide on polychloral CSP. Yamagishi et al. [63] resolved the enantiomers of menthol and binaphthol on CSP containing poly(*t*-butyl isocyanide). An optically active polyphenylacetylene derivative was coated onto silica gel and the developed CSP was tested for the chiral resolution of *trans*-stilbene oxide and Tröger's base [64]. In another study, Yashima et al. [65] also developed new polyacetylene-based CSPs and used them to resolve the enantiomers of *trans*-stilbene oxide, Tröger's base, and spiropyrans derivatives. Umeda et al. [66] chemically bonded polysaccharides like polyether to silica gel and the developed CSP was used for the chiral resolution of amino acids.

9.4.2 Condensation Polymers

Several polyamides and polyurethanes have been synthesized by condensation polymerization and used as the precursors for the preparation of different CSPs. These CSPs showed different chiral recognition properties than those of addition polymers. Allenmark and Andersson [67] prepared a CSP containing silica gel supported by *s*-triazine derivatives of L-valine isopropyl esters (Fig. 14) and the

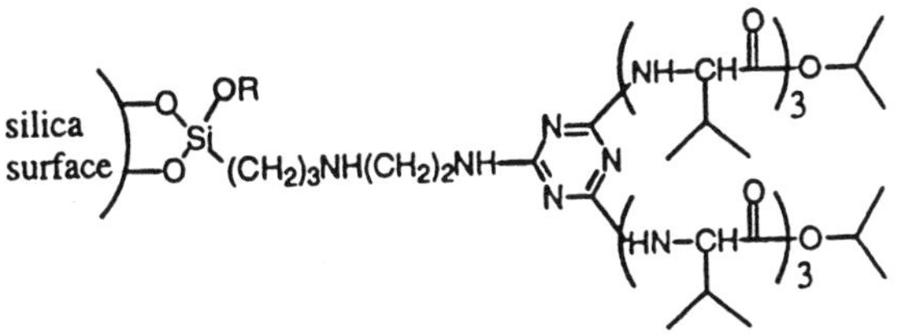

silica surface OR Si $(CH_2)_3NH(CH_2)_2NH$ N N N NH CH O O 3 HN CH O O 3

FIGURE 14 Structure of polyamide-based CSP (*s*-triazine derivative of L-valine isopropyl ester) (condensation polymer). (From Ref. 37.)

developed CSP was found suitable for the chiral resolution of amino acid derivatives. Hirayama et al. [68] resolved some amino acid derivatives on poly(L-leucine) or poly(L-phenylalanine) chemically bonded to poly(methyl acrylate) macropore beads. Poly(*N*-benzyl glutamine) chemically bonded to polystyrene beads also resolved the enantiomers of mandelic acid and hydantoin derivatives [69]. It has also been reported [70–74] that polyamides prepared from chiral diamine or chiral dicarboxilic acids or esters can resolve some polar racemic compounds having functional groups capable of hydrogen-bondings. The chiral recognition capacities depended on the structures of the chiral units and the groups binding the chiral units. For example, polyamides having methylene groups of even number in the diamine units show higher recognition than those with an odd number. Sinibaldi et al. [75] prepared monosized nonporous urea formaldehyde resin and used it for the chiral resolution of phenylalanine using potassium and copper(II) acetate buffer as the mobile phase. The chiral recognition capacities of polyurethanes and polyurea urethanes have also been evaluated [76–78].

9.4.3 Cross-Linked Gels (Template Polymerization)

Generally, cross-linked polymers recognize chiral resolution through chiral cavities. Some polymers were synthesized and used for chiral resolution purposes [79–82]. The applications of these CSPs include the chiral resolution of amino acids, β-blockers, and profens. Wulf et al. [83] observed that the chiral resolution capacity of these gels is significantly affected by the kind and amount of cross-linker agent.

9.5 MOLECULARLY IMPRINTED CSPs

Polysaccharides, proteins, and synthetic polymers are readily available polymers for the preparation of CSPs. A problem with these as well as other common CSPs is the limited predictability of elution orders and separability, making screening of stationary phase libraries a necessary step in the method development. Polymers imprinted with chiral templates promise to alleviate these drawbacks and offer a new generation of tailor-made CSPs with predictable selectivities [84–86]. Molecularly imprinted polymers (MIPs) can be prepared by a number of approaches that are different in the way the template is linked to the functional monomer and, subsequently, to the polymeric binding sites. The approaches to generate imprinted binding sites are shown in Figure 15. The first example of molecular imprinting of organic network polymers was introduced by Wulf [80]; it was based on a covalent attachment strategy (i.e., covalent monomer template and covalent polymer template). Most of the polymers used in the preparation of CSPs through the molecularly imprinted technique are synthesized by free-radical

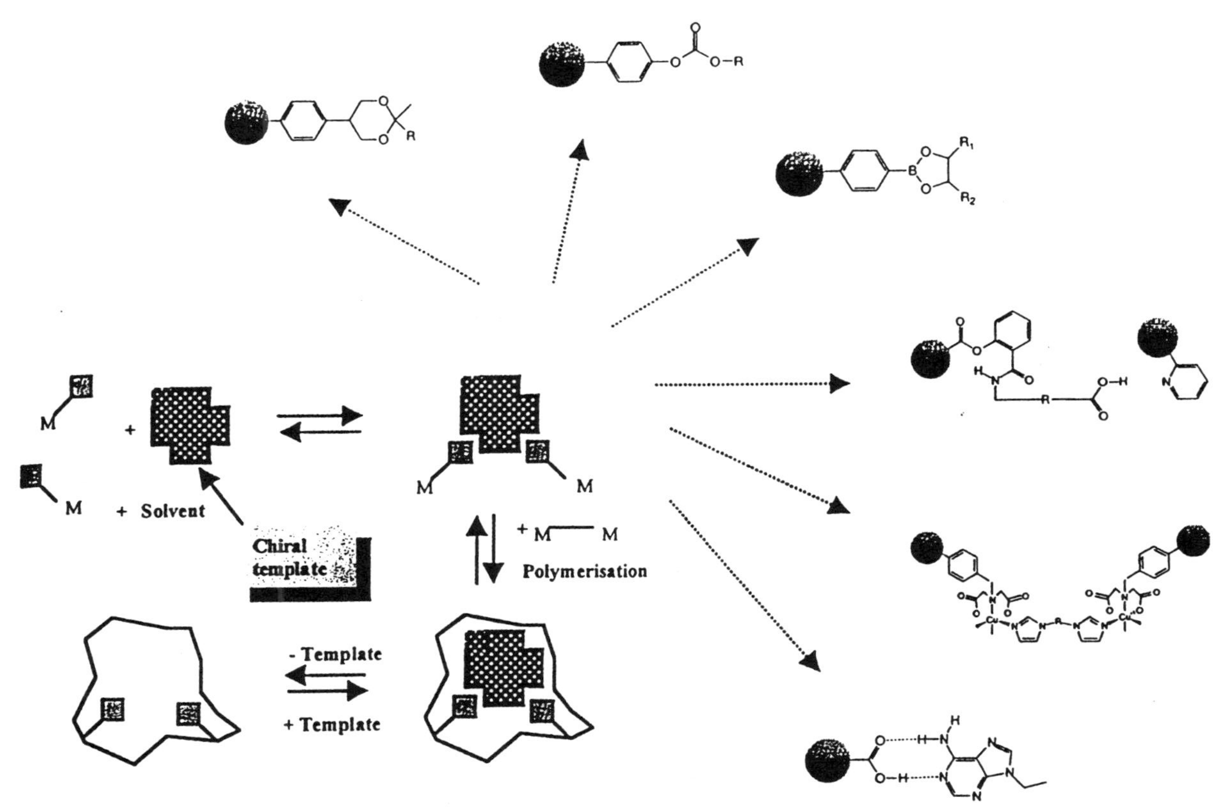

FIGURE 15 Different approaches to generate imprinted binding sites. (From Ref. 85.)

polymerization mechanisms involving monofunctional unsaturated monomers and an excess of cross-linking di- or triunsaturated monomers which resulted in porous organic network materials [87]. Most of the polymers prepared by this technique are based on vinylic, acrylic, and methacrylic monomers, which have already been discussed in Section 9.4. The most important applications of these CSPs are the chiral resolution of amino acids, amino acid derivatives, amines, carboxylic acids, peptides, β-blockers, profens, alkaloids, and other pharmaceuticals [85].

Kempe and Mosbach [88] improved the performance of MIPs by coating the surface of macroporous silica gel or by grafting the MIP onto polymer beads, and the resulted CSPs showed a moderate improvement in enantioselectivity. Generally, an increase in the temperature [89] and gradient elution [90] yielded a highly improved performance of MIP CSPs. Lin et al. [91] prepared L-aromatic amino acid-imprinted polymers using azobisnitrile as their photoinitiators or thermal initiators at temperature ranging from 4°C to 60°C. The resulting polymers were grounded and sieved to a 25-μm size and packed into the column. The developed CSP was used to resolve the enantiomers of amino acids. The authors also studied the effect of temperature on the chiral resolution of amino acids using this CSP (Fig. 16) and it was observed that chiral resolution improved at a high temperature. A typical example of the chiral resolution on a

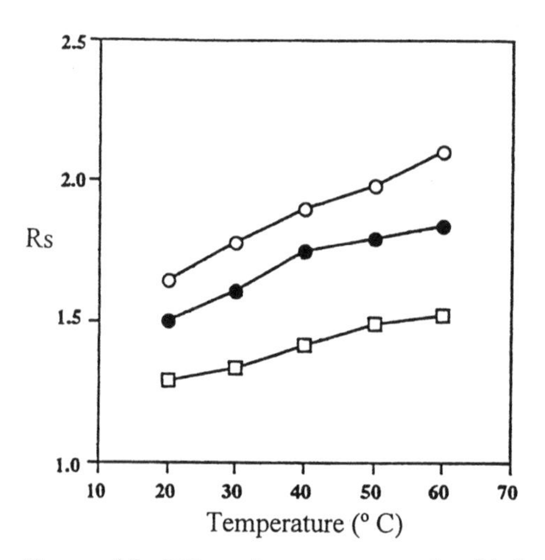

FIGURE 16 Effect of temperature on the chiral resolution of DL-phenylalanine on CSP prepared by imprinted L-phenylalanine molecule by capillary electrochromatography using acetonitrile–acetic acid (95 : 5, v/v). Imprinted polymers were prepared at (□) 60°C, (●) 40°C, and (○) 4°C. (From Ref. 91.)

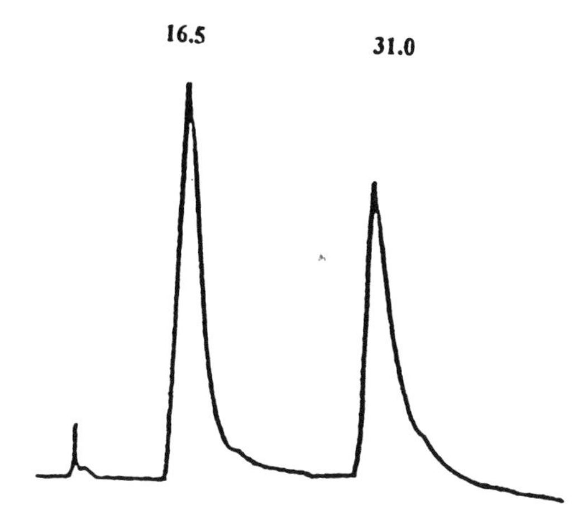

FIGURE 17 Chromatograms of the chiral resolution of *Z*-tyrosine-OH on *Z*-(*S*)-tyrosine-OH-imprinted poly(pentaerythritol triacrylate-*co*-methylmethacrylate) CSP using chloroform–acetic acid (66 : 4, v/v) as the mobile phase. (From Ref. 8.)

MIP is shown in Figure 17, representing the chromatograms of the chiral resolution of *Z*-tyrosine-OH on a *Z*-(*S*)-tyrosine-OH-imprinted poly(pentaerythritol triacrylate-*co*-methylmethacrylate) CSP.

9.6 MISCELLANEOUS CSP IN SFC

The above-discussed miscellaneous CSPs have also been used in supercritical fluid chromatography (SFC) for chiral resolution. Macaudiere et al. [92] carried out a comparison of the chiral resolution using HPLC and SFC on Chiralpak OT. The chromatographic behavior of this CSP was found to be quite different in HPLC and SFC. A CSP based on the adsorption of a chiral anthrylamine on porous graphitic carbon successfully resolved the enantiomers of tropic acid derivatives and anti-inflammatory agents in SFC [93]. Schleimer et al. [94] immobilized Chirasil nickel(II) bis[3-(heptafluorobutanoyl)-(1*R*)-10-methylene-camphorate to poly(dimethylsiloxane) and packed in the capillary. The developed CSP was used for the chiral resolution of 1-phenylethanol and camphor enantiomers. The retention and separation factors depended on the applied pressure, the density of the mobile phase, temperature, and flow rate. The structure of this CSP is shown in Figure 18 [94]. The effect of the temperature on the chiral resolution of 1-phenylethanol and camphor is presented in Figure 19. The elution profiles of 1-phenylethanol and camphor at different temperatures and pressures are shown in Figure 20 [94]. In addition, the effect of

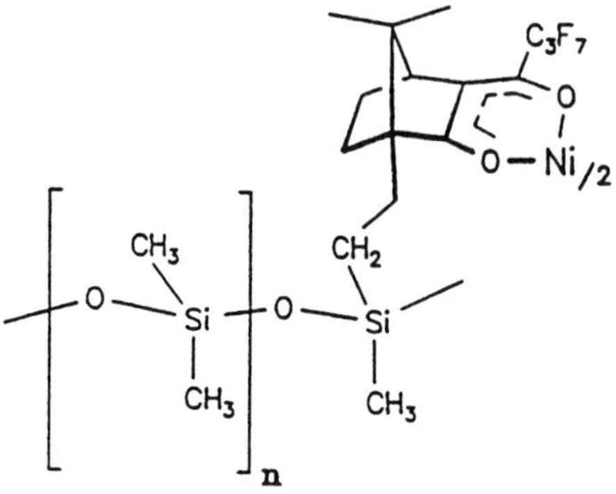

FIGURE 18 Structure of the Chirasil–nickel CSP. (From Ref. 94.)

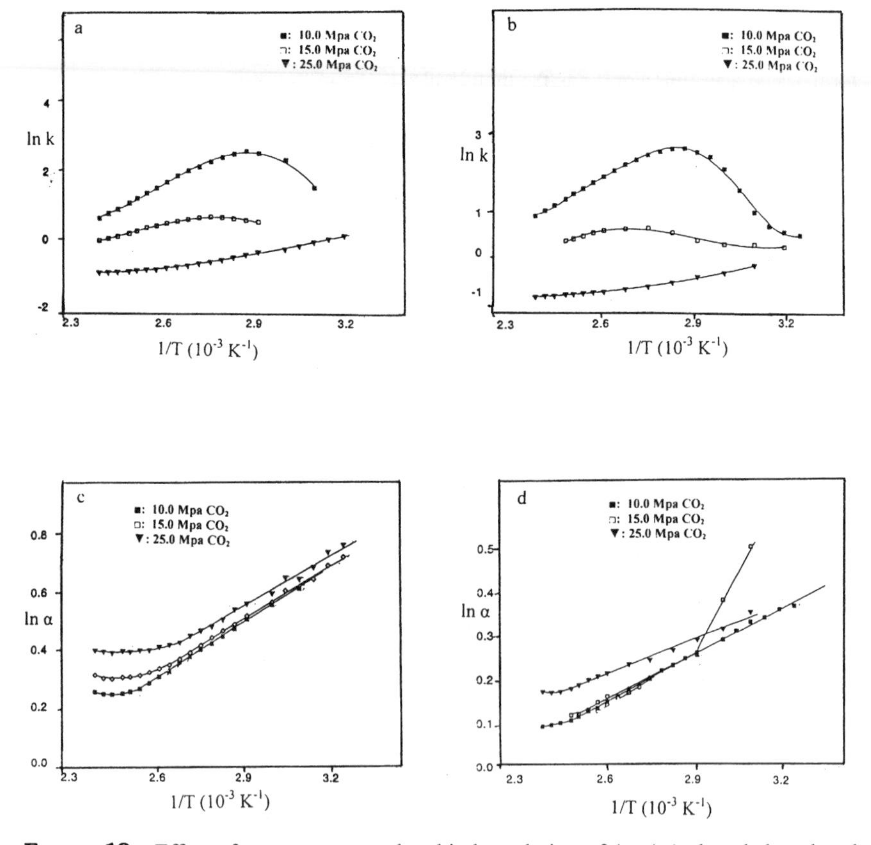

FIGURE 19 Effect of temperature on the chiral resolution of (a, c) 1-phenylethanol and (b, d) camphor on capillary containing the Chiralsil–nickel CSP (SFC). (From Ref. 94.)

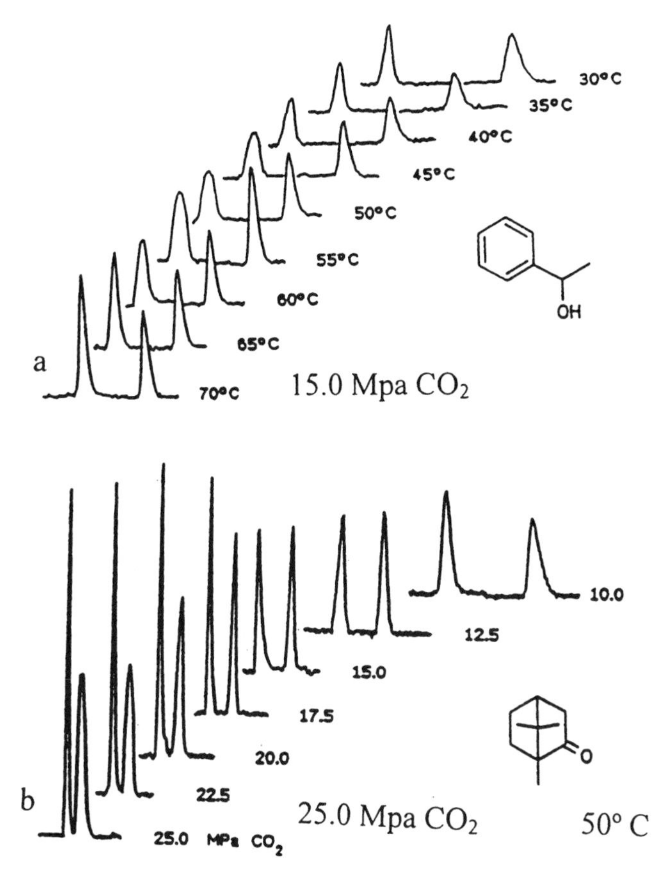

FIGURE 20 Elution profile of (a) 1-phenylethanol at different temperatures and (b) camphor at different pressures on a capillary containing a Chiralsil–nickel CSP (SFC). (From Ref. 94.)

flow rate on the chiral resolution of camphor is given in Figure 21. A perusal of all of these figures indicates that the chiral resolution depends on the different SFC parameters.

9.7 MISCELLANEOUS CSP IN CEC

Sinibaldi et al. [95] resolved the enantiomers of 2-arylpropionic acids (flobufens) and dansyl amino acids using a coated capillary of polyterguride using capillary electrochromatography (CEC). It was found that the analytes, in the range of the buffer pH between 2.5 and 4.0, were driven by anodic electro-osmotic flow

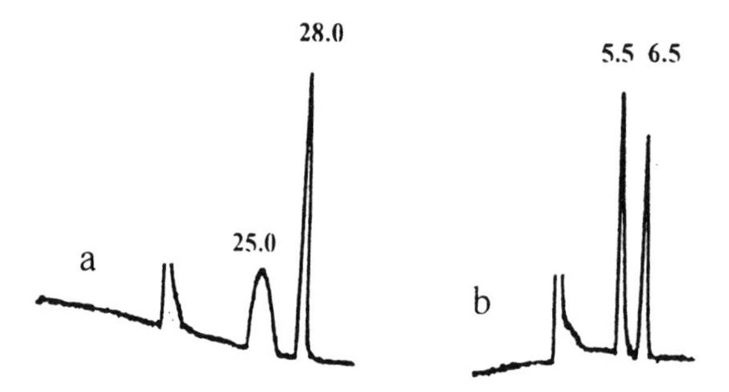

FIGURE 21 Chromatograms of the chiral resolution of camphor at different flow rates on the capillary containing the Chiralsil–nickel CSP (SFC): (a) $u_0 = 0.314\,\text{cm/sec}$, (b) $u_0 = 1.335\,\text{cm/sec}$. (From Ref. 94.)

originated by the positively charged moieties of the ergolinic skeleton and, only partially, by their anodic electrophoretic mobility. The *S*-enantiomer was better retained than *R*-enantiomer. The retention of the solutes was affected by the concentrations and composition of the eluents. The effect of pH (on the chiral resolution of flobufen) and acetonitrile concentration (on the chiral resolution of fenoprofen) is shown in Figure 22, which indicates high enantioselectivities at higher pH values and acetonitrile concentration [95]. Mandl et al. [5] used *O*-(*tert*-butylcarbamoyl) quinine as the chiral selector for the reversible enantiomerization of axially chiral 2-dodecycloxy-6-nitrophenyl-2-carboxylic acid by stopped-flow capillary electrochromatography (sfCEC). The authors compared the chiral resolution of 2-dodecycloxy-6-nitrophenyl-2-carboxylic acid by sfHPLC and sfCEC and reported the best resolution by sfCEC. Furthermore, the authors also studied the effect of temperature on the chiral resolution of 2-dodecycloxy-6-nitrophenyl-2-carboxylic acid. The effect of temperature is shown in Figure 23, which indicates that the chiral resolution decreased with the increase in temperature [6].

The helical chiral poly(diphenyl-2-pyridylmethylmethacrylate)-based CSP, in the form of a capillary, was used for the chiral resolution in aqueous and nonaqueous modes of CEC [96]. The contribution of the aminopropyl groups of the silica gel and the pyridyl groups of the chiral selector to the anodic electroosmotic flow (EOF) generated in these capillaries was evaluated. Furthermore, different nonaqueous BGEs (methanol, acetonitrile), the influence of the ammonium acetate concentration, the apparent pH of the BGE, and the composition of the aqueous–organic BGE were investigated with regard to the EOF and basic chromatographic parameters. The effect of the contributions of pressure-driven and electrokinetically driven flows and mass-transfer characteristics of the stationary phase on the plate heights and enantioseparations was also evaluated.

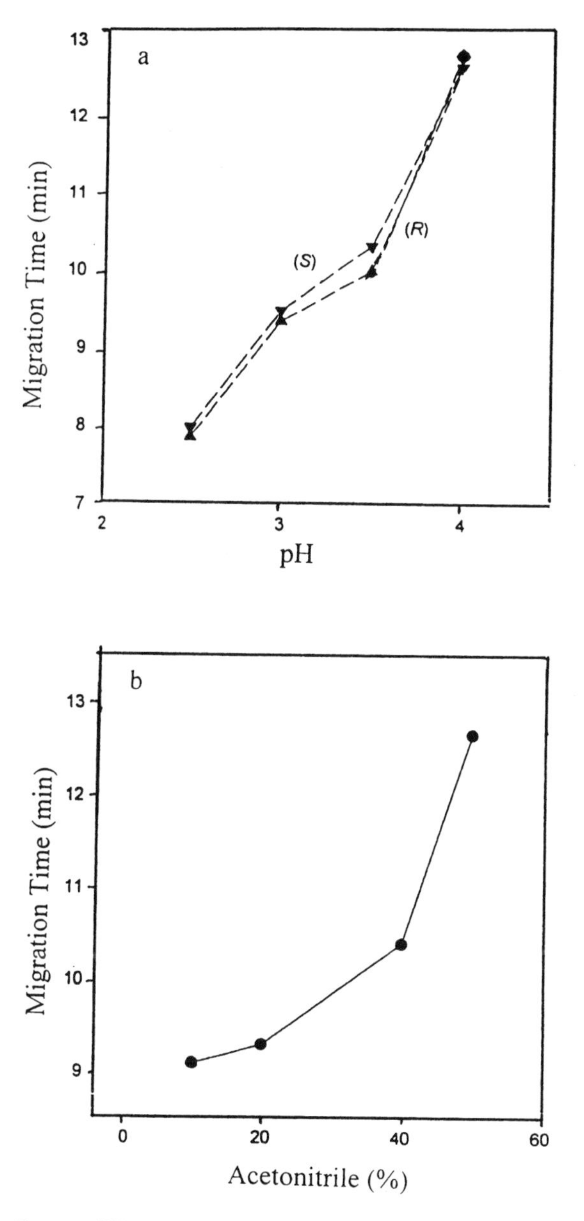

FIGURE 22 Effect of (a) pH on the chiral resolution of flobufen and (b) concentration of acetonitrile on the chiral resolution of fenoprofen on polyterguride CSP using a phosphate buffer (20 mM). (From Ref. 95.)

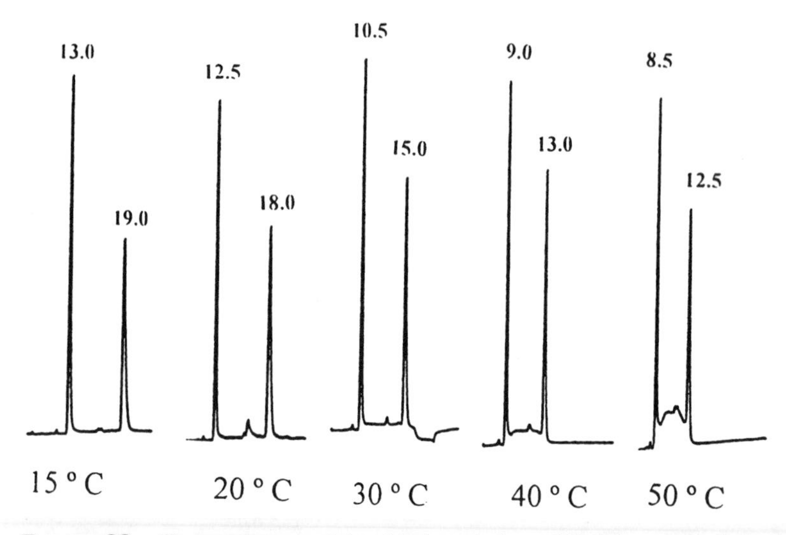

FIGURE 23 Chromatograms of the chiral resolution of 2′-dodecycloxy-6-nitrobiphenyl carboxylic acid on *O*-(*tert*-butylcarbomyl) quinine-based CSP at different temperatures using acetonitrile–methanol (80 : 20, v/v), containing 400 mM acetic acid and 4 mM triethylamine, as the mobile phase. (From Ref. 6.)

Peters et al. [97] developed several monolithic chiral selectors based on 2-hydroxyethyl methacrylate (*N*-L-valine-3,5-dimethylanilide) carbamate with ethylene dimethacrylate, 2-acrylamido-2-methyl-1-propanesulfonic acid, and butyl or glycidyl methacrylate and these chiral selectors were packed in capillaries. The developed CSPs were used for the chiral resolution of *N*(3,5-dinitrobenzoyl) leucine diallylamide enantiomers.

The CSPs prepared by the molecular imprint technique have also been used for chiral resolution by CEC [98–100]. Lin et al. [91] synthesized L-aromatic amino acid-imprinted polymers using azobisnitriles with either photoinitiators or thermal initiators at temperatures ranging from 4°C to 60°C. Methacrylic acid (MAA) was used as the functional monomer and ethylene glycol dimethacrylate (EDMA) was used as the cross-linker. The resulting polymers were ground and sieved to a particle size less than 10 μm, filled into the capillary columns, and used for enantiomeric separations of some amino acids at different temperatures. The relationships of separation factor and column temperatures were demonstrated to be linear between the logarithm of the separation factors and the inverse of the absolute temperature (Fig. 24). The authors also compared the obtained chiral resolution with the chiral resolution achieved by HPLC and reported the best resolution on CEC. The chromatograms of the chiral resolution of DL-

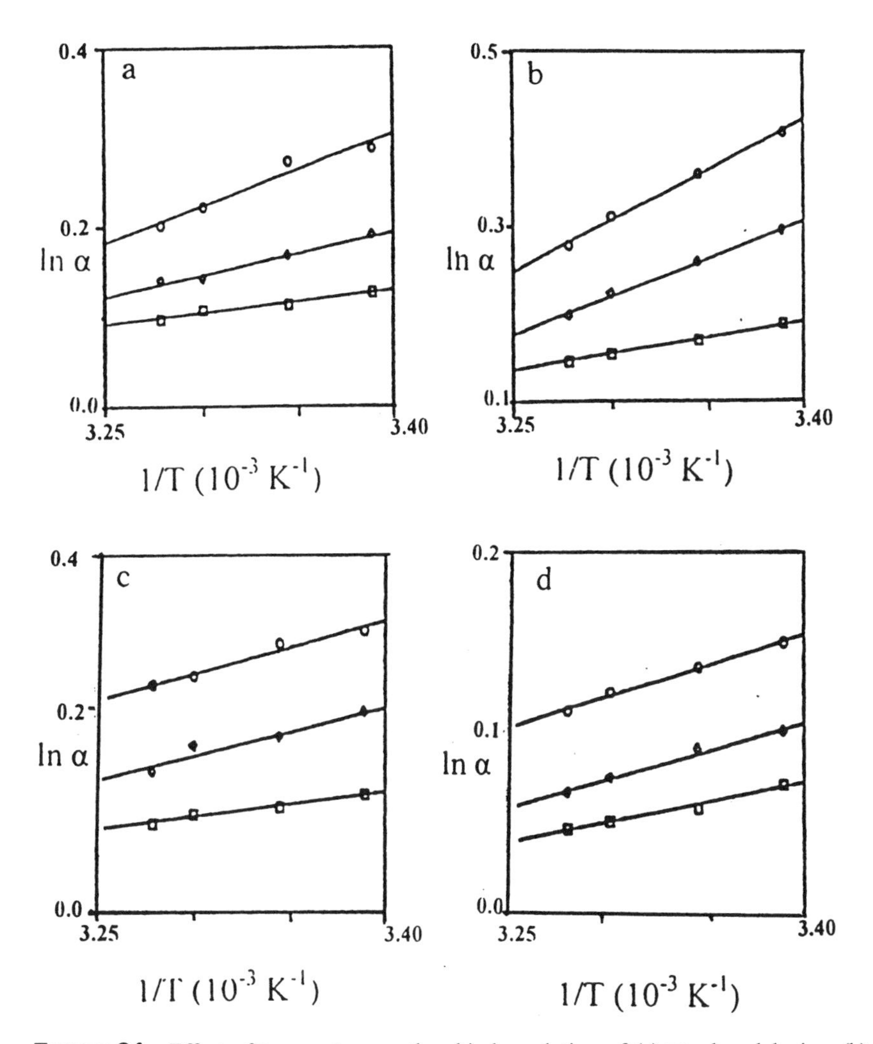

FIGURE 24 Effect of temperature on the chiral resolution of (a) DL-phenylalanine, (b) DL-phenylglycine, (c) DL-tyrosine, and (d) DL-3-(3,4-dihydroxyphenyl) alanine on CSP obtained by an imprinted L-phenylalanine molecule by CEC using acetonitrile–water–acetic acid (80 : 10 : 10, v/v/v). Imprinted polymers were prepared at (□) 60°C, (◇) 40°C, and (○) 4°C. (From Ref. 91.)

phenylalanine at different temperatures are shown in Figure 25. The chiral resolution of some amino acids on imprinted molecules is given in Table 3. The chiral resolution on different types of CSP using CEC is summarized in Table 4.

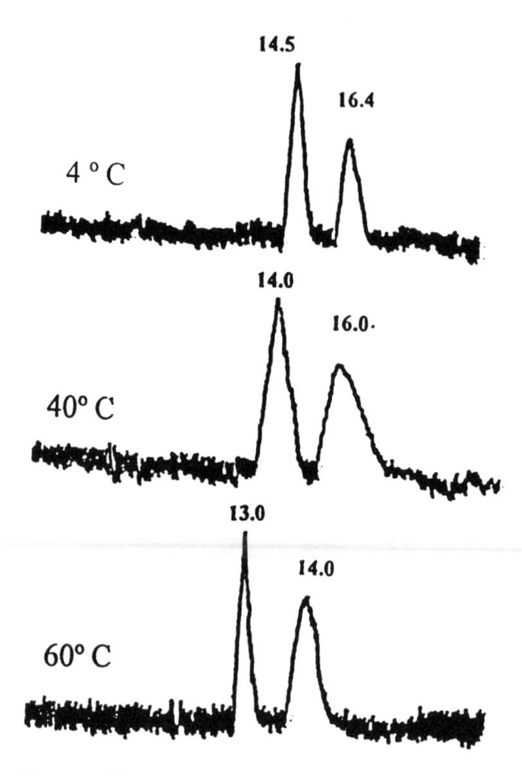

FIGURE 25 Chromatograms of the chiral resolution of DL-phenylalanine (at different temperatures) on CSP obtained by an imprinted L-phenylalanine molecule by CEC using acetonitrile–water–acetic acid (80 : 10 : 10, v/v/v). (From Ref. 91.)

TABLE 3 Chiral Resolution of Some Amino Acids on Molecularly Imprinted CSPs

Amino acid	Imprinted molecule	α	R_s
DL-Phe	L-Phe	1.15	1.74
DL-Phe	L-Phe-AN	1.13	1.60
DL-Try	L-Try	1.05	0.94
DL-DNS-Leu	DNS-L-Leu	1.05	1.04

Note: DNS = dansyl.
Source: Ref. 9.

TABLE 4 Chiral Resolution of Some Racemic Compounds on Different CSPs Using CEC

Racemic compound	CSP	Ref.
Flubufen, dansyl-DL-amino acids	Polyterguride	95
N-Derivatized amino acids	Cinchona types	101, 102
Bendroflumethiazide	Polyacrylamides	103
Benzoin, methylbenzoin, ethylbenzoin, isopropylbenzoin, benzoin oxime, cyclobutyldianilide carbamate, 1,1′-binaphthyl-2,2′-diol, *trans*-stilbene oxide, and Trögers's base	Polymethacrylate	104, 96
DL-Clomipramine	Lactone	105
3,5-Dinitrobenzoyl-DL-alanine	α-Naphthylethylamine	105
DL-Phenylalanine, dansyl-DL-leucine	L-Phenylalanine, Dansyl-Leucine	106
RS-Propranolol, *RS*-metoprolol	*R*-Propranolol	107
RS-2-Phenylpropionic acid	*S*(+)-2-Phenylpropionic acid	108

9.8 MISCELLANEOUS CSP IN TLC

The miscellaneous types of CSP were also used for the chiral resolution of racemic compounds by thin-layer chromatography (TLC). The CSPs for TLC are prepared by impregnating the thin layers by a suitable chiral selector. The various approaches of thin-layer impregnation include the mixing of the chiral selector with the adsorbent and preparation of thin layers, immersion of a thin-layer plain plate ascending or descending into the solution of chiral selector, exposing the thin layer to the vapors of the chiral reagent, spraying of the chiral reagent solution onto thin layer, and, sometimes, by the reaction of the inert material of the thin layer with the chiral reagent. Aboul-Enein et al. [109] and Bhushan and Martens [110] reviewed the chiral resolution on TLC and discussed the use of a variety of chiral selectors as impregnating reagents for the thin layers. Furthermore, they [109,110] described the use of these impregnating thin layers for the chiral resolution of some racemic compounds. A brief survey of the impregnating reagent and their application toward the chiral resolution of different racemic compounds are given in Table 5.

A combination of TLC and molecular imprinting technique was reported by Mosbasch in 1994 [122]. One approach of the technique involves noncovalent prearrangement of functional monomers in the presence of a print molecule prior to polymerization. After the removal of the print molecules from the resulting macroporous polymer matrix, the resulting polymer contains recognition sites which, because of the shape and the arrangement of the functional groups, have

TABLE 5 Chiral Resolution of Some Racemic Compounds on Different CSPs Using TLC

Racemic compound	CSP	Ref.
Amino acids	(−)-Brucine	111
	(1*R*,3*R*,5*R*)-2-Azabicyclo-[3,3,0]-octane-3-carboxylic acid	112
	Copper(II)-impregnated chitin and chitosan layers	113
Phenylthiohydantoin amino acids	(+)-Tartaric acids, (+)-ascorbic acid	114
Dansyl amino acids	(1*R*,3*R*,5*R*)-2-Azabicyclo-[3,3,0]-octane-3-carboxylic acid	115
	Poly-L-phenylalanineamide	116
	N,N-Di-*n*-propyl-L-alanine with copper(II) acetate	117
Atropine and colchicine	L-Aspartic acid	118
DL-Penicillamine	Copper(II) acetate	119
Lactic acid	Copper(II) acetate	120
2-Arylpropionic acid	(−)-Brucine	121

affinity for the print molecules. For example, the chiral resolution of DL-phenylalanine anilide on an imprinted chiral selector was reported by Kriz et al. [123]. The CSP was prepared by the molecularly imprinted technique L-phenylalanine anilide and impregnated onto thin-layer plates. Various concentrations of acetic acid in acetonitrile was used as the mobile phase.

9.9 CHIRAL RECOGNITION MECHANISMS

As discussed earlier, the miscellaneous types of CSP contain different types of structure containing various groups and atoms. There are only a few reports dealing with the determination of chiral recognitions mechanisms on these CSPs. However, Allenmark et al. [20] attempted to explain the chiral recognition mechanisms of the separation of amino alcohols, profens, β-blockers, benzodiazepinones, and benzothiadiazines on CSP based on *N,N*-diallyl-(*R,R*)-tartaric acid diamide. The authors reported the formation of various types of interaction such as hydrogen-bonding, dipole–dipole stacking, and charge-transfer complexes between the CSP and the racemic compounds. In another study, Franco et al. [7] synthesized nine new quinine carbamate dimers and immobilized them onto silica gel. The chiral recognition mechanisms were ascertained on these phases using Fourier transform infrared and x-ray analysis and it was observed that hydrogen-bonding and π–π interactions are responsible for the chiral resolution. Therefore, the chiral recognitions on these CSPs depend on the structures of

CSPs and racemic compounds. The chiral resolutions are controlled by different types of force and interaction such as hydrogen-bondings, π–π interactions, formation of charge-transfer and inclusion complexes, dipole interactions, and steric forces. Weak forces such as van der Waals and coordination bondings also play a crucial role in the chiral resolution on these miscellaneous types of CSP. The presence of a chiral moiety on the CSP is essential, as it provides the chiral environment to the racemic compounds. The racemic compounds fit on the chiral moiety in a different fashion and are stabilized, with different values of binding energies, through the above-cited forces. As a result of the flow of the mobile phase, the two enantiomers elute at different retention times and, hence, the chiral resolution occurs.

9.10 CONCLUSION

In spite of the development of more successful and reliable CSPs (Chaps. 2–8), these miscellaneous types of CSP have their role in the field of the chiral resolution also. The importance of these CSPs lies in the fact that they are readily available, inexpensive, and economic. Moreover, these CSPs can be used for some specific chiral resolution purpose. For example, the CSP based on the poly(triphenylmethyl methacrylate) polymer can be used for the chiral resolution of the racemic compounds which do not have any functional group. The CSPs based on the synthetic polymers are, generally, inert and, therefore, can be used with a variety of mobile phases. The development of CSPs based on the molecularly imprinted technique has resulted in various successful chiral resolutions. The importance and application of these imprinted CSPs lies in the fact that the chiral resolution can be predicted on these CSPs and, hence, the experimental conditions can be designed easily without greater efforts. Because of the ease of preparation and the inexpensive nature of these CSPs, they may be useful and effective CSPs for chiral resolution. Briefly, the future of these types of CSP, especially synthetic polymers and polymers prepared by the molecularly imprinted technique, is very bright and will increase in importance in the near future.

REFERENCES

1. Lämmerhofer M, Lindner W, J Chromatogr A 741: 33 (1996).
2. Piette V, Lämmerhofer M, Bischoff K, Lindner W, Chirality 9: 157 (1997).
3. Maier NM, Nicoletti L, Lämmerhofer M, Lindner W, Chirality 11: 522 (1999).
4. Lämmerhofer M, Maier NM, Lindner W, Am Lab 30: 71 (1998).
5. Mandl A, Nicoletti L, Lämmerhofer M, Lindner W, J Chromatogr A 858: 1 (1999).
6. Tobler E, Lämmerhofer M, Mancini G, Lindner W, Chirality 13: 641 (2001).
7. Franco P, Lämmerhofer M, Klaus PM, Lindner W, J Chromatogr A 869: 111 (2000).

8. Lämmerhofer M, Lindner W, Recent developments in liquid chromatographic enantioseparation, in Separation Methods in Drug Synthesis and Purification, Valko K (Ed.), Elsevier, Amsterdam, Vol. 1, p. 337 (2000).
9. Messina A, Girelli AM, Flieger M, Sinibaldi M, Sedmera P, Cvak L, Anal Chem 68: 1191 (1996).
10. Dondi M, Flieger M, Olsovska J, Polcaro CM, Sinibaldi M, J Chromatogr A 859: 133 (1999).
11. Olsovska J, Flieger M, Bachachi F, Messina A, Sinibaldi M, Chirality 11: 291 (1999).
12. Flieger M, Sinibaldi M, Cvak L, Castellani L, Chirality 6: 549 (1994).
13. Sinibaldi M, Flieger M, Cvak L, Messina A, Pichini A, J Chromatogr A 666: 471 (1994).
14. Padiglioni P, Polcaro CM, Marchese S, Sinibaldi M, Flieger M, J Chromatogr A 756: 119 (1996).
15. Blaschke G, Mempel G, Müller WE, Chirality 5: 419 (1993).
16. Dobashi A, Oka K, Hara S, J Am Chem Soc 102: 7122 (1980).
17. Dobashi Y, Hara S, J Org Chem 52: 2490 (1987).
18. Dobashi A, Dobashi Y, Kinoshita K, Hara S, Anal Chem 60: 1987 (1988).
19. Brain K, Rao KRN, Lloyd MJB, Novel chiral stationary phases, in Chiral Separations by HPLC, Krstulovic AM (Ed.), Ellis Horwood, Chichester (1989).
20. Allenmark SG, Andersson S, Möller P, Sanchez D, Chirality 7: 248 (1995).
21. Okabe H, Itabashi Y, Ota T, Kuksis A, J Chromatogr A 829: 81 (1998).
22. Abou-Basha LI, Aboul-Enein HY, Biomed Chromatogr 10: 69 (1999).
23. Bisel P, Schlauch M, Weckert E, Sin KS, Frahm AW, Chirality 13: 89 (2001).
24. Gasparrini F, Misiti D, Villani C, Chirality 4: 447 (1992).
25. Gasparrini F, D'Acquarica I, Villani C, Cimarelli C, Palmieri G, Biomed Chromatogr 11: 317 (1997).
26. Vaton-Chanvrier L, Peulon V, Combert JC, Chromatographia 46: 613 (1997).
27. Vaton-Chanvrier L, Combert Y, Combert JC, Chromatographia 54: 31 (2001).
28. Iuliano A, Salvadori P, Felix G, Tetrahedron: Asymm 10: 3353 (1999).
29. Gross B, Schurig V, Lamparth I, Hirsch A, J Chromatogr A 791: 65 (1997).
30. Alamo AD, Sally S, Tiritan ME, Matlin S, Chirality 11: 416 (1999).
31. Akelah A, Sherrington DC, Chem Rev 81: 557 (1981).
32. Farina M, Topics Spectrochem 17: 1 (1987).
33. Okamoto Y, Yashima E, Prog Polym Sci 15: 263 (1990).
34. Okamoto Y, Nakano T, Chem Rev 94: 349 (1994).
35. Yashima E, Maeda K, Okamoto Y, Nature 399: 449 (1999).
36. Allenmark SG, Chromatographic Enantioseparations. Methods and Applications, Ellis Horwood, Chichester (1988).
37. Nakano T, J Chromatogr A 906: 205 (2001).
38. Okamoto Y, CHEMTECH 177 (1987).
39. Okamoto Y, Hatada K, J Liq Chromatogr 9: 369 (1986).
40. Okamoto Y, Honda S, Okamoto I, Yuki H, Murata S, Noyori R, Takaya H, J Am Chem Soc 103: 6971 (1981).

41. Schmid R, Antoulas S, Rüttimann A, Schmidt M, Vecchi M, Helv Chim Acta 73: 1276 (1990).
42. Sedo J, Ventosa N, Ruiz-Molina D, Mas M, Molins E, Rovira C, Veciana J, Angew Chem Int Ed Engl 37: 330 (1998).
43. van Es JJGS, Biemans HAM, Meijer WE, Tetrahedron: Asymm 8: 1825 (1997).
44. Harada N, Saito A, Koumura N, Roe DC, Jager WF, Zijlstra RWJ, de Lange BL, J Am Chem Soc 119: 7249 (1997).
45. Reetz MT, Merk C, Naberfeld G, Rudolph J, Grebenow N, Goddard R, Tetrahedron Lett 38: 5273 (1997).
46. Irurre J, Santamari J, Gonzalez-Rego MC, Chirality 7: 154 (1995).
47. Okamoto Y, Mohri H, Nakamura M, Hatada K, J Chem Soc Jpn 435 (1987).
48. Araki A, Inada K, Shinkai S, Angew Chem Int Ed Engl 35: 72 (1996).
49. Okamoto Y, Yashima E, Ishikura M, Hatada K, Polym J 19: 1183 (1987).
50. Nakano T, Taniguchi K, Okamoto Y, Polym J 29: 540 (1989).
51. Ren C, Chen C, Xi F, Nakano T, Okamoto Y, J Polym Sci 31: 2721 (1993).
52. Mohri H, Okamoto Y, Hatada K, Polym J 21: 719 (1989).
53. Tamai Y, Qian P, Matsunaga K, Miyano S, Bull Chem Soc Jpn 65: 817 (1992).
54. Blaschke G, Angew Chem Int Ed Engl 19: 13 (1980).
55. Blaschke G, Bröeker W, Frankel W, Angew Chem Int Ed Engl 25: 830 (1986).
56. Blaschke G, J Liq Chromatogr 9: 341 (1986).
57. Hasegawa S, Sakuri Y, 52nd National Meeting of the Chemical Society of Japan, Abstract 2J36 (1986).
58. Yoshizako K, Hosoya K, Kimata K, Araki T, Tanaka N, J Polym Sci 35: 2747 (1997).
59. Pino P, Ciardelli F, Lorenzi GP, Natta G, J Am Chem Soc 82: 1487 (1962).
60. Kunieda N, Chakihara H, Kinoshita M, Chem Lett 317 (1990).
61. Kakuchi T, Takaoka T, Yokota K, Polym J 22: 199 (1990).
62. Ute K, Hirose K, Hatada K, Vogl O, Polym Prepr Jpn 41: 41 (1992).
63. Yamagishi A, Tanaka I, Taguchi M, Takahashi M, Chem Commun 1113 (1994).
64. Yashima E, Huang S, Okamoto Y, Chem Commun 1811 (1994).
65. Yashima E, Matsushima T, Nimura T, Okamoto Y, Korea Polym J 4: 139 (1996).
66. Umeda S, Satoh T, Satoh K, Yokota K, Kakuchi T, J Polym Sci 36: 901 (1998).
67. Allenmark SG, Andersson S, J Chromatogr A 666: 167 (1994).
68. Hirayama C, Ihara H, Tanaka K, J Chromatogr 450: 271 (1988).
69. Doi Y, Kiniwa H, Nishikaji T, Ogata N, J Chromatogr 396: 395 (1987).
70. Okamoto Y, Nagamura Y, Fukumoto T, Hatada K, Polym J 23: 1197 (1991).
71. Saigo K, Prog Polym Sci 17: 35 (1992).
72. Saigo K, Nakamura N, Adegawa Y, Noguchi S, Hasegawa M, Chem Lett 337 (1988).
73. Saigo K, Shiwaku T, Hayashi K, Fujioka K, Sukegawa M, Chen Y, Yonezawa N, Hasegawa M, Hashimoto T, Macromolecules 23: 2830 (1990).
74. Tamai Y, Matsuzaka Y, Oi S, Miyano S, Bull Chem Soc Jpn 64: 2260 (1991).
75. Sinibaldi M, Castellani L, Federici F, Messina A, Girelli AM, Lentini A, Tesarova E, J Liq Chromatogr 18: 3187 (1995).
76. Kobayashi T, Kakimoto M, Imai Y, Polym J 25: 969 (1993).

77. Chen Y, Lin JJ, J Polym Sci 30: 2699 (1992).
78. Chen Y, Tseng HH, J Polym Sci 31: 1719 (1988).
79. Wulf G, Angew Chem Int Ed Engl 28: 21 (1989).
80. Wulf G, Angew Chem Int Ed Engl 34: 1812 (1995).
81. Remcho VT, Tam ZJ, Anal Chem 72: 248A (1999).
82. Kriz D, Ramstrom O, Mosbach K, Anal Chem 69: 345A (1997).
83. Wulf G, Kemmer R, Vietmeier J, Poll HG, Nouv J Chim 6: 681 (1982).
84. Bartsch RA, Maeda M, (Eds.), Molecular and Ionic Recognition with Imprinted Polymers, ACS Symposium Series No. 703, Oxford University Press, Washington DC (1998).
85. Sellergren B, J Chromatogr A 906: 227 (2001).
86. O'Brien TP, Snow NH, Grinberg N, Crocker L, J Liq Chromatogr Relat Technol 22: 183 (1999).
87. Cowie JMG, Polymers—Chemistry and Physics of Modern Materials, Blackie, Glasgow (1991).
88. Kempe M, Mosbach K, J Chromatogr A 694: 3 (1995).
89. Sellergren B, Shea KJ, J Chromatogr A 690: 29 (1995).
90. Kempe M, Anal Chem 68: 1948 (1996).
91. Lin JM, Nakagama T, Uchiyama K, Hobo T, Biomed Chromatogr 11: 298 (1997).
92. Macaudiere P, Caude M, Rosset R, Tambute A, J Chromatogr Sci 27: 583 (1989).
93. Wilkins SM, Taylor DR, Smith RJ, J Chromatogr A 697: 587 (1995).
94. Schleimer M, Fluck M, Schurig V, Anal Chem 66: 2893 (1994).
95. Sinibaldi M, Vinci M, Federici F, Flieger M, Biomed Chromatogr 11: 307 (1997).
96. Krause K, Chankvetadze B, Okamoto Y, Blaschke G, J Microcol Sep 12: 398 (2000).
97. Peters EC, Lewandowski K, Petro M, Svec F, Frechet MJ, Anal Commun 3: 83 (1998).
98. Lin JM, Nakagama T, Wu XZ, Uchiyama K, Hobo T, Fresenius J Anal Chem 357: 130 (1997).
99. Lin JM, Uchiyama K, Hobo T, Chromatographia 47: 625 (1998).
100. Shibukawa A, Bunseki 131 (1998).
101. Lämmerhofer M, Lindner W, J Chromatogr A 829: 115 (1998).
102. Lämmerhofer M, Tobler E, Lindner W, J Chromatogr A 887: 421 (2000).
103. Krause K, Girod M, Chankvetadze B, Blaschke G, J Chromatogr A 837: 51 (1999).
104. Krause K, Chankvetadze B, Okamoto Y, Blaschke G, Electrophoresis 20: 2772 (1999).
105. Pesek JJ, Matyska MT, Menezes SJ Chromatogr A 853: 151 (1999).
106. Schweitz L, Andersson LI, Nilsson S, Anal Chem 69: 1179 (1997).
107. Schweitz L, Spegel P, Nilsson S, Analyst 125: 1899 (2000).
108. Brüggemann O, Freitag R, Whitcombe MJ, Vulfson EN, J Chromatogr A 781: 43 (1997).
109. Aboul-Enein HY, El-Awady MI, Heard CM, Nicholls PJ, Biomed Chromatogr 13: 531 (1999).
110. Bhushan R, Martens J, Biomed Chromatogr 15: 155 (2001).
111. Bhushan R, Ali I, Chromatographia 23: 141 (1987).

112. Bhushan R, Martens J, Thiong'o GT, J Pharm Biomed Anal 21: 1143 (2000).
113. Malinowska I, Rozylo JK, Biomed Chromatogr 11: 272 (1997).
114. Bhushan R, Ali I, J Chromatogr 392: 460 (1987).
115. Bhushan R, Martens J, Walbaum S, Joshi S, Parshad V, Biomed Chromatogr 11: 286 (1997).
116. Sinibaldi M, Messina A, Girelli AM, Analyst 113: 1245 (1988).
117. Weinstein S, Tetrahedron Lett 25: 985 (1984).
118. Bhushan R, Ali I, Chromatographia 35: 679 (1993).
119. Bhushan R, Ali I, HPLC-93, 17th International Symposium of Liquid Chromatography (1993).
120. Cecchi L, Malaspina P, Anal Biochem 192: 219 (1991).
121. Bhushan R, Thiong'o GT, Biomed Chromatogr 13: 276 (1999)
122. Mosbach K, Trends Biochem Sci 19: 9 (1994).
123. Kriz D, Berggen C, Andersson LI, Mosbach K, Anal Chem 66: 2636 (1994).

10

Enantiomeric Separation by the Chiral Mobile Phase Additives

There are two approaches for the direct chiral resolution by liquid chromatography: on chiral stationary phases, discussed in Chapters 2–9 and using the chiral mobile phase additives (CMPAs). In the latter type of approach, the chiral resolution is carried out by mixing a suitable chiral selector into the mobile phase. The CMPAs, which vary greatly, offer the advantage of using less expensive conventional achiral columns. As compared to the chiral columns, the achiral conventional columns are more rugged, and more efficient and have higher capacities. Generally, the chiral resolution using CMPAs is carried out on reversed-phase columns and, therefore, a variety of mobile phases including acids, bases, and organic solvents can be used without any problem. The chiral selectors are often identical to those used for covalently bonded CSPs. The most commonly used chiral selectors of the CMPAs are cyclodextrins, ligand exchangers, proteins, macrocyclic antibiotics, and other miscellaneous types of chiral molecule. The best CMPAs are those having good solubility in the mobile phase and low ultraviolet (UV) absorbance. Moreover, the chiral selectors are selected arbitrarily, which can form diastereoisomers with the racemic compounds to be resolved. Briefly, in spite of the development of suitable and successful CSPs, few reports on the chiral resolution by this mode of chromatography exist. In view of this, the present chapter describes the art of chiral resolution using CMPAs in various models of liquid chromatography.

10.1 CYCLODEXTRIN MOBILE PHASE ADDITIVES

Cyclodextrins (CDs) are cyclic and nonreducing oligosaccharides and obtained from starch. The structures and properties of these molecules were discussed in detail in Chapter 3. These molecules are soluble in aqueous mobile phases and, hence, most of the chiral resolution was carried out under the reversed-phase mode. Therefore, cyclodextrins were used frequently as CMPAs for the chiral resolution of a wide variety of racemic compounds. The nontoxicity, nonvolatile, poor UV absorbance, stability over a wide range of pHs, and inexpensive natures of cyclodextrins make them superb CMPAs.

The first enantiomeric separation of mandelic acid using β-cyclodextrin as CMPA was reported by Debowski et al. [1]. Since then, many enantiomers have been resolved using cyclodextrins as CMPAs and a few review articles have been published on this issue [2–6]. Gazdag et al. [7,8] used α-, β-, and γ-cyclodextrins for the chiral resolution of norgestrel, hydrocortisone acetate, triamcinolone acetonide, prednislone, and hydrocortisone racemic compounds. The mobile phase used was water–methanol with LiChrosorb RP-18, Nucleosil 5 C_{18}, Hypersil ODS, and Ultrasphere ODS columns. Rizzi and Plank [9] described the chiral resolution of alanine, glutamine, threonine, valine, leucine, phenylalanine, ethylphenylhydantoin, hexabarbitals, oxazepam, and nomifensine using β-cyclodextrin as the CMPA. Reepmeyer [10] reported the enantiomeric resolution of thalidomide on Hypersil C_{18}, Altima C_{18} and Novapak C_{18} columns using β-cyclodextrin as the CMPA. Ameyibor and Stewart [11] reported the chiral resolution of some β-blockers and profens using β-cyclodextrin as the CMPA. Similarly, β-cyclodextrin was used as the mobile phase additive for the chiral resolution of a variety of racemic compounds [12–18]. The application of α-, β-, and γ-cyclodextrins as CMPAs for the chiral resolution of different compounds is summarized in Table 1. To show the nature of the chiral resolution by this model of chromatography, the chromatograms of the chiral resolution of thalidomide on Prodigy C_{18} column using β-cyclodextrin as the mobile phase additive are shown in Figure 1.

The chiral resolution using cyclodextrins as mobile phase additives is also controlled by a number of chromatographic parameters as in case of CSPs. The chiral resolution occurred by the formation of diastereoisomeric inclusion complex formation and, hence, the composition of the mobile phase, pH, concentration of cyclodextrins, and temperature are the most important controlling parameters.

The chiral resolution is optimized by adjusting the concentrations of buffers and organic modifiers [7–11]. Ameyibor and Stewart [11] studied the chiral resolution of some drugs using different ratios of 0.1% trifluoroacetic acid and acetonitrile. The results of this finding are summarized in Table 2. It is clear from this table that the 98 : 2 ratio of 0.1% trifluoroacetic acid and acetonitrile is the

TABLE 1 Chiral Resolution of Some Racemic Compounds Using Cyclodextrins as the Mobile Phase Additives

Racemic compound	Cyclodextrins	Ref.
Aminomethyl benzodiaxane derivative	Carboxymethyl β-CD	19
Barbituric acid derivatives	β-, DM-β-, TM-β-CDs	20, 21
Benzoin	β-, TM-β-CDs	22–24
1,1-Binaphthyl-2-2′-diyl hydrogen phosphate	β-, TM-β-CDs	25
Brompheniramine	β-CD	24
Budesonide	β-, γ-CDs	8, 26, 27
5-Butyl-1-methyl-5-phenylbarbituric acid	TM-β-CD	23
Camphene	α-CD	28
Chlorpheniramine hydrogen maleate chlorthalidone	Methylated β-CD	29
Chlorthalidone	β-CD	9, 30, 31
Cyclobarbital	β-, DM-β-CDs	32
Dansyl amino acids	β-, TM-β-, γ-CDs	9, 25, 33–35
Ethotoin	β-CD	36
Ephedrine	CE-β-CD	19
Ethyl mandelate	DM-β-, TM-β-, AC-β-CDs	22, 23, 32
1-Ferrocenyl-ethanol	β-, TM-β-CDs	25, 37
Flumecinol	β-, γ-CDs	7
Glutethimide	β-, TM-β-CDs	22, 38
Glycyl-DL-phenylalanine	β-CD	29
Heptobarbital	β-, DM-β-CDs	32
Hexobarbital	CE-β-, CM-β-, DM-β-, TM-β-CDs	9, 19, 22, 23, 36
Hydantoins	β-, DM-β-CDs	15, 21
5-(*m*-Hydroxy phenyl)-5-phenylhydantoin	β-CD	36
5-(*p*-Hydroxy phenyl)-5-phenylhydantoin	β-CD	36, 39
Isosalsoline	β-CD	40
Laudanosoline	β-CD	40
Lorazepam	β-CD	41
Mandelic acid	β-CD	1, 32
Mephenytoin	β-, TM-β-CDs	38
Methyl mandelate	β-, AC-β-, DM-β-, TM-β-CDs	22, 23, 32, 42
Methylphenobarbital	β-CD	22, 23, 36, 38
1-Methyl-5-phenyl-propylbarbital	TM-β-CD	22

(*continued*)

TABLE 1 Continued.

Racemic compound	Cyclodextrins	Ref.
N-Methylsalsolinol	β-CD	43
Morsuxamide	β-, TM-β-CDs	38
Nomifensine	β-CD	24
Norgesterl	γ-CD	7, 8, 26
Oxazepam	β-CD	30, 41
α-Pinene	α-CD	18, 28, 44
β-Pinene	α-CD	28
Propranolol	β-CD	29
Pseudoephedrine	β-CD	45
Salsolinol	β-CD	43
trans-Sobrerol	β-CD	46
Terbutaline	β-CD	30, 31
Tetrahydro-papaveroline	β-CD	40
Thalidomide	β-CD	10
Thiamylal	β-CD	36
Trimeprazine	β-CD	24, 47
β-Blocker and profens	β-CD	11
Trimipramine	β-CD	12
Doxazosin	β-, CM-β-CDs	13
Chlorthalidone	β-CD	14
Oxazolidone	α-, β-, γ-CDs	16
Naproxen	Methyl-β-CD	17

Abbreviations: AC, peracetylated; CE, carboxyethyl; CM, carboxymethyl; DM, heptakis(2,6-di-*O*-methyl); TM, heptakis(2,3,6-tri-*O*-methyl).

best combination for the chiral resolution of all drugs studied. Moreover, it can be concluded from Table 2 that the resolution decreased by increasing the amount of acetonitrile. The effect of organic modifiers on the chiral resolution was also studied [7–11]. The most important modifiers used were methanol, ethanol, 2-propanol, acetonitrile, and tetrahydrofuran. A very interesting study on the effect of these organic modifiers was carried out by Gazdag et al. [7] and the results are shown in Figure 2. It may be concluded from this figure that the resolution increased by increasing the amount of these organic modifiers. The same authors [8] also studied the effect of the ionic strength of sodium perchlorate on the chiral resolution of norgestrel, budesonide, and flumecinol (Fig. 3). Figure 3 indicates the small effect of the ionic strength on the chiral resolution.

The concentration of cyclodextrins is also very important for optimizing the chiral resolution. Some reports have been published on this issue [7–14,45]. Mularz et al. [45] studied the effect of the concentration of β-cyclodextrin on the chiral resolution of ephedrine and pseudoephedrine. The authors reported

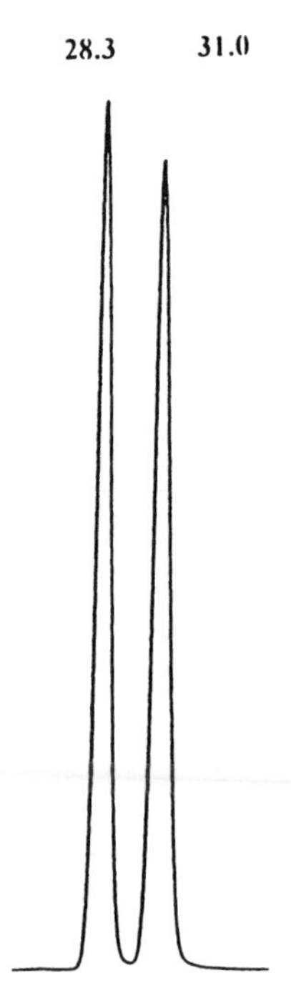

FIGURE 1 Chromatograms of the chiral resolution of thalidomide on the Prodigy C_{18} column using phosphate buffer (10 mM, pH 4.0)–ethanol (95 : 5, v/v) as the mobile phase containing β-cyclodextrin (20 mM) as the CMPA. (From Ref. 10.)

an increase in enantioselectivity by increasing the concentration of β-cyclodextrin. Similarly, Reepmeyer [10] studied the effect of the concentration of β-cyclodextrin on the chiral resolution of thalidomide enantiomers and the author reported an increase in chiral resolution by increasing the amount of cyclodextrin. In one of the studies, Gazdag et al. [7] carried out the chiral resolution of norgestrel, hydrocortisone, hydrocortisone acetate, triamcinolone acetonide, and prednislone. The authors reported the decrease in the chiral resolution of these drugs by increasing the concentration of cyclodextrin. The effect of the concentration of the cyclodextrin is shown in Figures 4a and 4b.

The effect of pH on the chiral resolution in this type of chromatography is also studied [8,13,14]. The different trends of pH on the chiral resolution were observed. For example, the effect of pH on the chiral resolution of oxazosin [13] and chlorthalidone [14] is shown in Figures 5a and 5b. It is clear from these figures that the chiral resolution first increases and then decreases.

TABLE 2 Chiral Resolution of Some Drugs Using Different Ratios of 0.1% Trifluoroacetic Acid (TFA) and Acetonitrile (ACN) Containing β-Cyclodextrin as the CMPA on an Octadecylsilane Column

	Mobile phase (v/v)			
Racemic compound	TFA	ACN	k	R_s
Lorazepam	98	2	6.04	1.05
	96	4	6.60	1.05
	94	6	7.37	1.02
	92	8	6.54	1.00
	90	10	7.13	0.90
Temazepam	98	2	8.58	1.34
	96	4	8.25	1.25
	94	6	7.23	1.14
	92	8	7.42	1.10
	90	10	7.37	1.00
Oxazepam	98	2	3.79	1.97
	96	4	4.66	1.90
	94	6	4.19	1.74
	92	8	4.60	1.74
	90	10	4.11	1.65
Chlorthalidone	98	2	5.79	1.78
	96	4	nr	nr
	94	6	nr	nr
	92	8	nr	nr
	90	10	nr	nr

Note: nr = not resolved.
Source: Ref. 11.

10.2 LIGAND-EXCHANGER MOBILE PHASE ADDITIVES

The ligand exchangers were also used as mobile phase additives for the chiral resolution in liquid chromatography [48,49]. For the first time, LePage et al. [50] used L-2-alkyl-4-octyldiethylenetriamine with Zn(II) as the mobile phase additive for the chiral resolution of dansyl amino acids on the octylsilyl (C_8)-bonded phase. Furthermore, the same group [51] extended their work for the chiral resolution of amino acids and amine using different chiral mobile phase additives such as L-propyl-*N*-octylamine with Ni(II). Later, Gil-Av et al. [52,53] used ligand exchangers [D- or L-proline-Cu(II) complex] as the CMPAs for the enantiomeric separation of amino acids. Since then, the chiral resolution of a

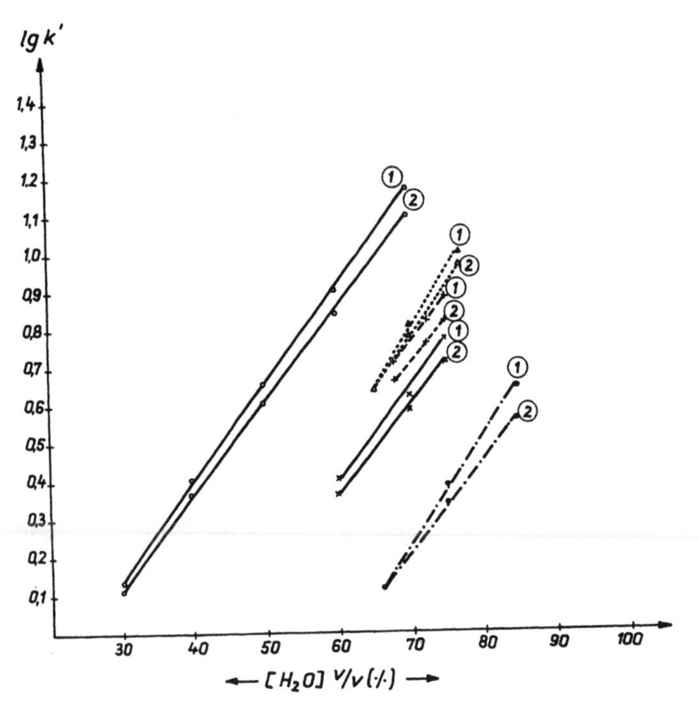

FIGURE 2 Effect of the organic solvents on the enantioselectivity of norgestrel [(1) L- and (2) D-enantiomers]: (○) methanol, (χ) ethanol, (●) 2-propanol, (+) acetonitrile, and (△) tetrahydrofuran. (From Ref. 7.)

variety of racemic compounds was carried out using this approach [54–58]. The application of ligand exchangers as the mobile phase additives for the chiral resolution is summarized in Table 3. The typical chromatograms of the chiral resolution of dansyl amino acids using the Cu(II) complex of (*S*,*S*)-*N*,*N*-2-hydroxypropyl-phenylalaninamide as the mobile phase additive are shown in Figure 6.

The effect of various chromatographic parameters, as in the case of cyclodextrin mobile phase additives, on the chiral resolution of racemic compounds using ligand exchangers as CMPAs has also been studied. Galaverna et al. [55] studied the effect of the pH of the mobile phase and the concentration of mobile phase additives (copper complexes of amino acid amides). These findings are shown in Figure 7 (pHs) and Figure 8 (concentration of CMPAs). It can be concluded from these figures that the chiral resolution is affected by changing the pH and the concentration of the mobile phase additive. Similarly, Galaverna et al. [57,58] studied the effect of pH, concentration of CMPA, ionic strength of the mobile phase, and the eluent polarity on the chiral resolution of

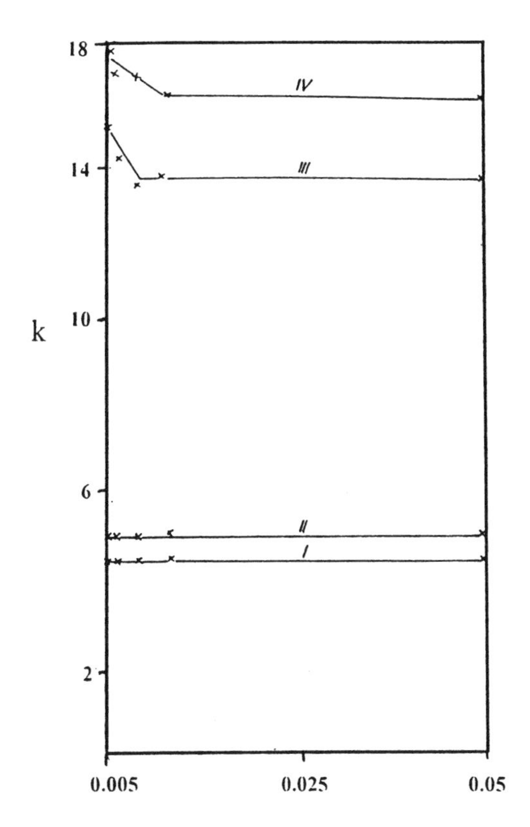

FIGURE 3 Effect of sodium perchlorate on the chiral resolution of norgestrel [(I) D- and (II) L-enantiomers] and budenoside [(III) *R*- and (IV) *S*-enantiomers] on Hypersil ODS column using methanol–water (1 : 1, v/v) as the mobile phase containing γ-cyclodextrin (10 mM) as the CMPA. (From Ref. 8.)

amino acids. The authors reported different trends of the chiral resolution of various amino acids. In a very interesting study, Marchelli and co-workers [56] studied the effect of the structures of CMPA. The authors studied the effect of *S,S* and *R,S* configurations of *N,N*-2-hydroxypropyl-phenylalaninamide (Table 4). The authors reported the best separation with the *S,S* configuration as the CMPA.

10.3 PROTEIN MOBILE PHASE ADDITIVES

Proteins have also been used as CMPAs for chiral resolution by liquid chromatography. Bovine serum albumin (BSA) and α_1-acid glycoproteins (AGP) were investigated as CMPAs. Allenmark et al. [87] used BSA protein as the CMPA for

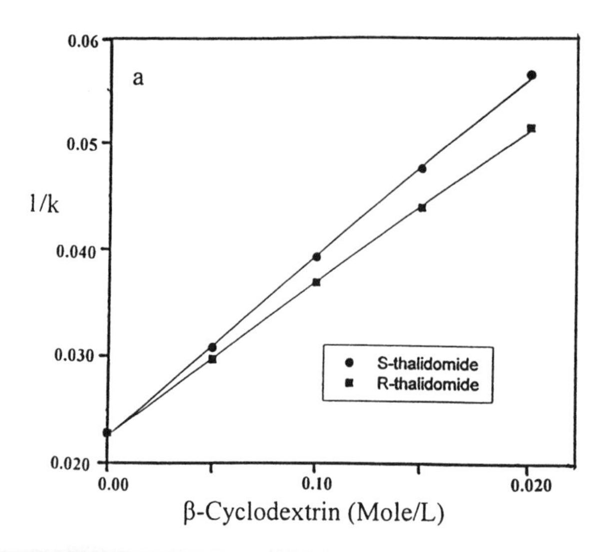

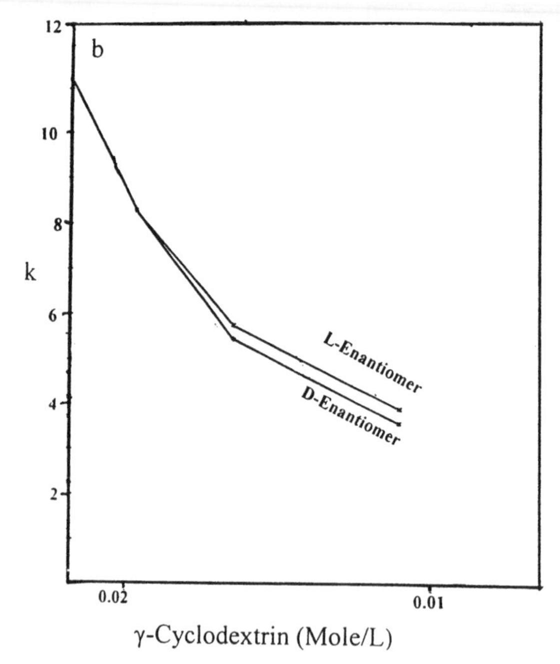

FIGURE 4 Effect of the concentrations of cyclodextrins CMPAs (a) on the chiral resolution of thalidomide on the Prodigy C_{18} column using phosphate buffer (10 mM, pH 4.0)–ethanol (95 : 5, v/v) as the mobile phase (from Ref. 10) and (b) on the chiral resolution of norgestrel on the LiChrosorb RP-18 column using water–methanol (3 : 4, v/v) as the mobile phase (from Ref. 7).

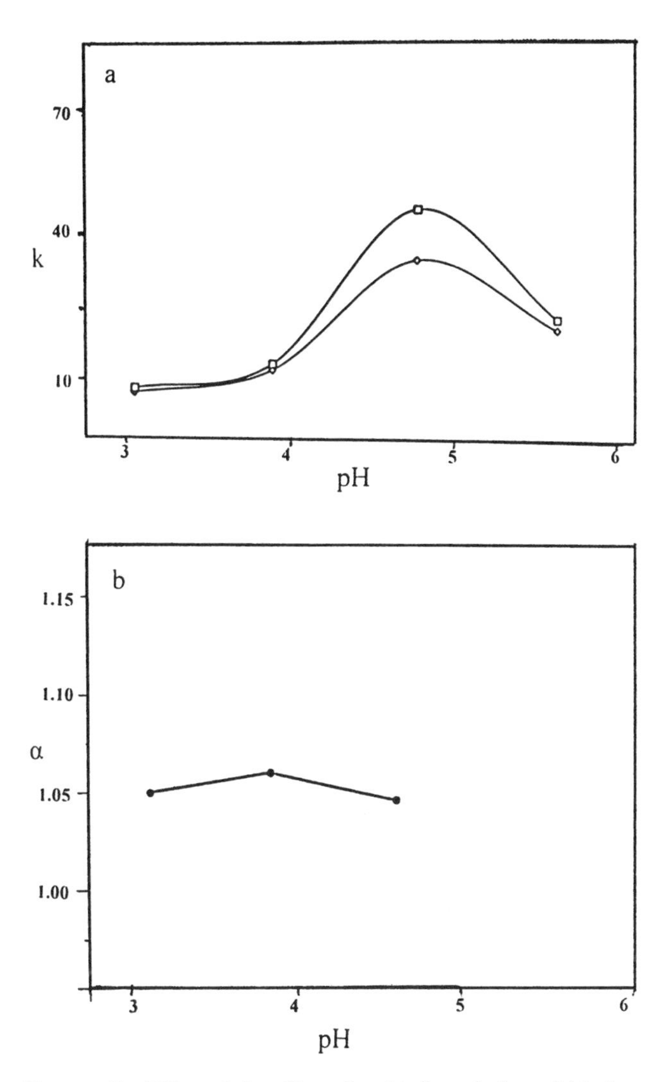

FIGURE 5 Effect of the pH on the chiral resolution of (a) doxazosin on Nova-Pak C_8 column using methanol–20 mM sodium dihydrogen phosphate (15.3 : 50, v/v) as the mobile phase containing carboxymethyl β-cyclodextrin (15 mM) as the CMPA (from Ref. 13), (◇) first and (□) second eluted enantiomers, and (b) of chlorthalidone on the LiChrosphere C_{18} column using methanol–0.1 M sodium phosphate buffer (25 : 75, v/v) as the mobile phase containing β-cyclodextrin (10 mM) as the CMPA and triethylamine (10 mM) as the mobile phase modifier (from Ref. 14).

TABLE 3 Chiral Resolution of Some Racemic Compounds Using Ligand Exchangers as the Mobile Phase Additives

Racemic compound	Ligand exchangers	Ref.
Amino acids	Cu(II)–L- or D-proline	52, 53
	Cu(II)–L-phenylalanine	59
	Cu(II)–L-aspartyl-L-phenylalanine methyl ester	60
	Cu(II)–L-aspartylcyclohexyl amide	61–63
	Cu(II)–*N*-(*p*-toluenesulphonyl)-L-phenylalanine	64, 65
	Cu(II)–*N*-(*p*-toluenesulphonyl)-D-phenylglycine	66, 67
	Cu(II)–*N*,*N*-dialkyl-L-amino acids	68
	Cu(II)-Di-*n*-propyl-L-alanine	69–71
	Cu(II)–*N*,*N*,*N*,*N*-tetramethyl-(*R*)-propane1,2-diamine	72
	Cu(II)-*N*-methyl-L-phenylalanine, *N*,*N*-dimethyl-L-phenylalanine	73
	Cu(II)–(*R*,*R*)-tartaric acid mono-*n*-octylamine	74
	Cu(II)–amino acid amides	55
	Cu(II)–terdentate ligands	57
	Cu(II)–(*S*,*S*)-*N*,*N*-bis(phenylalanyl)-ethanediamine, (*S*,*S*)-*N*,*N*-bis(methylphenylalanyl)-ethanediamine	58
Dansyl amino acids	Zn(II), Cd(II), Ni(I), Cu(II), Hg(II) complexes of L-2-alkyl-4-*n*-octyldiethylene triamine	51, 75
	Ni(II)–L-propyl-*n*-octylamine, L-propyl-*n*-dodecylamide	75, 76
	Cu(II)–L-proline	77–79
	Cu(II)–L-arginine	78
	Cu(II)–L-histidine, L-histidine methyl ester	78, 80–82
	Cu(II)–*n*-(*R*)- or *N*-(*S*)-2-hydroxypropyl(*S*)-phenylalaninamide	56
	Cu(II)–terdentate ligands	57
Hydroxy acids	Cu(II)–L-phenylalanine	83, 84
	Cu(II)–*N*-methyl-L-phenylalanine *N*,*N*-dimethyl-L-phenylalanine	73
	Cu(II)–(*S*)-phenylalaninamide	85
	Cu(II)–L-proline	84
Amino alcohols	Cu(II)–(*R*,*R*)-tartaric acid mono-*n*-octylamide	74
	Cu(II)–9-(3,4-dihydroxybutyl)guanine	86

the chiral resolution of acidic drugs. Hermansson resolved the enantiomers of disopyramide using AGP as the CMPA [88]. Hedeland et al. [89] used cellobiohydrolase I as CMPA for the chiral resolution of propranolol enantiomers. The high absorbance of the proteins in the UV region makes them unpopular and, therefore, the proteins could not be used frequently as the CMPAs.

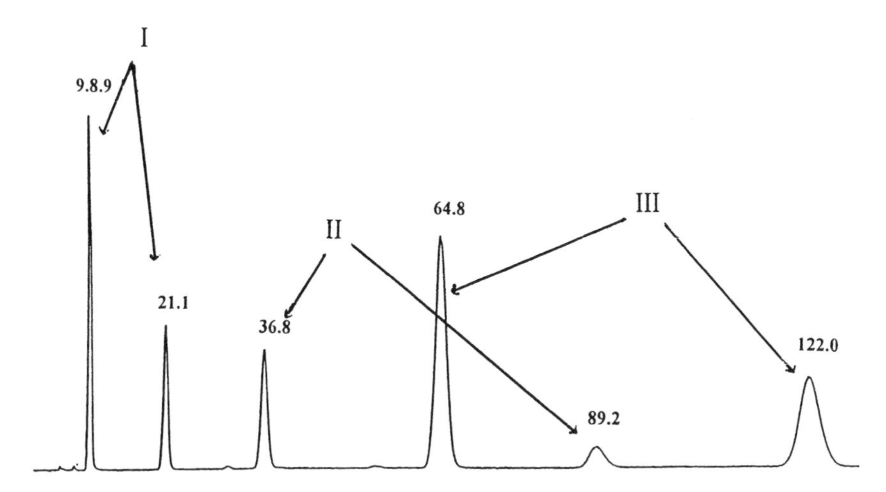

FIGURE 6 Chromatograms of the chiral resolution of dansyl amino acids on the Radialpak C_{18} column using water–acetonitrile (80 : 20, v/v) as the mobile phase containing the Cu(II) complex of (*S*,*S*)-2-hydroxypropyl-(*S*)-phenylalaninamide as the CMPA. I: DNS–Glu; II: DNS–α-Nbu; III: DNS–Ser. (From Ref. 56.)

10.4 MACROCYCLIC GLYCOPEPTIDE ANTIBIOTIC MOBILE PHASE ADDITIVES

Although the macrocyclic glycopeptide antibiotic CSPs are very effective for the chiral resolution of many racemic compounds, their use as chiral mobile phase additives is very limited. Only a few reports are available on this mode of chiral resolution. It is interesting to note that these antibiotics absorb UV radiation; therefore, the use of these antibiotics as the CMPAs is restricted. However, Armstrong et al. used vancomycin as the CMPA for the chiral resolution of amino acids by thin-layer chromatography, which will be discussed in Section 10.7.

10.5 MISCELLANEOUS CHIRAL MOBILE PHASE ADDITIVES

Other chiral molecules were also used as CMPAs. Lämmerhofer and Lindner [90] resolved the enantiomers of N-derivatized amino acids (e.g., 3,5-dinitrobenzoyl,

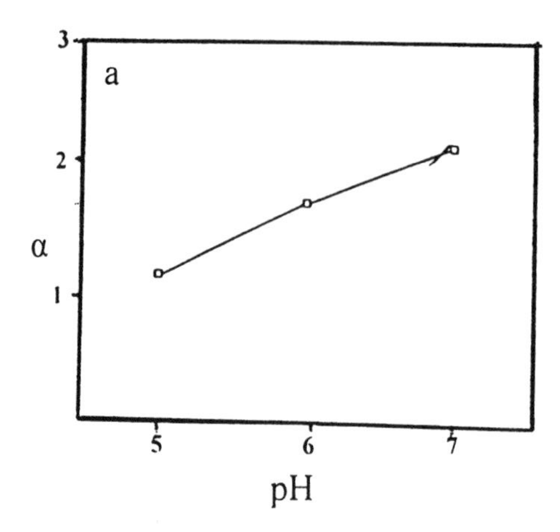

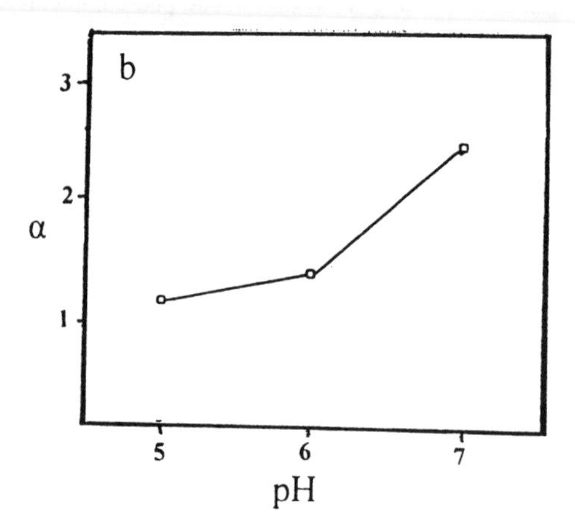

FIGURE 7 Effect of pH on the chiral resolution of α-aminobutyric acid on the Spherisorb ODS-2 column using water as the mobile phase containing (a) Cu(II)–L-proline (4 mM) and (b) Cu(II)–L-methylphenyl alanine as CMPAs separately and respectively (from Ref. 55).

3,5-dinitrobenzyloxycarbonyl, 2,4-dinitrophenyl, and 9-fluorenylmethoxycarbonyl amino acids) on the Hypersil ODS-3 column in combination with a quinine carbamate-type chiral ion-pair agent which was added to aqueous and non-aqueous buffered mobile phases, respectively.

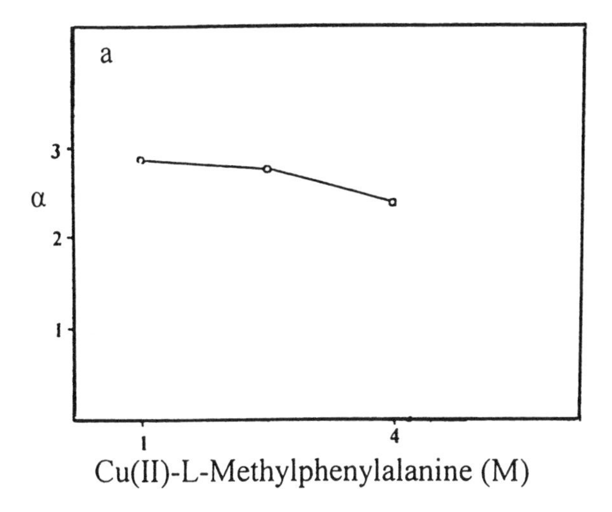

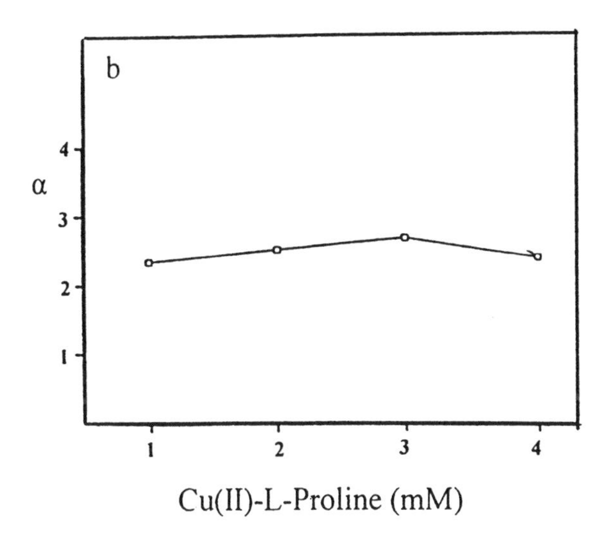

FIGURE 8 Effect of the concentration of (a) Cu(II)–L-methylphenyl alanine and (b) Cu(II)–L-proline on the chiral resolution of α-aminobutyric acid on the Spherisorb ODS-2 column using water as the mobile phase (from Ref. 55).

TABLE 4 Effect of the Structure of *S,S* and *R,S* Configurations of *N,N*-2-hydroxy-propyl-phenylalaninamide CMPA on the Chiral Resolution of Some Dansyl Amino Acids

	S,S Configuration		*R,S* Configuration	
Dansyl amino acid	k	α	k	α
Glutamine	1.28	1.62	0.29	2.41
Aspragine	2.21	1.48	0.32	2.38
Serine	7.28	1.41	1.65	2.10
Threonine	6.93	1.77	1.58	1.97
Valine	14.07	1.37	5.03	0.73
nor-Valaine	15.28	1.31	8.88	0.71
Leucine	20.10	1.82	11.97	0.84
nor-Leucine	25.93	1.41	16.38	0.66
Methionine	29.14	1.43	7.76	0.87
Phenylalanine	32.14	1.09	17.44	0.80

Source: Ref. 56.

10.6 CHIRAL MOBILE PHASE ADDITIVES IN SFC AND CEC

In addition to high-performance liquid chromatography (HPLC), the chiral resolution using CMPAs was also carried out by supercritical fluid chromatography (SFC) [91] and capillary electrochromatography (CEC) [92–98]. Salvador et al. [91] used dimethylated β-cyclodextrin as the mobile phase additive on porous graphite carbon as the solid phase for the chiral resolution of tofizopam, warfarin, a benzoxazine derivative, lorazepam, flurbiprofen, temazepam, chlorthalidone, and methyl phehydantoin by SFC. The authors also studied the effect of the concentration of dimethylated β-cyclodextrin, the concentration of the mobile phase, the nature of polar modifiers, outlet pressure, and the column temperature on the chiral resolution.

Wang and Porter [92] resolved the enantiomers of oxazepam, lorazepam, and temazepam using β-cyclodextrin as the CMPA by CEC. The authors varied separation parameters such as voltage and mobile phase. Wei et al. [93] resolved the enantiomers of phenylephrine and synephrine by varying the concentration of β-cyclodextrin (CMPA), pH, electrolyte concentration, and temperature. Lelievre et al. [99] separated the enantiomers of chlorthalidone using hydroxypropyl β-cyclodextrin as the CMPA. Lämmerhofer and Lindner [90] resolved the enantiomers of N-derivatized amino acids (e.g., 3,5-dinitrobenzoyl, 3,5-dinitro-benzyloxycarbonyl, 2,4-dinitrophenyl, and 9-fluorenylmethoxycarbonyl amino

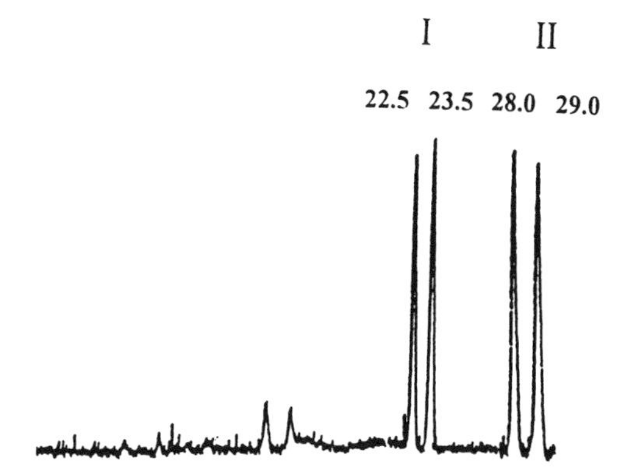

FIGURE 9 Electropherograms of the chiral resolution of (I) 1-cyanobenz[*f*]isoindole (CBI)–selenomethionine and (II) CBI–selenoethionine enantiomers using a mixture of boric acid (10 mM) and phosphate buffer (30 mM, pH 7) as the mobile phase containing β-cyclodextrin (30 mM) and taurodeoxycholic acid (50 mM) as CMPAs with sodium dodecyl sulfate (50 mM) as the mobile phase modifier (from Ref. 97).

acids) by enantioselective ion-pair formation and packed CEC using RP-18 silica particles (Hypersil ODS-3 μm packed capillary column, 335 mm) using a quinine carbamate-type chiral ion-pair agent which was added to aqueous and nonaqueous buffered mobile phases respectively.

Ban et al. [94] used β-cyclodextrin as the CMPA for the chiral resolution of racemorphan enantiomers by micellar electrokinetic chromatography (MEKC). The authors investigated the effect of buffer pH, concentration of mobile phase additives, and the detection wavelength. Similarly, Lucangioli et al. [95] resolved the enantiomers of sertraline hydrochloride and related substances using β-cyclodextrin as the CMPA by MEKC. Garcia-Ruiz et al. [96] used carboxymethylated and permethylated β-cyclodextrins as CMPAs for the chiral resolution of polychlorinated biphenyls by electrokinetic chromatography (EKC). For a reference, the chromatograms of the chiral resolution of seleno amino acid derivatives by MEKC are shown in Figure 9.

10.7 CHIRAL MOBILE PHASE ADDITIVES IN TLC

The approach of CMPAs has also been used in thin-layer chromatography (TLC) for the chiral resolution of a variety of racemic compounds [100–110]. Lepri et al. [104,105] used BSA as a mobile phase additive for the chiral resolution of dansyl amino acids and other drugs by TLC. Armstrong et al. [101,102] used underivatized and hydroxyethyl and hydroxypropyl β-cyclodextrins for the chiral resolution of dansyl amino acids, alkaloids, and other compounds. Aboul-Enein et al. [103] used β-cyclodextrin for the chiral resolution of aminoglutethimide, acetylaminoglutethimide, and dansyl aminoglutethimide. Armstrong and Zhou [106] used vancomycin as CMPA for the chiral resolution of dansyl amino acids, 6-aminoquinolyl-*N*-hydroxysuccinimidyl carbamate-derivatized amino acids, and other racemic drugs on TLC under the reversed-phase mode. To stabilize the TLC plates, sodium chloride was added into the mobile phase. Chemically bonded diphenyl-F reversed-phase plates were used and obtained from Whatman Chemical Separation Division Inc. The mobile phase comprised different ratios of acetonitrile, sodium chloride (0.6 M), and 1% triethyl ammonium acetate buffer (pH 4.1).

The effect of various TLC parameters such as concentration of β-cyclodextrins, the pH of the mobile phase, and the concentration of organic modifiers were also studied by different workers [100–103]. Armstrong et al. [101,102] observed an increase in the enantioselectivity by increasing the amount of β-cyclodextrins (Figs. 10a and 10b). Similar behavior of the chiral resolution of aminoglutethimide, acetylaminoglutethimide, and dansyl aminoglutethimide enantiomers was observed by increasing the concentration of β-cyclodextrins in mobile phase by Aboul-Enein et al. [103] (Fig. 10c). Armstrong et al. [101] studied the effect of the concentration of acetonitrile on the chiral resolution of dansyl serine amino acid (Fig. 11a). Figure 11a indicates that no chiral resolution was observed at lower and higher concentrations of acetonitrile. However, the best resolution occurred using 20–40% acetonitrile. Aboul-Enein et al. [103] reported no marked effect of the concentration of acetonitrile on the chiral resolution of aminoglutethimide. However, the authors reported the best resolution of this compound using 70% methanol as the organic modifier (Fig. 11b). In the same study, the authors [103] also studied the effect of pH on the chiral resolution of aminoglutethimide. The effect of pH on the chiral resolution of aminoglutethimide is presented in Figure 12. It is interesting to observe from this figure that the higher resolution occurred at pH 2.5 and 9.0, whereas the lower resolution was observed at pH 7.0 and 12.0.

10.8 CHIRAL RECOGNITION MECHANISMS

It is a well-known fact that the chiral environment is essential in liquid chromatography for the chiral resolution. In this modality of chromatography, the chiral situation is provided by the CMPAs. Basically, the chiral recognition mechanisms on this type of liquid chromatography are similar to those on CSPs. The differences between the chiral recognition mechanisms on these models (i.e., CMPAs and CSPs) lies in the fact that the diastereoisomeric complexes of the racemic compounds are formed in the mobile phase (CMPAs), whereas these complexes are formed on stationary phase in case of CSPs mode. These

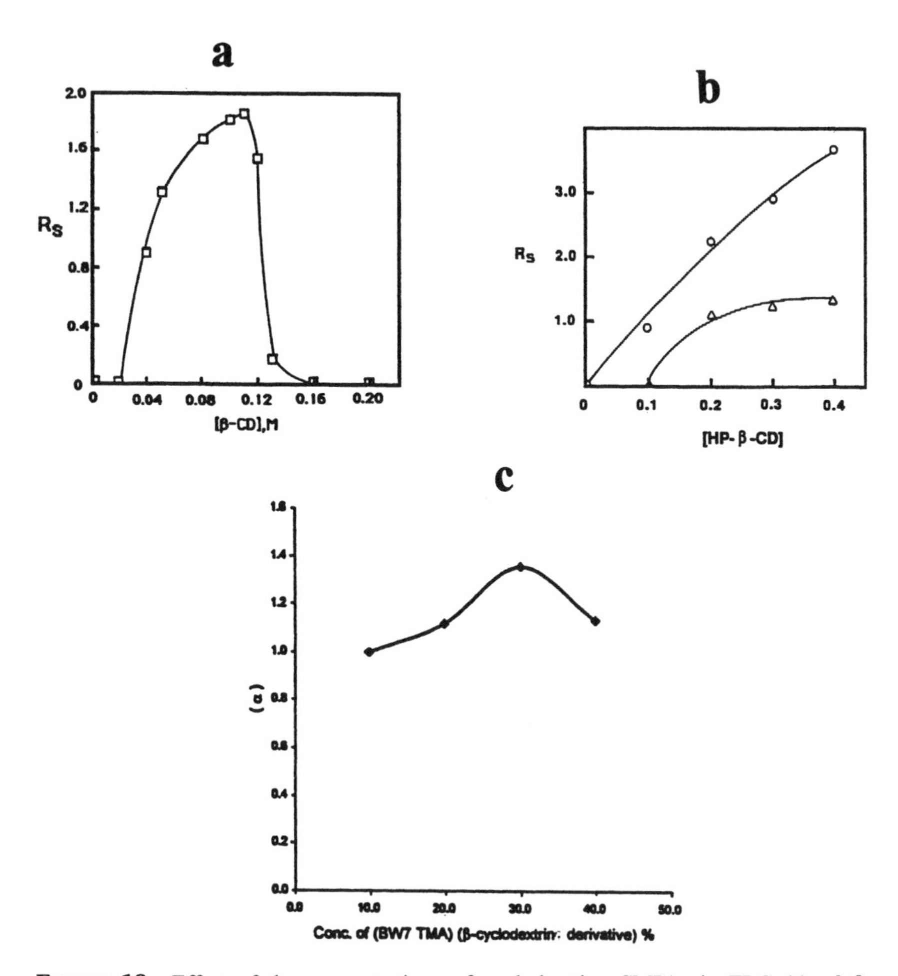

FIGURE 10 Effect of the concentrations of cyclodextrins CMPAs in TLC (a) of β-cyclodextrin on the chiral resolution of dansyl-DL-glutamic acid using acetonitrile–water (30 : 70, v/v, saturated with urea) as the mobile phase, (○) D- and (●) L-enantiomers (from Ref. 101), (b) of hydroxypropyl-β-cyclodextrin on the chiral resolution of dansyl-DL-leucine (circles) and dansyl-DL-valine (triangles) using aqueous triethylamine as the mobile phase (from Ref. 102), and (c) of hydroxy trimethylpropyl ammonium-β-cyclodextrin on the chiral resolution of *RS*-aminoglutethimide using water and acetonitrile mixture as the mobile phase (from Ref. 103).

diastereoisomeric complexes possess different physical and chemical properties and, therefore, are separated on some solid support (achiral column). The formation of diastereoisomeric complexes depends on the type and nature of the CMPAs and analytes. In the case of cyclodextrins, inclusion complexes are

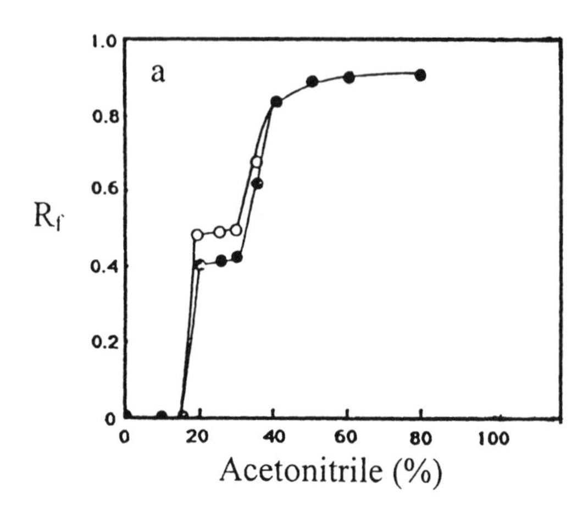

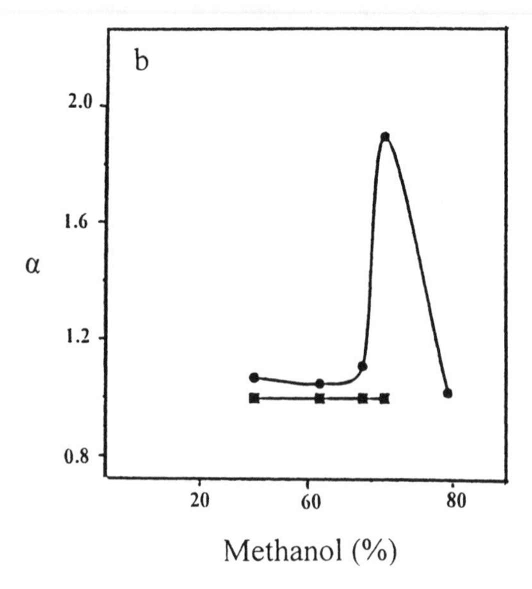

FIGURE 11 Effect of the concentrations of organic modifiers in TLC (a) of acetonitrile on the chiral resolution of dansyl-DL-serine, (○) D- and (●) L-enantiomers (from Ref. 101), and (b) of methanol on the chiral resolution of *RS*-aminoglutethimide (from Ref. 103).

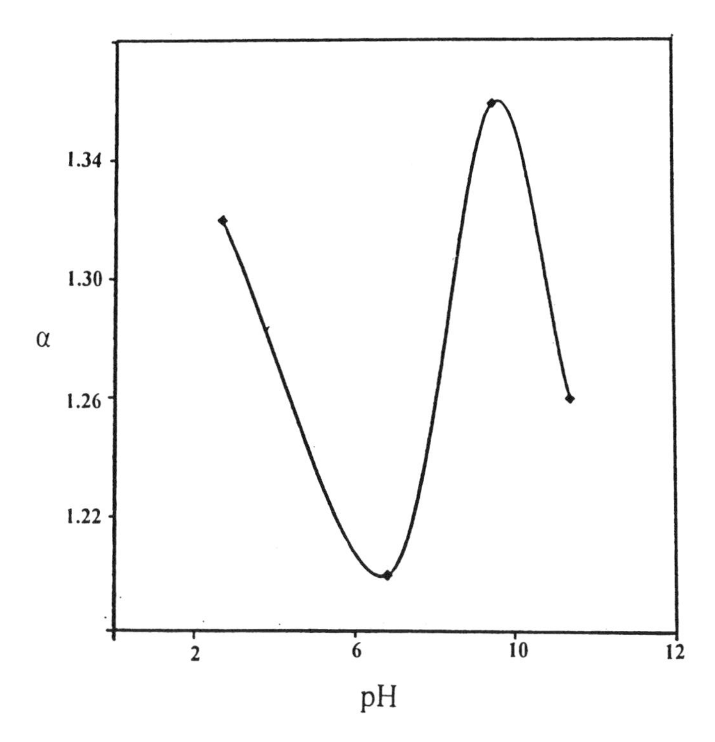

FIGURE 12 Effect of pH on the chiral resolution of *RS*-aminoglutethimide by TLC using water and acetonitrile mixture as the mobile phase containing hydroxy trimethylpropyl ammonium-β-cyclodextrin as the CMPA (from Ref. 103).

formed, whereas in case of other CMPAs, simple chiral diastereoisomeric complexes are formed. Again, as in case of CSPs, the diastereoisomeric complexes formation is controlled by a number of interactions such as π–π complexation, hydrogen-bondings, dipole–dipole interactions, ionic bindings, and steric effects.

Lämmerhofer and Lindner [90] explained the chiral resolution of N-derivatized amino acids by CEC. The authors explained the formation of the transient diastereomeric ion-pairs between negatively charged analyte enantiomers and a positively charged chiral selector by multiple intermolecular interactions which might be differentially adsorbed to the ODS stationary phase. Furthermore, they claimed that the enantioseparation was achieved because of different observed mobilities of the analyte enantiomers originating from different ion-pair formation rates of the enantiomers and/or differential adsorption of the diastereoisomeric ion-pairs to the ODS stationary phase [90].

10.9 CONCLUSION

It is interesting to note that the first report on the chiral resolution was published using the chiral mobile phase additive approach. Later, the CSPs are developed and, because of certain advantages, this approach is common nowadays. Of course, the chiral resolution by this approach is inexpensive, as it involves the use of inexpensive achiral columns and, therefore, it is considered an effective method of the chiral resolution. The main advantage of this method lies in the fact that a wide variety of mobile phases can be used and hence the chiral resolution can be obtained easily. However, this approach is not attractive because of some serious drawbacks and limitations. The serious weakness of this method is the wastage of the costly chiral selectors. Sometimes the CMPAs, especially proteins, absorb in the UV range, creating problems in detection. Moreover, the resolved enantiomers are obtained in the form of diastereoisomeric complexes and cannot be used directly for further pharmacological studies. Although this approach is not attractive yet, it can be used as a mean of an inexpensive method for chiral resolution.

REFERENCES

1. Debowski J, Sybilska D, Jurczak J, J Chromatogr 237: 303 (1982).
2. Sybilska D, Zukowski J, Cyclodextrin additives, in Chiral Separations by HPLC: Applications to Pharmaceutical Compounds, Krstulovic AM (Ed.), Ellis Horwood, New York (1989).
3. Camilleri P, Biomed Chromatogr 5: 128 (1991).
4. Husain N, Warner IM, Am Lab 80 (1993).
5. Han SM, Biomed Chromatogr 11: 259 (1997).
6. Jakubetz H, Juza M, Schurig V, GIT Labor Fachzeitschr 42: 479 (1998).
7. Gazdag M, Szepesi G, Huszar L, J Chromatogr 351: 128 (1986).
8. Gazdag M, Szepesi G, Huszar L, J Chromatogr 436: 31 (1988).
9. Rizzi AM, Plank C, J Chromatogr 557: 199 (1991).
10. Reepmeyer JC, Chirality 8: 11 (1996).
11. Ameyibor E, Stewart JT, J Liq Chromatogr 20: 855 (1997).
12. Ameyibor E, Stewart JT, J Liq Chromatogr 20: 3107 (1997).
13. Owens P, Fell AF, Coleman MW, Berridge JC, Chirality 9: 184 (1997).
14. Herraez-Hernandez R, Campins-Falco P, J Chromatogr B 740: 169 (2000).
15. Yao TW, Zeng S, Biomed Chromatogr 15: 141 (2001).
16. Lepri L, Boddi L, Del Bubba M, Cincinelli A, Biomed Chromatogr 15: 196 (2001).
17. Healy LO, Murrihy JP, Tan A, Cocker D, McEnery M, Glennon JD, J Chromatogr A 924: 459 (2001).
18. Bielejewska A, Duszczyk K, Sybilska D, J Chromatogr A 931: 81 (2001).
19. Szeman J, Ganzler K, J Chromatogr A 668: 509 (1994).
20. Zukowski J, Sybilska D, Bojarski J, Szejtli J, J Chromatogr 436: 381 (1988).
21. Zukowski J, Sybilska D, Bojarski J, J Chromatogr 364: 225 (1986).

22. Pawlowska M, J Liq Chromatogr 14: 2273 (1991).
23. Pawlowska M, Chirality 3: 136 (1991).
24. Cooper AD, Jefferies TM, Gaskell RM, Anal Proc 29: 258 (1992).
25. Hu R, Takeuchi T, Jin JY, Miwa T, Anal Chim Acta 259: 173 (1994).
26. Gazdag M, Szepesi G, Huszar L, J Chromatogr 371: 227 (1986).
27. Szepesi G, Gazdag M, J Pharm Biomed Anal 6: 623 (1988).
28. Zukowski J, Pawlowska M, J High Resolut Chromatogr 16: 505 (1993).
29. Clark BJ, Mama JE, J Pharm Biomed Anal 7: 1883 (1989).
30. Walhagen A, Edholm LE, Chromatographia 32: 215 (1991).
31. Wannman H, Walhagen A, Erlandsson P, J Chromatogr 603: 121 (1992).
32. Zukowski J, Nowakowski R, J Liq Chromatogr 12: 1545 (1989).
33. Shimada K, Hirakata K, J Liq Chromatogr 15: 1763 (1992).
34. Takeuchi T, Nagae N, J Chromatogr 595: 121 (1992).
35. Takeuchi T, Miwa T, Anal Chim Acta 292: 275 (1994).
36. Eto S, Noda H, Noda A, J Chromatogr 579: 253 (1992).
37. Takeuchi T, Miwa T, J Chromatogr A 666: 439 (1994).
38. Sybilska D, Bielejewska A, Nowakowski R, Duszczyk K, Jurczak J, J Chromatogr 625: 349 (1992).
39. Eto S, Noda H, Minemoto M, Noda A, Mizukami Y, Chem Pharm Bull 39: 2742 (1991).
40. Stammel W, Woesle B, Thomas H, Chirality 7: 10 (1995).
41. Fell AF, Noctor TAG, Mama JE, Clark BJ, J Chromatogr 434: 377 (1988).
42. Pawlowska M, Zukowski J, J High Resolut Chromatogr 14: 138 (1991).
43. Deng Y, Maruyama W, Yamamura H, Kawai M, Dostert P, Naoi M, Anal Chem 68: 2826 (1996).
44. Zukowski J, J High Resolut Chromatogr 14: 361 (1991).
45. Mularz EA, Cline-Love LJ, Petersheim M, Anal Chem 60: 2751 (1988).
46. Italia A, Schiavi M, Ventura P, J Chromatogr 503: 266 (1990).
47. Cooper AD, Jefferies TM, J Pharm Biomed Anal 8: 847 (1990).
48. Nimura N, Ternary complexation, in Chiral separations by HPLC: Applications to Pharmaceutical Compounds, Krstulovic AM (Ed.), Ellis Horwood, New York, p. 107 (1989).
49. Clark BJ, Separations of enantiomeric compounds by chiral selectors in the mobile or solvent phase, in A Practical Approach to Chiral Separations by Liquid Chromatography, Subramanian G (Ed.), VCH, Weinheim, p. 311 (1994).
50. LePage J, Lindner W, Davies G, Seitz D, Karger BL, J Chromatogr 185: 323 (1979).
51. LePage J, Lindner W, Davies G, Seitz D, Karger BL, Anal Chem 51: 433 (1979).
52. Hare PE, Gil-Av E, Science 104: 1226 (1979).
53. Gil-Av E, Tishbee A, Hare P, J Am Chem Soc 102: 5115 (1980).
54. Davankov VA, Kurganov AA, Ponomareva TM, J Chromatogr 452: 309 (1988).
55. Galaverna G, Corradini R, de Munari E, Dossena A, Marchelli R, J Chromatogr A 657: 43 (1993).
56. Marchelli R, Corradini R, Bertuzzi T, Galaverna G, Dossena A, Gasparrini F, Galli B, Villani C, Misti D, Chirality 8: 452 (1996).
57. Galaverna G, Corradini R, Dossena A, Chiavaro E, Marchelli R, Dallavalle F, Folesani G, J Chromatogr A 829: 101 (1998).

58. Galaverna G, Corradini R, Dallavalle F, Folesani G, Dossena A, Marchelli R, J Chromatogr A 922: 151 (2001).
59. Oelrich E, Preusch H, Wilhelm E, J High Resolut Chromatogr 3: 269 (1980).
60. Gilon C, Lesham R, Tapuhi Y, J Am Chem Soc 101: 7612 (1979).
61. Gilon C, Lesham R, Greushka E, Anal Chem 52: 1206 (1980).
62. Gilon C, Lesham R, Greushka E, J Chromatogr 203: 365 (1981).
63. Greushka E, Lesham R, J Chromatogr 255: 41 (1983).
64. Nimura N, Suzuki T, Kasahara Y, Kinoshita T, Anal Chem 53: 1380 (1981).
65. Nimura N, Toyama A, Kinoshita T, J Chromatogr 234: 482 (1982).
66. Nimura N, Toyama A, Kasahara Y, Kinoshita T, J Chromatogr 239: 671 (1982).
67. Nimura N, Toyama A, Kinoshita T, J Chromatogr 316: 547 (1984).
68. Weinstein S, Angew Chem Suppl 96: 425 (1982).
69. Weinstein S, Engel M, Hare P, Anal Biochem 121: 370 (1982).
70. Weinstein S, Weiner S, J Chromatogr 303: 244 (1984).
71. Weinstein S, Grinberg N, J Chromatogr 318: 117 (1985).
72. Kurganov A, Davankov V, J Chromatogr 218: 559 (1981).
73. Wernicke R, J Chromatogr Sci 23: 39 (1985).
74. Lindner WF, Hirschbock I, J Liq Chromatogr 9: 551 (1986).
75. Lindner WF, LePage J, Davies G, Seitz D, Karger BL, J Chromatogr 185: 323 (1979).
76. Tapuchi Y, Miller N, Karger BL, J Chromatogr 205: 325 (1981).
77. Lam S, Chow F, J Liq Chromatogr 3: 1579 (1980).
78. Lam S, Chow F, Karmen A, J Chromatogr 199: 295 (1980).
79. Lam S, J Chromatogr 234: 483 (1982).
80. Lam S, Karmen A, J Chromatogr 239: 451 (1982).
81. Lam S, Karmen A, J Chromatogr 289: 339 (1984).
82. Lam S, J Chromatogr Sci 22: 416 (1984).
83. Klemisch W, von Hodenberg A, Vollmer K, J High Resolut Chromatogr 4: 535 (1981).
84. Horikawa R, Sakamoto H, Tanimura T, J Liq Chromatogr 9: 537 (1986).
85. Galaverna G, Panto F, Dossena A, Marchelli R, Bigi F, Chirality 7: 331 (1995).
86. Forsman U, J Chromatogr 303: 217 (1984).
87. Allenmark S, Bromgren B, Boren H, J Chromatogr 237: 473 (1982).
88. Hermansson J, J Chromatogr 316: 537 (1984).
89. Hedeland M, Isaksson P, Pettersson C, J Chromatogr A 807: 297 (1998).
90. Lämmerhofer M, Lindner W, J Chromatogr A 839: 167 (2001).
91. Salvador A, Herbreteau B, Dreux M, Karlsson A, Gyllenhaal O, J Chromatogr A 929: 101 (2001).
92. Wang S, Porter MD, J Chromatogr A 828: 157 (1998).
93. Wei W, Luo G, Xiang R, Yan C, J Microcol Sep 11: 263 (1999).
94. Ban E, Choi S, Lee JA, Lho DS, Yoo YS, J Chromatogr A 853: 439 (1999).
95. Lucangioli SE, Hermida LG, Tripodi VP, Rodriguez VG, Lopez EE, Rouge PD, Carducci CN, J Chromatogr A 871: 207 (2000).
96. Garcia-Ruiz C, Martin-Biosca Y, Crego AL, Marina ML, J Chromatogr A 910: 157 (2001).

97. Mendez SP, Gonzalez EB, Sanz-Medel A, Biomed Chromatogr 14: 8 (2000).
98. Fanali S, Catarcini P, Blaschke G, Chankvetadze B, Electrophoresis 22: 3131 (2001).
99. Lelievre F, Yan C, Zare RN, J Chromatogr A 723: 145 (1996).
100. Debowski J, Grassini-Strazza G, Sybilska D, J Chromatogr 349: 131 (1985).
101. Armstrong DW, He FY, Han SM, J Chromatogr 448: 345 (1988).
102. Armstrong DW, Faulkner JR, Han SM, J Chromatogr 452: 323 (1988).
103. Aboul-Enein HY, El-Awady M, Heard CM, J Liq Chromatogr Relat Technol 23: 2715 (2000).
104. Lepri L, Coas V, Desideri PG, Pettini L, J Planar Chromatogr Mod TLC 5: 175 (1992).
105. Lepri L, Coas V, Desideri PG, J Planar Chromatogr Mod TLC 5: 364 (1992).
106. Armstrong DW, Zhou Y, J Liq Chromatogr 17: 1695 (1994).
107. Huang MB, Li HK, Li GL, Yan CT, Wang LP, J Chromatogr 742: 289 (1996).
108. Tivert AM, Backman A, J Planar Chromatogr Mod TLC 2: 472 (1989).
109. Tivert AM, Backman A, J Planar Chromatogr Mod TLC 6: 216 (1993).
110. Duncan JD, Armstrong DW, J Planar Chromatogr Mod TLC 3: 656 (1990).

Index